Studies of Brain Function, Vol. 15

Coordinating Editor
V. Braitenberg, Tübingen

Editors
H. B. Barlow, Cambridge
T. H. Bullock, La Jolla
E. Florey, Konstanz
O.-J. Grüsser, Berlin-West
A. Peters, Boston

Studies of Brain Function

Uwe Windhorst

How Brain-like is the Spinal Cord?

Interacting Cell Assemblies in the Nervous System

With 60 Figures

Springer-Verlag Berlin Heidelberg New York
London Paris Tokyo

Professor Dr. Uwe Windhorst
Zentrum Physiologie und Pathophysiologie
der Universität Göttingen
Abteilung Neuro- und Sinnesphysiologie
Humboldtallee 23
3400 Göttingen, FRG

ISBN 978-3-642-51122-6   ISBN 978-3-642-51120-2 (eBook)
DOI 10.1007/978-3-642-51120-2

Library of Congress Cataloging-in-Publication Data.
Windhorst, Uwe, 1946– How brain-like is the spinal cord? (Studies of brain
function ; vol. 15) Includes index. 1. Spinal cord. 2. Nerves, Spinal. 3. Neural
circuitry. 4. Neurology-Philosophy. I. Title. II. Series: Studies of brain function ; v. 15. [DNLM: 1. Interneurons-physiology. 2. Motor Neurons-physiolo-
gy. 3. Spinal Cord-physiology. W1 ST937KF v. 15 / WL 400 W764h]
QP371.W56   1988   599'.0188   87-28781

Softcover reprint of the hardcover 1st edition 1988

Media conversion: Daten- und Lichtsatz-Service, Würzburg
Offsetprinting and binding: Konrad Triltsch, Graphischer Betrieb, Würzburg
2131/3130-543210

To my father,
my sister, Lisbeth and Siggi

Preface

"Theorizing about brain functions is often considered slightly disreputable and anyhow a waste of time – perhaps even 'philosophical'"

P. S. CHURCHLAND[1]

At present there are no unanimously accepted general concepts of brain operation and function. This is especially the case with regard to so-called "higher" functions such as perception, memory or the coupling between sensory input and motor output. There are a number of different reasons for this. Some may be related to experimental limitations allowing the simultaneous recording of the activities of only a restricted number of neurones. But there are also conceptual difficulties hindering the transition from "single-neurone" schemes, in which neurones are assigned relatively specific tasks (such as feature detection), to more complex schemes of nerve cell assemblies (for a discussion of some of the difficulties see Abeles 1982; von der Malsburg 1981; Krüger 1983). Whilst much is known about the basic properties and functions of single neurones, whose operations we hope to understand in the foreseeable future, this does not hold true in the same way for the working of large assemblies of neurones.

Much of this book is concerned with theories and hypotheses about the workings of large assemblies of neurones. Is this a reasonable project in the light of the difficulties involved? Churchland (1986) has reviewed some of the pros and cons about theorising on brain function. She starts by listing the cons, which are often heard from experimental neuroscientists: (1) At present, since not enough is known about the basic structure of the brain, any theory about its working is premature. (2) Available theory of brain function is too abstract,

[1] Churchland PS (1986) Neurophilosophy. Toward a unified science of the mind/brain. The MIT Press, Cambridge (Mass.), p 403

mostly untestable and irrelevant to experimental neuroscience. (3) It is difficult to get funding for theoretical research. These misgivings, particularly the first, contribute to the gap between experimental neuroscientists and theorists and may be one important reason for the tendency toward reductionism. At present, this tendency is supported by the success of basic research, for example, in membrane biophysics. By contrast, the work on integrated circuits in complicated nervous tissues is impeded by our inability to grasp the wiring diagrams which are being elucidated with anatomical and physiological techniques.

Churchland's arguments for theory in neuroscience may be summarised as follows. (1) Good theory motivates and organises experimental research. Without a theoretical background, data gathering often becomes a quasi-random process solely determined by some newly mastered experimental technique. (The underlying hope is that one day the collected data will fit into a theory: "inductivist fallacy".) On the contrary, theory ought to provide the questions which experiments are to answer. (2) The idea that experiments can be done without any theoretical background is a misconception. Experimenters plan their experiments on the basis of a – perhaps unconscious – evaluation of their worth. This background is made up of a conglomerate of assumptions, intuitions and prejudices. It should be clarified into a rich, coherent and well-ramified theory rather than be ignored. (3) Theory does not usually emerge from raw data on its own. For example, if the function of a neural assembly in coordinating movements is to be explained, this cannot be achieved by simply looking at the properties of the participating neurones, because the interactions between the neurones are nonlinear. The whole is more than the sum of its parts, and a theory of the whole is an interpretation of the interaction of the components. It is in this spirit that this book is written.

Having occupied myself with correlations between discharge patterns of neurones (more specifically between those of motoneurones and muscle spindle afferents) for some years and having grown rather pessimistic about ever being able to make much physiological sense of them (see Windhorst 1978a, b), three near-synchronous events suddenly refreshed my interest and ultimately induced me to write this book. Moshe Abeles' book on *Local cortical circuits* (1982) took a bold step towards a brain theory based upon synchronisation of neuronal firing. In conversations with Christoph von der Malsburg I became aware of his *Correlation Theory of Brain Function* (1981), which in many respects moves on the same tracks as Abeles' ideas. And finally, Peter H. Ellaway organised a symposium on correlations between neuronal discharge patterns (London, 1982), which recently found a successful sequel (Oxford, 1985). These three events encouraged me to collect material distributed throughout the literature, originally intended as a basis for further discussion, in the form of a review paper. After much stimulating criticism from various people I finally presented an incomplete draft of the manuscript to Valentino Braitenberg, asking for advice as to its possible future. He immediately responded positively and suggested publication in this series.

I share Moshe Abeles' view "that the nervous system is essentially a statistical machine" (1982; p. 91); and I share Christoph von der Malsburg's view that information in this statistical machine is encoded not only in the number of active nerve cells and their mean firing rates ("frequency code"), but also in interactions between neuronal and neuromuscular elements of cell assemblies expressed in correlations of their firing patterns. Without these convictions, the undertaking would have been fatuous or useless. Much of the book therefore deals with interneuronal correlations and their causes and consequences. This might appear to be a very specialised and narrow topic, but it is not. Correlations are a fundamental aspect of information processing in terms of redundancy, security of information transfer, and organisation of neuronal assemblies. They also have to do with codes and thus with an important and little understood aspect of function. I hope to be able to make this point. The other point to be emphasised is the locus of such more sophisticated information processing. I argue that the "lower" spinal cord does use this type of encoding in much the same way as the "higher" central nervous system, particularly the cortex. To make this point sufficiently clear, one would have to present a theoretical framework of the function of both structures, within which the more sophisticated information processing by correlations could be understood. Needless to say, at present no generally accepted theories are at hand. All I can do is to offer some fragments of such frameworks, "piecemeal" as it were. The provocative question in the book's title will therefore not be answered conclusively.

The book is organised like variations on a theme, so that some topics are discussed several times, but from varying viewpoints. The plan evolves from a discussion of correlations between neuronal firing patterns at supraspinal levels, related topics of synaptic modifiability, and theories of brain function built upon them (Chap. 1). Of course, these aspects cannot be dealt with exhaustively in the present monograph (for a recent anthology of brain theories see Palm and Aertsen 1986). The first chapter serves as a frame for the following main discussion which is principally concerned with spinal cord networks and functions. By implicitly and explicitly referring back to ideas outlined in the "supraspinal" part, correspondences of functional principles between spinal and supraspinal levels hopefully emerge, thus yielding fragments of answers to the main question.

More specifically, Chap. 1 is concerned with correlations between discharge patterns of different supraspinal neurones and those events which can be considered as their causes and consequences. Correlations between neuronal firings are ubiquitous in the nervous system, from cortex via brainstem and spinal cord to the periphery. The forms and types of correlation are surprisingly similar at all levels. But functional interpretations of them are rather dissimilar. Whereas correlations, synchrony in particular, between firings of nerve cells at the supraspinal level have stimulated the development of intricate theories of brain function, nothing comparable has happened with regard to similar findings of firing correlations at the spinal or peripheral levels. Examples of brain theories based on correlations will be presented in the initial sections (1.3.1 to 1.3.3), which include descriptions of the "synfire chain" hypothesis developed by Abeles (1982), the "correlation theory of brain function" (von der Malsburg 1981, 1983, 1985), and related experimental results on neuronal plasticity. The brain stem section (1.4) shows how correlations between neuronal discharges could also play an important functional role at this level. The spatio-temporal cross-correlation model for acoustic pitch detection of Loeb et al. (1983) is a telling example. But also the functional organisation of activity in the lower brainstem reticular formation may depend significantly on the way neuronal discharges are correlated (Schulz et al. 1985).

Chapter 2 on the spinal cord discusses the types of correlations observed between the firings of spinal neurones, as well as the origins and consequences of these correlations. As indicated above, such correlations have not led to hypotheses or theories of spinal cord function as they have at higher levels. At the lower level, correlations between parallel neuronal elements have been used rather as an indirect means of studying the input organisation of the neurones recorded. This forbearance to develop more intricate theories based on correlations may be partially due to the fact that the spinal cord is commonly regarded as a reflex apparatus producing motor output. Correlations, in particular synchrony, between discharge patterns of α- and γ-motoneurones are usually regarded as detrimental, since they may increase the amplitude of physiological tremor. This may however be a one-sided view if generalised to all neurones involved in spinal circuits including peripheral muscle receptors.

Since the stability problem is an important issue in connexion with neuronal correlations *at all levels of the nervous system*, Chap. 3 is devoted to correlations between entire networks of different neurones which play a significant role in different forms of physiological tremor. It is argued that these forms of tremor reflect different "functional states" of the spinal cord which bear certain similarities to corresponding states at other levels of the nervous system. Spatially ordered (topographical) and "species specific" connexion patterns are discussed as a means of organising different neural activity states.

Where the third chapter is predominantly concerned with the conventional stretch reflex loop and its constituents, Chap. 4 makes a step towards integrating an intraspinal recurrent system into a more elaborate motor control scheme by proposing a parallelism of organisation between spinal recurrent circuitry and proprioceptive feedback. It is suggested that recurrent inhibition and the muscle spindle system represent two "reference models" for the execution of posture and movement in a parameter-adaptive control system.

Finally, Chap. 5 outlines some current problems in spinal cord physiology. Complexity is emphasised in order to shake simple views and thereby encourage the search for alternatives. Within this context more complex neuronal activity patterns including correlations attain an elevated significance.

A good part of the material collected here is speculative although based on a sound body of experimental data. Many *ideas* and *concepts* are taken over from workers in various fields. To further the reader's orientation I have included summaries which give an overview of the material but also contain some discussion, particularly following longer presentations of experimental results.

I am very grateful to the many people who assisted me in one or the other way in writing this book. First of all, I would like to thank my co-workers C. N. Christakos, W. Koehler, J. Meyer-Lohmann, U. Niemann, R. Rissing and C. Schwarz for their cooperation and endurance in stressful experiments and lengthy data evaluation. Mrs. B. Müller-Fechner typed much of the manuscript. I am grateful for encouragement, critical comments and suggestions by V. Braitenberg, P. H. Ellaway, T. M. Hamm, Z. Hasan, H.-D. Henatsch, P. A. Kirkwood, G. E. Loeb, F. J. R. Richmond, and D. G. Stuart. Particular thanks are due to S. Cleveland and M. Hulliger who took great pains to improve the manuscript.

Part of the work presented was supported by grants from the Deutsche Forschungsgemeinschaft, Sonderforschungsbereich 33 ("Nervensystem und biologische Information", Göttingen). Thanks are due to Elsevier, The Journal of Neurophysiology, The Journal of Physiology, Raven Press, and Springer-Verlag, and to the respective authors, for the permission to reproduce figures.

Göttingen, February 1988   UWE WINDHORST

Contents

"A few months ago Werner Heisenberg and Wolfgang Pauli believed that they had made an essential step forward in the direction of a theory of elementary particles. Pauli happened to be passing through New York, and was prevailed upon to give a lecture explaining the new ideas to an audience which included Niels Bohr. Pauli spoke for an hour, and then there was a general discussion during which he was criticized rather sharply by the younger generation. Finally Bohr was called on to make a speech summing up the argument. 'We are all agreed,' he said, 'that your theory is crazy. The question which divides us is whether it is crazy enough to have a chance of being correct. My own feeling is that it is not crazy enough.'

The objection that they are not crazy enough applies to all the attempts which have so far been launched at a radically new theory of elementary particles ... When the great innovation appears, it will almost certainly be in a muddled, incomplete and confusing form. To the discoverer himself it will be only half-understood; to everybody else it will be a mystery. For any speculation which does not at first glance look crazy, there is no hope."

F. J. Dyson[2]

"... our ignorance may result from the lack of an adequate theoretical framework in the light of which to order and interpret the relevant facts, a problem that has in more or less degree retarded progress in fields as diverse as particle physics, pure mathematics, the study of cancer, and the functioning of the central nervous system."

J. Kendrew[3]

---

[2] Dyson FJ (1958) Innovation in physics. Sci Am 199:74–82, pp 79–80
[3] Kendrew J (1977) Introduction. In: Duncan R, Weston-Smith M (eds) The encyclopaedia of ignorance. Pergamon, Oxford, pp 205–207; p. 206

1   Correlations Between
Spike Trains of Supraspinal Neurones

> "... there is a pessimistic note in microelectrode neurophysiology when one considers that the relevant chunks of experience (things, persons, attitudes, sentences) are most likely not represented within the brain by single neurons, but by sets of neurons too numerous to be observed systematically by means of arrays of microelectrodes"
>
> V. Braitenberg [4]

It is most difficult to explain, in neurophysiological terms, the highest brain functions, which are considered to take place in the phylogenetically young neocortex or, at most, in the cerebellum. However, this must not distract us from the fact that many "lower" functions executed by phylogenetically older parts of the central nervous system (CNS), such as spinal cord and brainstem, are not yet well understood, either; and, perhaps surprisingly, there may be reasons similar to those existing for the "high" levels. Indeed, it is not at all clear to what extent the operational principles, i.e. methods by which the CNS solves problems, are different at different levels of the hierarchically organised nervous system. This is not meant to imply that these various levels could not be concerned with different tasks or different aspects of a common task. But, at least to a certain extent, hierarchically higher and phylogenetically newer levels have had to make use of, and thus adapt to, functional principles governing the working of the lower levels. Conversely, the latter may have partially adapted to new demands resulting from the new possibilities which have been opened by the evolution of the higher levels. Because of this mutual dependency, it is highly improbable that high and low levels have developed totally disjoint principles of operation.

It would thus appear justified and perhaps promising to look for common features in brain and spinal cord function. This will be attempted here by starting from a *seemingly* very specific feature, namely correlations between discharges of neurones and their possible role. But, by definition, such correlations have to do with the interaction of neurones and, hence, with the working of assemblies or networks of nerve cells. Thereby immediate relations to integrative function are brought into focus. The main emphasis will be put on neural assemblies in and around the spinal cord. Cortical and

---

[4] Braitenberg V (1977) On the texture of brains. Springer New York, p 3

cerebellar networks will only be mentioned in passing (for recent reviews see monographs by Creutzfeldt 1983, and Ito 1984).

1.1 Correlations Between Discharge Patterns of Cortical Neurones

With the advent of modern, computer-aided data processing the determination of neuronal interactions by means of statistical cross-correlation computations has become commonplace (see Sect. 1.2 and App. A), and has produced a huge amount of data showing such interdependencies between firing patterns of different nerve cells at all levels of the nervous system. (No complete reference list can be given here. For a review of early work see Moore et al. 1966. Further single examples are, for the retina: Arnett and Spraker 1981; Mastronarde 1983 a, b, c; for the cerebral cortex: recent review: Krüger 1983; Holmes and Houchin 1966; Toyama 1978; Toyama et al. 1981; Burns and Webb 1979; Abeles 1982; Schneider et al. 1983; Metherate and Dykes 1985; Wyler 1985; for the cerebellum: Bell and Grimm 1969; Bell and Kawasaki 1972; Ebner and Bloedel 1981 b; for the brainstem: Madden and Remmers 1982; Schulz et al. 1985; Hilaire et al. 1984; for spinal neurones see below).

Usually the appearance of such correlations, for example, more-than-random tendencies towards synchronous firing, depends very much on the momentary state of the animal preparation, e.g. depth and kind of anaesthesia, state of attention (wakefulness and sleep), experimental manipulations such as stimulations and lesions, etc. (see, for instance, Holmes and Houchin 1966; Milner-Brown et al. 1975; Abeles 1982; Ebner and Bloedel 1981 b; Ellaway and Murthy 1982; Kirkwood et al. 1982 b; Kirkwood et al. 1984; Schneider et al. 1983; Schulz et al. 1985). The same applies to data obtained in man. These complicated relations cannot be reviewed here in any completeness. Indeed, they are so confusing that they have much contributed to the uncertainty as to the physiological significance of cross-correlated discharges of groups of neurones. Thus, the only somewhat safe conclusion which may be drawn from the muddle is that if correlations between neurones do at all play any significant role for the normal functioning of the nervous system they do so as *dynamically changing patterns*. This general statement would in turn explain part of the difficulties in getting hold of the above role since the statistical techniques of determining cross-correlated firings commonly require at least quasi-stationary conditions, thereby nearly precluding the investigation of dynamic changes. It would be of great help if the at present available techniques could be further refined in such a direction as to take account of dynamic changes. Another shortcoming of the present statistical methods is that cross-correlations are mostly computed for *pairs* of cells (Sect. 1.3 and App. A) or, at most, for triples of cells (App. B). The disadvantages of this restriction were made plain by Abeles (1982) who compared cross-correlograms computed for three simultaneously recorded cortical neurones (using the method of Perkel et al. 1975; App. B) with results obtained from conventional cross-correlograms for (the three) *pairs* of cells. The three-neurone technique revealed so much more information about the interaction of neurones that Abeles was led to propose an extended hypothesis on brain function (Sect. 1.3.1). There have been some attempts to apply conventional cross-correlation techniques under more dynamic conditions. For example, Gerstein and Michalski (1981) studied the interaction of several cortical neurones recorded simultaneously in response to regularly repeated stimulus presentations (e.g. sweeps of a light bar over the receptive fields of a visual cortical neurone). They found more cross-correlated firing between the neurones (during each single stimulus presentation) than could be accounted for by the stimulus alone. They acknowledge that the question whether stimulus-related synchrony of firing patterns among several neurones is used as a code cannot be ultimately determined. But they feel that at least the *possibility* for coding by synchronous firing is demonstrated by their data from the visual and the auditory system. "Such coding schemes would allow the short processing time to detection that is behaviorally observed" (Gerstein and Michalski 1981; p. 101). Similar results were reported by Schneider et al. (1983). As rightly pointed out by Gerstein and Michalski, cross-correlation analysis is to a large extent, at least implicitly and tacitly, concerned with the problem of neural code, and not only with the detection of connexions between the neurones under study. (Ebner and Bloedel 1981a, calculated auto–correlations of cerebellar Purkinje cell simple spikes in temporal relation to climbing fibre input and concluded that the precise temporal patterning of spikes within a single train may contain information above and beyond that contained in the mean discharge rate.)

1.2 The Basic Method of Cross-Correlating Spike Trains

Although the notions of auto- and cross-correlation are discussed more extensively and rigorously in App. A, it appears appropriate for a better understanding of the following to give here a qualitative and practical introduction to conventional cross-correlations between two spike trains before indulging in specific examples. In principle, the computation of cross-correlograms is an intuitively straightforward matter.

Assume two spike trains A and B are recorded, and the upper two traces in Fig. 1 represent a segment of the data. In order to find out whether the probability of firing of cell B is temporally related to spike occurrences of cell A, a *cross-correlogram* is conventionally constructed as follows. One of the two spike trains is chosen as "reference" train, in this case train A. In the

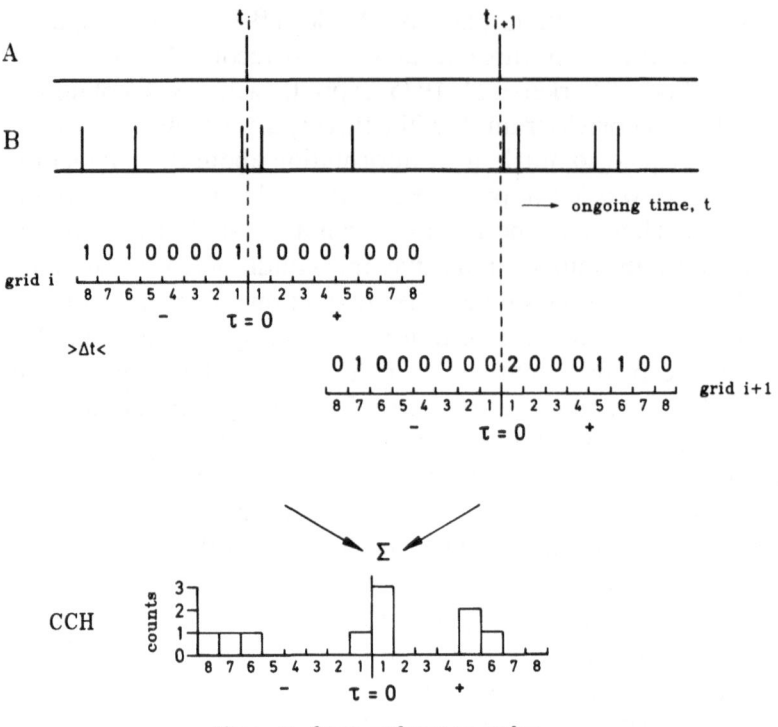

Fig. 1. Scheme of the basic method of cross-correlating two spike trains. The *upper two traces* show a segment of two spike trains labelled *A* and *B*, of which *A* is assumed to be the train yielding the reference events. Below the *upper traces*, representing original data, are shown two "*grids*" which are centred around t_i and t_{i+1}, the instants of spike occurrence in train *A*. These *grids* are used to temporally locate spikes in train *B* (to within the bin width Δt) with respect to spikes in train *A*. Note that spikes in *B* can be captured (and counted) twice (or more times) by successive *grids* which may overlap. The lower cross-correlation histogram (CCH) is the linear sum (as represented by a Σ) of counts within corresponding bins in *grids* i and $i + 1$. The number of events in the bins is represented by *bars*. For fuller description see text

small segment displayed, cell A discharges twice, at t_i and t_{i+1}. Aligned with these instants, temporal grids are projected onto spike train B. That is, from each "reference event" in A (at t_i and t_{i+1}), relative time τ, which extends into backward (negative) and forward (positive) direction, is divided into bins of selectable width Δt. Here, each grid (centred around t_i or t_{i+1}) is made up of 8 backward bins (labelled by negative numbers) and 8 forward bins (positive numbers). The length of a grid corresponds to the analysis time. Now, if a spike in train B occurs within a bin of the i-th grid, it is represented as a count labelled 1 assigned to that particular bin; non–occurrence of a spike is here symbolised by a zero. The same procedure is followed with the grid i+1 centred around t_{i+1} and further grids centred around all the other spikes occurring in train A. Thus, to each spike occurrence in train A is assigned an array of numbers consisting of the counts within bins -8 to $+8$. These vectors are then added, as illustrated for the two specimens assigned to t_i and t_{i+1} in Fig. 1 (lower row). The number of counts per bin in this provisional cross-correlation histogram (CCH), collected for only two reference events, is represented by the height of bars spanning the bins. To intuitively under-stand the meaning of such a cross-correlogram, collected for many reference events in train A, it is helpful to imagine how it would look like if the two spike trains A and B were independent of each other. Then, the temporal positioning of spikes in train B would bear no relation to the instants t_i, t_{i+1}, and so forth, at which spikes in train A occur. That is, the probability of discharge of cell B is the same at all relative times τ. The cross-correlogram would therefore be flat apart from statistical fluctuations from bin to bin (due to limited sampling). Any excess of bin counts above or below these fluctu-ations would point to some dependency of the spike occurrences in both trains.

It is worthwhile to note the following. (1) Whereas the abscissa of the upper two traces in Fig. 1 (representing the original data) is labelled "ongoing time, t", the abscissa of any grid or cross-correlogram is labelled "time τ from reference pulse" because its zero time is always reset to the instant t_i of occurrence of reference spikes in A; the time axis of a correlogram thus represents time relative to the reference events and is therefore denoted by another symbol (τ instead of t). Although different authors often use different terminology, for an understanding of cross-correlation it is essential to un-derstand the difference in the notions of time. (2) In this representation of the cross-correlogram (Fig. 1, bottom), the counts within each bin are natural numbers. These counts depend on the "local" firing rate of cell B, on the bin width chosen, and on the number of reference spikes of cell A (which in turn depends on the mean firing rate of A and the length of recording). In order to get rid of the incidental dependency on the number of reference events, the following normalisation may be performed. Dividing the bin contents (counts) by the number of reference events used to collect the histogram yields the *relative frequency* of the occurrence of spikes in train B per spike in A. This relative frequency is still dependent on bin width. This in turn can be circumvented by dividing the relative frequency values by the bin width, which leaves us with the *rate of firing* of cell B relative to a spike in A. These different measures have indeed all been used by different authors. (3) If the reference pulse train is a stimulus train, the cross-correlation histogram becomes a peristimulus-time histogram (PSTH). If reference (A) and corre-lated (B) spike trains are identical, the result is an auto-correlation histogram, which shows interdependence of spike occurrences within a single spike train. (4) The choice of bin width Δt depends on a number of practical factors, such as desired temporal resolution of cross-correlogram features, discharge rates of the two cells and length of recording, strength of correlation etc. Also, the analysis time (length of grids) is chosen according to practical considerations (see Fig. 10). There are no absolute rules that can be generally recommended. (5) The interpretation of a cross-correlogram is usually more difficult than its construction because possible rhythms in the two spike trains mirror themselves in the cross-correlogram (see Perkel et al. 1967b; Glaser and Ruchkin 1976). These intricacies cannot be treated here, but do not play a prominent role in the following. (6) The grids can also be designed slightly differently in that a central bin (labelled 0) is centred around $\tau = 0$, i.e., $\tau = 0$ is assigned to the middle of the zero bin. This would render it easier to detect precise synchrony of discharge.

1.3 Correlations Between Discharge Patterns of Cortical Neurones and Their Role in Some Recent Models of Brain Function

1.3.1 Synfire Chains

Abeles (1982) extensively studied the interactions of cortical neurones (in cats and monkeys) by means of cross-correlation computations. The sort of results obtained from recordings of cortical cells in unanaesthetised, muscle-relaxed cats that were not explicitly subjected to any specific stimulation is shown in Fig. 2 (his Fig. 8), which displays cross-correlograms or "cross-renewal densities". In the underlying experiment three neurones were recorded simultaneously with the same electrode, and all possible combinations of cross-correlograms between the three pairs of cells are shown (A to C). On the abscissa the relative time from the firings of one cell (yielding the reference events at zero time) is plotted, and on the ordinate the firing rate of the second cell correlated with respect to the reference firings of the first is plotted. The range within which 99% of the firing rate values of the second cell should be contained if the two cells are uncorrelated is indicated by the two horizontal lines.

The type of correlation found most often (in 45% of the neurone pairs studied) in the preparation chosen is illustrated by Fig. 2C. The curve is more or less flat except for random fluctuations.

The second type is shown in Fig. 2A. This plot shows a peak around zero time that extends up to about 30–50 ms on either side. This means that the correlated cell is more likely to discharge around the occurrence of a spike in the reference cell than further before or later. Abeles (1982) considers two possible causes of this type of cross-correlation: (1) Neurone 5 excited neurone 6 and vice versa (reciprocal excitation). (2) Both neurones were excited from a common source (common input), which would simultaneously increase the firing rate of both cells 5 and 6. Abeles argues that, of the two alternatives, the common input hypothesis is more attractive, because the

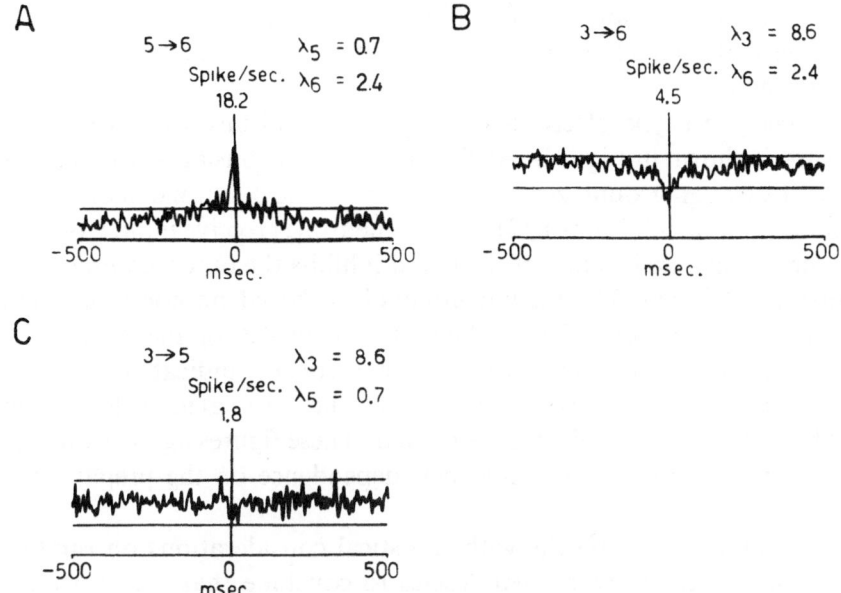

Fig. 2 A–C. Relations between three cortical neurones. The spike activities of three neurones were simultaneously recorded with the same electrode. All possible cross-correlations between the three pairs of cells are shown. The cells are labelled by *numbers* on top. The *arrows* between them indicate the direction of correlation; for example, $5 \rightarrow 6$ (in **A**) means that the discharge of cell No. 5 was correlated with respect to spikes (as reference events) of cell No. 6. The λ's indexed by the cell numbers give their mean discharge rates. **A** Cells 5 and 6 probably share a common input that either excites or (less likely) inhibits both of them; **B** Cells 3 and 6 probably share a common input which excites one of them and inhibits the other; **C** Cells 3 and 5 fire independently of each other. (With permission from Abeles 1982; his Fig. 8)

probability that the two cells excited each other is low and the width of the cross-correlogram peak of Fig. 2A is too large for reciprocal excitation. Moreover, in the visual cortex of the cat, the cross-correlograms of most pairs switched from a flat curve into a curve with a peak around zero delay when the cat went to sleep (Burns and Webb 1979). Such changes could be explained by presuming that the appearance of a peak during sleep results from the synchronising influences of subcortical nuclei. Although the foregoing interpretation is based for an excitatory common input, it holds true also for a common input inhibiting both neurones, although in this case the peak would probably be smaller and wider. Of the neurone pairs 28% were found to show such signs of common input. But this figure may apply only to the special type of preparation used in Abeles' studies (unanaesthetised, muscle-relaxed cat), because signs of common input proved to disappear and reappear as the animal changed its state of arousal (Abeles 1982).

There is another type of cross-correlogram in which the peak is not convolved around zero time but is displaced slightly from it by a certain latency. This relation, which might indicate an excitatory synaptic connexion between the reference and the correlated cells, occurred in 12.5% of the neurone pairs.

The third type of cross-correlogram, found in only 2% of the neurone pairs, is shown in Fig. 2 B. Abeles (1982) again considers two possible causes for the trough around zero delay: (1) Neurone 3 inhibits neurone 6 and vice versa (mutual inhibition). (2) The two neurones receive input from a common source which excites one of them and inhibits the other (common input with opposite effects). The second mechanism, based on common input, again appears to be more likely (Abeles 1982). In 8% of the neurone pairs the trough followed the reference zero time which is indicative of an inhibitory connexion between the reference and the correlated cells. According to Abeles 4.5% of the pairs were irregular. These figures again are subject to the same reservations concerning their dependence on the preparation as outlined above.

Comparing his results with statistical considerations on randomly connected cortical networks led Abeles to conclude that "... in summary our data support the view that each cortical cell receives and makes many local connections. They indicate that neurons in close proximity can make strong contacts, while remote neurons make weaker contacts" (1982; p. 38). Thus, although the pattern of intracortical connexions between cortical neurones may show some randomness of organisation, they also exhibit a spatial gradient. This was recently confirmed by deRibaupierre et al. (1985) who investigated the presence and types of correlations between neurones in the cat auditory cortex. They rarely found signs of direct excitatory or inhibitory synaptic connexions which, if existent, were limited to adjacent units (4/24). Signs of common inputs occurred frequently for adjacent units (79%), but their frequency decreased progressively with distance, reaching 7% at 5 mm. In about 40% of such cases of common inputs, these were modified by the presentation of an acoustic stimulus (deRibaupierre et al. 1985). Similar results were reported by Toyama et al. (1981) for the cat visual cortex and by Metherate and Dykes (1985) for the cat somatosensory cortex, with the specification that correlations indicative of intracortical excitatory or inhibitory cell interactions existed for neurone pairs belonging to the same ocularity stripes or surroundings with the same types of rapidly or slowly adapting cutaneous input, respectively. Common inputs from other nonlocal sources may also be supposed to correlate neighbouring neurones in a stronger way than remote neurones. The spatial organisation of this input would be subjected to a number of different rules (topographical and various functional mappings; see Creutzfeldt 1983). The spatial distribution of correlation patterns is therefore important and appears to be a common feature elsewhere, too, as will be amply demonstrated in the following.

The basic types of cross-correlogram described above were also seen and very precisely analysed in retinal ganglion cells by Mastronarde (1983 a, b, c; see also Chap. 2).

In addition to the commonly used (conventional) cross-correlograms for *pairs* of cells, Abeles (1982) employed the display technique of Perkel et al. (1975) (see also Abeles 1983) to reveal time relations between firings of three simultaneously recorded neurones. This technique is explained in App. B. In most cases, Abeles (1982), by using this method, did not detect more complex interactions between the three cells than could also be detected by the conventional two-cell cross-correlograms, "as could be expected from the finding that most cells have their own way of responding to their input and that interactions, when they exist, are usually quite weak" (p. 62). An interesting exception is illustrated in Fig. 3, which exemplifies relations between the discharge patterns of three neurones found in 8% (10 out of 129) of the three-cell groups. The three conventional cross-correlograms for the three pairs of neurones at the top again suggest the presence of a common input for each pair. In Fig. 3D two dark bins can be recognised (enclosed by rectangles), one near the centre and another near $T_4 - T_1 = 375$ ms. The latter implies that neurone 5 fired frequently within 25 ms after the firing of neurone 4, but only when neurone 1 fired 375 ms earlier. In all such cases of special time relations between three neurones, the time delays seen were quite long, tens to hundreds of milliseconds. At first sight, these complex firing patterns might be discarded as chance events, but in fact the occurrence of these relations by mere chance is highly improbable. But then, if they cannot be discarded, this poses a difficult problem. The long time delays suggest the involvement of a large number of synapses within a long chain of neurones, which would have to be traversed by signals with almost no failure. Abeles (1982) thinks it unlikely that such a long chain, in which the activity of one neurone is strictly coupled to that of another exists in the cortex, because he never detected in his cross-correlograms such strong connexions at short or long delays. Thus an alternative arrangement that could result in good time-locking after many synapses should be looked for.

Abeles (1982) discussed several possible arrangements among which the most likely one was termed "synfire chain". This chain consists of a set of neurones that converges on a subsequent set, which in turn converges on another set and so forth (see Fig. 4 B). When the neurones in the first set fire in near-synchrony, each of the cells in the second set receives near-synchronous synaptic inputs which synchronously excite the subsequent next set, and so forth. Synchronous presynaptic cell discharges are so important because "it seems that coincident firing is about 10 times more efficient than asynchronous integrated activity" (Abeles 1982; p. 68). (One may question, however, whether the broad-peak correlations extending over several tens of

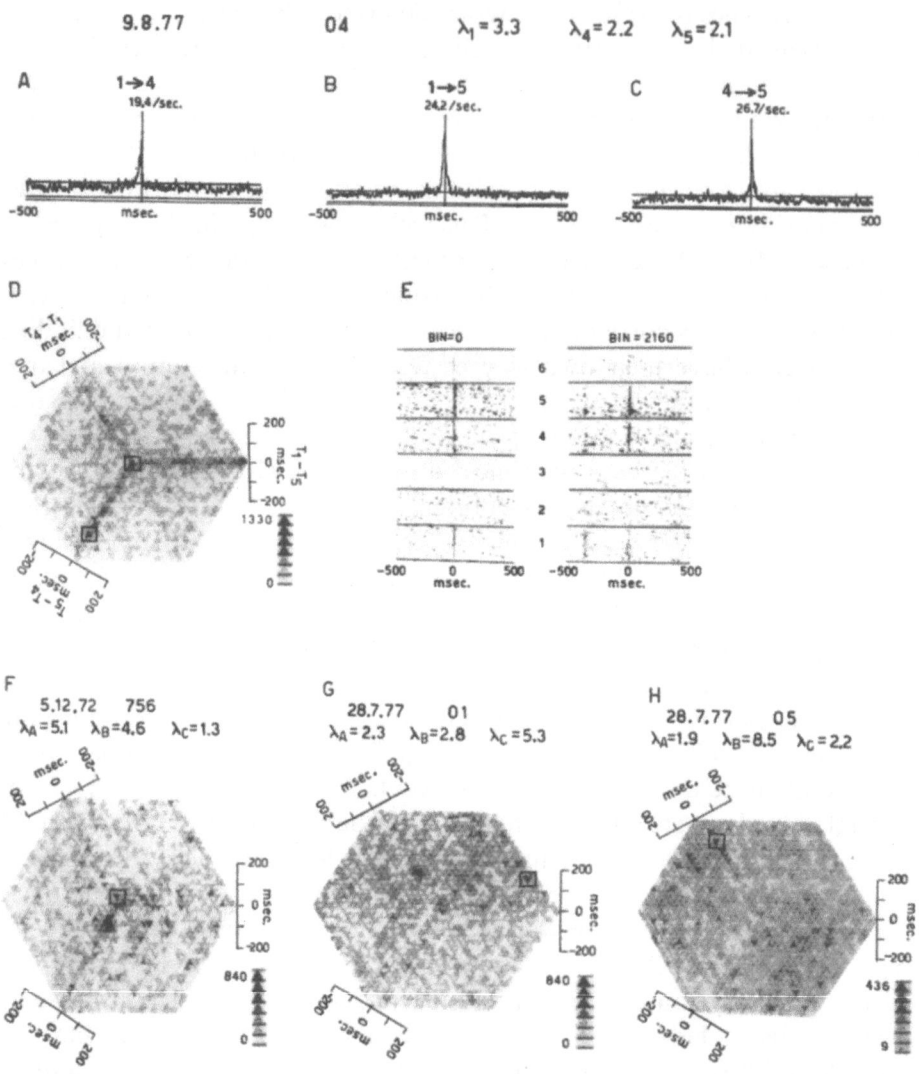

Fig. 3 A–H. Spatio-temporal organisation of activity. **A, B** and **C** The three possible cross-correlograms obtained from cells 1, 4, and 5 in a group of six cells. No exceptional relations are seen; **D** The three-cell density ("snowflake") indicates that besides the three common inputs there is also a preferred arrangement in time, as indicated by the two dark bins enclosed by *rectangles*; **E** A dot display of the firing times of all six cells is triggered on every occurrence of the preferred sequence. *On the left* are seen the cases in which cell 1 fired at high rate 25 ms after cells 4 and 5 fired together. *On the right* are shown cases in which cell 5 fired at high rate shortly after cell 4 fired if cell 1 fired 375 ms earlier; **F, G,** and **H** Other examples of complex spatio-temporal sequences of firing. Significant bins are enclosed by *rectangles*. (With permission modified from Abeles 1982; his Fig. 22)

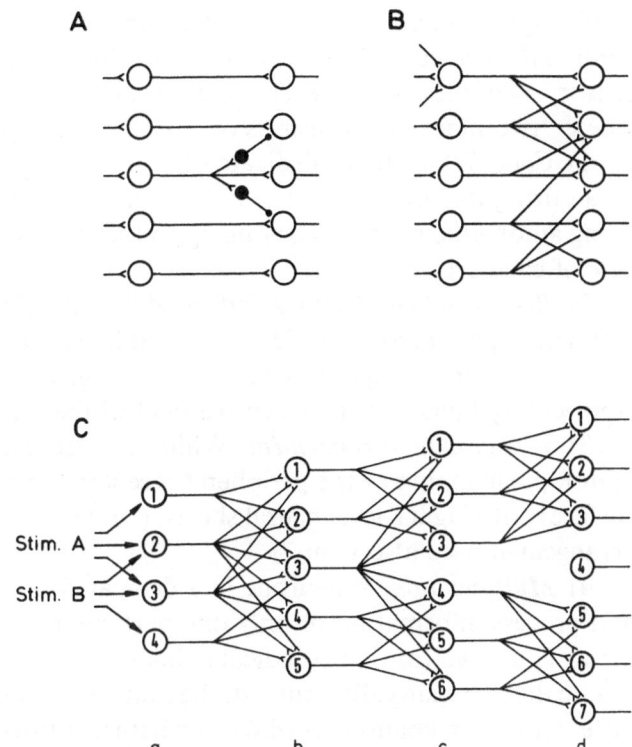

Fig. 4 A–C. Different types of neural chains. **A** Dedicated lines established by one-to-one connexions with superimposed lateral inhibition; **B** "Synfire-chain" with different numbers of synaptic connexions from pre- to postsynaptic neurones; **C** Separation of stimuli affecting overlapping sets of neurones. (With permission modified from Abeles 1982; **A** and **B** from his Fig. 23 B and D; **C** from his Fig. 31)

ms, as often observed by Abeles, comply with the requirement of relatively close synchrony which would make a "synfire chain" operative; for this problem see Sect. 2.8.2). Abeles (1982) envisages his "synfire chains" as very flexible dynamic functional entities, because very probably such a chain of synchronously firing neurones is activated only occasionally, whereas the cells are normally excited from other sources. The chain is therefore no rigid structure which is always traversed by a fixed pattern of activity, but a dynamic entity that may be turned on or off according to the spatio-temporal pattern of activity in the cortical network. "This concept of dynamic formation and dissolution of functional chains is similar to the 'cell assemblies' postulated by Hebb thirty years ago, except for the requirement that the transmission along the chain is secured by the synchronous firing of sets of cells" (Abeles 1982; p. 65).

From this starting point, Abeles (1982) constructs a far-reaching concept that will be reproduced here in some detail. To emphasise the novel features of his "synfire chain" hypothesis, he compares them with those of the conventional "dedicated line" concept (see Fig. 4 A) in which, for example, sensory neurones follow each other from the periphery up to the cortex. These "two modes of transmission" are now compared.

1) *Code*. Whilst in the dedicated-line arrangement the simplest code is the mean firing rate of a neurone, the code in the synfire chain is synchrony of firing among the neurones belonging to the same set (i.e. *not the mean firing rate* of each cell).

2) *Receptive field*. Along a dedicated line, the size of the peripheral receptive field is preserved or could be reduced by lateral inhibition (see Fig. 4 A). By contrast, in the synfire chain, the divergence-convergence structure of connectivity blurs out the receptive field of the individual neurone.

3) *Topographic arrangement*. Whilst the dedicated lines preserve the somatotopic mapping of the peripheral receptor surface (retina, cochlea, body surface) onto higher-level neural sheets, the synfire chain tends to mix up this arrangement (see also item 7).

4) *Multifunction*. A neurone in a dedicated line can be used to transmit and process information of only one particular type, e.g. about hair movement. Thus, two (or more) dedicated lines must not cross over if they are not to lose their specialty. By contrast, the same neurone may participate in many different synfire chains according to whether it fires synchronously with one set of neurones or with another (see Fig. 4 C). This is a very important property of neurones in synfire chains, because many neurones in the central nervous system receive polymodal input from various sources (see Sects. 2.9.1, 4.12 and 5.6).

5) *Immunity to damage*. If, in a given dedicated line, an individual neurone dies, then the entire line becomes useless, whilst the synfire chain is not so sensitive to loss of some neurones here and there.

6) *Parallel processing*. Since the dedicated lines must not cross over, a separate line must exist for each type of processing, which enables the simultaneous activation of many processes with little interference between the processes. The synfire chains can be activated in parallel only if they employ separate sets of neurones.

7) *Localisation*. Whilst sensory information in the periphery is transmitted through subsequent stages, each being localised in different anatomical structures, sets of neurones in a synfire chain are likely to be distributed in a quasi-random mode.

Abeles (1982) points out that, although *dedicated lines* and *synfire chains* were opposed as distinct categories, a whole spectrum of intermediate structures may exist. Thus, adding an increasing number of collaterals to a dedicated line structure (Fig. 4 A) eventually yields a synfire chain (Fig. 4 B).

Later Abeles (1982) discusses synfire chains with respect to some further key words such as "versatility" and "limitless variety" of the functional cell assemblies constituted by correlations, "memory traces" and "association", "interference" and "fine (sensory) discrimination" associated with synfire chains. The last feature is of some interest for the discussion later on. Figure 4 C schematically shows how two stimulus patterns excite a group of four cells (cell layer a), two cells being excited by both stimuli (polymodality). The stimuli may be separated (discriminated) in subsequent cell layers (b through d) by appropriate connexions and neuronal discharge patterns even if both stimuli are presented together. If each stimulus causes synchronised discharge in the cells affected by it, but in a way desynchronised with the synchronised pattern induced by the other stimulus, the synchronisations take different routes through the subsequent stages and are hence separated.

Although the basic elements of Abeles' scheme, i.e. the "synfire chains", appear simple or even simplistic, the synthesis of higher brain operations from these basic elements is not. This might erect barriers to intuitive understanding, in much the same way as the conceptually difficult transition from single-cell concepts to those of higher brain functions. But the starting point has changed because Abeles' basic elements are no longer single cells whose functions are more or less rigidly defined *anatomically* by their topographical positions and their input-output organisation, but rather versatile, dynamically changing assemblies of cells which are defined *functionally* through their spatio-temporal activity patterns. Certainly these assemblies are also connected by (at least indirect) synaptic contacts, but these connexions are thought to be modifiable through processes of synaptic learning which may depend exactly on those synchronisations which functionally define the given synfire chain.

1.3.2 A Correlation Theory of Brain Function: Synaptic Modulation

Christoph von der Malsburg (1981, 1983, 1985) has developed a theory in which the fine structure of neuronal discharge patterns, in particular synchronous firing of different neurones, plays a key role. The starting point was a critique of some current brain theories. In his 1985 paper, von der Malsburg writes that, according to many current views, nervous signals are interpreted in the code of mean discharge rate (see Abeles' item 1). The ensuing signal then varies on the time-scale of tenths of a second which is typical of global brain processes. Over short times, however, the structure of the neural network is commonly considered as fixed. On these views, the instantaneous state of the brain is therefore completely described by the array of firing rates of all neurones. Von der Malsburg then continues by suggesting that brain states may be considered as semantic symbols. On the currently dominant view of brain function, these symbols can be decomposed, without loss of structure, into semantic atoms, which are supposed to correspond either to single neurones or to small disjoint sets of them. That would imply that each neurone would bear a bit of symbolic meaning. On this view, then, the dynamically changing semantic symbol in the brain is built up additively from the basic symbols represented by the activity of all the active neurones.

One of the problems with the above semantic symbols is that they are represented by non–ordered sets of neurones which are not hierarchically structured. Thus, there is an infinite number of possibilities of decomposing the semantic symbols into subsymbols represented by subsets of cells. This causes ambiguity as to the specific arrangement of subsymbols in symbols. For example, a particular pattern in the visual field can and is usually decomposed into features which have a specific spatial organisation making up the pattern. This syntactical organisation is missing in the above way of representing semantic symbols by non–ordered sets of neurones. Von der Malsburg (1981, 1985) proposed to reduce or eliminate this ambiguity by introducing syntactical structure by way of additional variables representing the binding or grouping of elementary symbols into subsets. This may be achieved by way of temporal correlations between neural activity patterns. Von der Malsburg (1985) argues that correlations (synchrony of discharge) "... have a strong influence on nervous dynamics, since neurons essentially are coincidence detectors" (1985; p. 704), because synchronous inputs more effectively excite a postsynaptic cell than uncorrelated inputs. Correlations could therefore be processed by neural networks "as signals in their own right". They could be initiated in the network from an external source and then be propagated as suggested by Abeles (1982).

For the correlations to be controlled and stabilised, they should have an influence on the strengths of synaptic transmission between neurones. For this purpose, von der Malsburg proposes a mechanism that appears to be truly new because it has a fast time course: "*synaptic modulation*". He emphasises that this synaptic modulation must not to be confounded with synaptic plasticity, which in his definition represents processes, which act on a much longer time scale (see also below, Sect. 1.3.3). In synaptic modulation, the excitatory synaptic coupling between a pre- and postsynaptic cell is envisaged to change rapidly depending on the pre- and postsynaptic signals. This scheme, though applied to synaptic modulation, harks back to the idea of Hebbian modification, which is otherwise used in "conjunction theories of learning" (see Kohonen 1984).

The strength of a monosynaptic connexion is assumed to be modifiable, on two different time scales, by correlated firings of the pre- and postsynaptic cells. On a long-term scale (decades or permanent) the synaptic efficacy can be strengthened by such correlated activity; on a short-term scale (of the order of seconds or less), the synaptic efficacy can be in a resting state when there are no pre- and postsynaptic signals, or it is strengthened by correlated signals, or weakened by uncorrelated signals. Thus, correlations between pre-

and postsynaptic signals which naturally arise from the synaptic connexion reinforce each other in a positive feedback loop, "until a globally stable structure is reached" (von der Malsburg 1981; p. 25). Synaptic modulation is conceived as a basis for rapid pattern formation in the brain. It results from an instability inherent in the positive feedback loop between correlations and synaptic strengths. If two cells are linked with each other via different synaptic links, for instance a direct excitatory synapse and an indirect excitatory link (via interneurones), these two pathways may cooperate with each other. However, this cooperation ceases if the two pathways produce delays too different to ensure near-synchronous pre- and postsynaptic activity. In order to control the positive feedback, i.e. to ensure global stability, competitive and inhibitory mechanisms are needed. These processes then permit that the globally stable structures, which can be interpreted as "topological networks" or "topological connexion patterns" between neurones, can rapidly be (re-)constructed "on the fast time scale of thought processes" (von der Malsburg 1981; p. 29). These dynamics may drive two cells and their signals "away from a structureless or uncorrelated state" (1981; p. 25).

These notions are fairly abstract as long as they are not visualised, e.g. by computer simulations. These have partially been performed (von der Malsburg 1985) or are still underway. It is also useful to explain their significance in regard to some well-known problem, e.g. of sensory physiology. Von der Malsburg (1985) outlined the potential power of his theory on an unsolved problem, viz. the invariant pattern recognition problem. This problem is the following. Any object can be projected onto the retina in a huge variety of ways, characterised by different size, position, orientation, and so forth. If this object is always to be recognised as the same, the brain must organise an invariant representation of it. As pointed out by von der Malsburg (1985), invariant pattern recognition is an outstanding unsolved problem not only for the biological sciences but also for engineering.

Von der Malsburg then investigates two attempts at solving the problem: template matching and extraction of invariant features. The first is immediately discarded as impractical (too many templates would be required). The feature extraction model is explained on a simplified three-layer version of Rosenblatt's perceptron. This scheme (see Fig. 5) consists of three layers, S, $A^{(1)}$, and $A^{(2)}$, where S represents the sensory surface of the retina, and $A^{(2)}$ corresponds to a layer of cells specifically responding to certain subpatterns or features in S irrespective of their location on the retina. The projection between S and $A^{(2)}$ is accomplished by a complicated network located in $A^{(1)}$. This projection is such that a cell in $A^{(2)}$ responding to a feature α receives its excitatory input from a class of corresponding cells in $A^{(1)}$ responding to the same feature but in fixed locations. Hence, when a specific pattern is projected onto S, all those cells in $A^{(2)}$ are activated which respond to particular features of the pattern. In this way, the response in $A^{(2)}$ becomes invari-

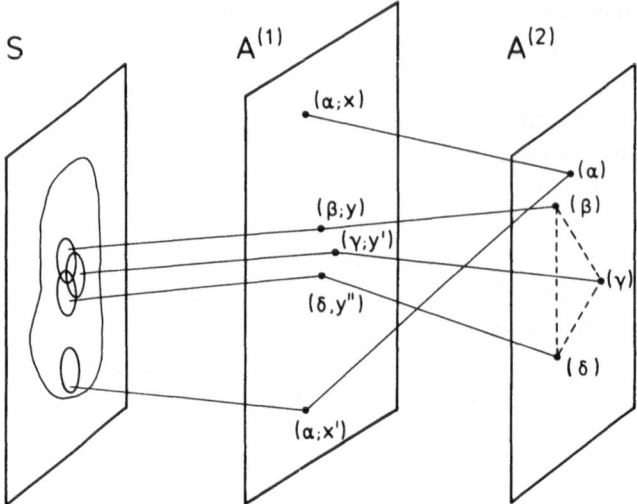

Fig. 5. How to disambiguate Rosenblatt's model for invariant pattern recognition. Cells in layer $A^{(2)}$ respond to features irrespective of their position on the sensory surface S. For each feature cell α there is a class of cells (α, x) in $A^{(1)}$, one cell for each position x. Cell α receives connexions from all cells (α, x), and can be fired by one of them. In this way, an activity pattern in S activates a set of feature cells in $A^{(2)}$. This set, by construction, is invariant with respect to the position of the sensory pattern. However, it is ambiguous. The ambiguity can be reduced if cells in $A^{(1)}$ encode their neighbourhood relationships by signal correlations (produced by common retinal origin or by short-range connexions). The correlations are propagated to $A^{(2)}$, where they activate a connexion pattern (*dashed lines*) which encodes the topology of $A^{(1)}$. (With permission from von der Malsburg 1985; his Fig. 8)

ant to translation. But "unfortunately this response is ambiguous, because many patterns of different structure may be composed of the same set of local features, arranged differently" (von der Malsburg 1985; p. 709).

The above approach suffers from the fact that all the information about spatial relationships between the features of the pattern is lost along with position information on the way from $A^{(1)}$ to $A^{(2)}$. Von der Malsburg (1985) then resumes his earlier proposal that the syntactical structure of neighbourhood relationships within $A^{(1)}$ could be encoded as temporal correlations between neural signals. For instance, the set of cells labelled $(\beta; y)$, $(\gamma; y')$ and $(\delta; y'')$ in Fig. 5 corresponds to a set of features (at a roughly specified location in S), whose spatial relationship is encoded in temporal correlations among the corresponding cells. These correlations could then be transmitted to $A^{(2)}$, where they would modulate the connexions between the corresponding cells, i.e. those labelled (β), (γ) and (δ) and connected by a dashed line. The pattern of synaptic connexions activated in $A^{(2)}$ would soon mirror similar connexions among cells in $A^{(1)}$ (but without position information) and would hence reflect neighbourhood relationships within $A^{(1)}$ and within

S, whereby the ambiguity present before in $A^{(2)}$ could be reduced or even eliminated.

Whether this really functions appears to depend on some conditions, the network needs to fulfil. First of all, the projection from $A^{(1)}$ to $A^{(2)}$ must be injective. Otherwise, if there existed many cell groups in $A^{(2)}$, which responded to the same, or even only similar, feature constellation, ambiguity would again arise and would have to be disambiguated by yet another layer. This situation might evolve if the CNS started off from a random and widespread connectivity between the layers which would not be specified during ontogenetic development. Hence, this situation could be avoided by the development of specific connexions (see next paragraph and Sect. 1.3.3). Secondly, the features need to be defined wide enough. Otherwise their neighbourhood relationship alone would not suffice to recognise an object, which may combine features not only in various positions and sizes, but also in arrangements differently distorted by varying perspectives. For the sake of a unifying theory, it would be desirable to clarify this feature issue with respect to Marr's (1982) "primitives". It would also be desirable to relate the three-layer model, as outlined by von der Malsburg, to known anatomical structures, in order to enable experimental testing, but this appears to remain wishful thinking for some time.

As argued by von der Malsburg (1983), some further problems could also be solved by the above theory of dynamical networks: figure – ground separation and combinatoric buildup of feature detectors of high levels from those of low levels. But also on the *long-term time scale of ontogenetic development* of, say, retinotectal or retinocortical projections, correlated or synchronised firing patterns of neurones are supposed to play an important part (von der Malsburg 1983). These topographical projections are proposed to result from two basic mechanisms. A global mechanism provides for the gross projection of retinal fibres to the correct general region of the tectum; in this process, gradients of chemical markers could play a role in guiding the fibres. The second mechanism is local and establishes the precise topological projection. It would work in almost the same way as the short-term establishment of neural assemblies outlined above, supplemented by short-range interconnexions between neighbouring cells in the retina as well as between neighbouring cells in the tectum, or the visual cortex. Retinotectal or -cortical fibres originating from neighbouring retinal cells would cooperate with each other because – due to common inputs – their firing patterns would be correlated. They would tend to make strong connexions with neighbouring target cells which in turn would be correlated, firstly by the correlated input, and secondly by their mutual interconnexions. This idea is consistent with some recent experimental results reviewed in the section on synaptic plasticity (1.3.3).

It is merely mentioned here in passing that Francis Crick (1984) has used some of von der Malsburg's ideas and notions to develop his "searchlight hypothesis". This outlines a concept of the function of the thalamic reticular complex in attentive scanning of visual inputs for particular combinations of features. Here, too, "Malsburg synapses", modifiable on a short time-scale by coincident neuronal activity, are basic elements in establishing neural cell assemblies whose constituent subsets represent different features.

It is also noted in passing that further, more or less abstract, concepts have been proposed, which implicitly or explicitly rest on the assumption of synchronised, or otherwise correlated, discharge among sets of neurones, which may even reverberate in closed loops (Eccles 1973; Sejnowski 1976; Edelman 1979). The idea of self-excitatory reverberatory circuits has been attractive and influential, particularly as a possible basis for short-term memory (see Tsukahara 1981). But the available experimental evidence for their existence is sparse, an exception being circuits between deep cerebellar and pre-cerebellar nuclei (Tsukahara 1981). Sejnowski (1976) presented a general mathematical theory in which he forwarded the "hypothesis that information is coded as correlations between spike trains and processed in a collection of neurons by correlations between membrane potentials" (p. 203). This form of information coding and processing was called "ensemble correlation". It may have several advantages: Firstly, the information capacity of available channels is used more efficiently; secondly, detailed sensory information can be coded more reliably by sensory afferents in apparently noisy background activity.

1.3.3 Synaptic Plasticity

In the models of both Abeles (1982) and von der Malsburg (1981, 1985) changes in the strength of synaptic connexions play a key role.

Synaptic modifiability has been intensively investigated at the neuromuscular synapse and at central synapses. A number of different processes acting at different time-scales have been distinguished. From studies on frog and mammalian neuromuscular junctions, the *enhancement* of synaptic efficacy can be divided into a few components with different decay time constants, i.e. *facilitation, augmentation* and *potentiation*. Facilitation can be described by two time constants of the order of 35–50 ms and 250–300 ms (Mallart and Martin 1967; Magleby 1973), augmentation by a time constant of the order of several (3–8) seconds (Magleby 1973; Magleby and Zengel 1975, 1976a, b), and potentiation by a time constant of several (2–3) minutes (Magleby and Zengel 1975, 1976a, b). Similar time constants for augmentation and potentiation were obtained for the post-tetanic decline of the rate of miniature endplate potentials (Rahamimoff et al. 1980). Synapses of the perforant pathway on granule cells of the rat fascia dentata also have decay time constants for augmentation between about 3 and 7 s, and for potentiation between about 30 and 80 s (McNaughton 1978, 1980). (Interestingly, nearly identical time constants were found at synapses of Aplysia L10 neurones; see Kretz et al. 1984.) Moreover, McNaughton (1980) found facilitation of brief duration. In intact preparations, the facilitation lasted for a time of the order of 100–200 ms when the lateral perforant path was stimulated, synapses of the medial perforant path showing only depression with a "relative facilitation" superimposed on the depression for short intervals (McNaughton 1980; his Fig. 5 A). In addition, in the hippocampus and other central nervous structures, there is a *long-term potentiation* (LTP; lasting for hours to permanent), which apparently is not found in synapses originating from single presynaptic fibres, but requires synchronous activation of several to many parallel synapses at the same cell (see McNaughton 1980). Whereas, at present, facilitation is the process, which von der Malsburg's modulation is most akin to (although it does not require postsynaptic discharge), LTP is commonly assumed to be one possible basis for medium- or long-term memory (see Lynch and Baudry 1984).

Only a brief overview of synaptic plasticity related to the importance of (correlated) neuronal firing patterns is given here (reviews: Tsukahara 1981; Thompson et al. 1983). It exists in the cerebral cortex, particularly the hippocampus (brief review: Bliss 1979; Eccles 1979, 1983; Voronin 1983), the cerebellar cortex (e.g. Ito et al. 1982, with further references; but see Racine et al. 1986) and other structures, such as the interpositus and vestibular nuclei (long-term potentiation: Racine et al. 1986). According to the site of action, the plastic phenomena may be divided into four categories: (a) homosynaptic and transsynaptic; (b) homosynaptic and presynaptic; (c) heterosynaptic and transsynaptic; (d) heterosynaptic and presynaptic. These notions may best be explained by examples. It should be noted from the outset that protagonists of these mechanisms tend to commit themselves to one or the other, whilst actually several mechanisms might be operative in combination.

Many examples of long-term potentiation (LTP) have been reported, particularly in the hippocampus (see above). This plastic change may be observed within *one synaptic channel*, which is usually activated experimentally by stimulating one particular input system (almost always consisting of many presynaptic fibres). This type of synaptic plasticity is commonly referred to as "homosynaptic" (see Saggau and ten Bruggencate 1987). In some cases, postsynaptic discharge was reported to be necessary for the effect to appear (e.g. Baranyi and Fehér 1981, working on the cat motor cortex). The temporal contiguity of activity in pre- and postsynaptic cells is consistent with Hebb's concept of synaptic modification (Hebb 1949; see also below). This type of plasticity would then fall under category (a) (preceding paragraph). Another important point is that it appears to be collateral-specific, i.e. occurs specifically at the terminals of some but not all of the collaterals of a neurone (N. McNaughton and Miller 1986). There are manifold possible molecular mechanisms that could underly Hebbian synaptic modification (for a brief overview see, e.g. Kohonen 1984; Wigström and Gustafsson 1985; see below). Such mechanisms are hypothesised to play a role in "conjunction (or correlation) theories of learning". Following Kohonen (1984), the importance of correlations between pre- and postsynaptic neuronal firing patterns can be understood along the following lines.

The presynaptic and postsynaptic neuronal activities are regarded as stochastic processes. The activity of the postsynaptic (output) neurone, if it exceeds a certain intensity level, is defined as an event Y with a probability of occurrence $P(Y)$; similarly, the activity of one presynaptic input line, if it exceeds a certain level, is defined as a stochastic event X_i with probability $P(X_i)$. The probability of both events happening concurrently is given by the joint probability $P(X_i, Y)$. For simplicity, Kohonen now makes the important assumption that X_i and Y are – for practical purposes – statistically independent, because the output Y depends on many other input events; thus

$$P(X_i, Y) = P(X_i) P(Y). \tag{1}$$

In the case of strong connexions between the pre- and postsynatic cells, this assumption would not be warranted, of course.

Kohonen then continues to formulate a mathematical description of the conjunction hypothesis of synaptic plasticity in statistical terms. It may be assumed that the expectation value of the synaptic efficacy, μ_i, denoted by $E(\mu_i)$, changes in time in proportion to $P(X_i, Y)$, and to a free parameter α'_i:

$$dE(\mu_i)/dt = \alpha'_i P(X_i) P(Y) \tag{2}$$

A simplified notation may be used as follows. If the input activity is denoted by ξ_i and the output activity by η, and if μ_i is written for $E(\mu_i)$, Eq. (1) may be rewritten as:

$$d\mu_i/dt = \alpha_i \xi_i \eta, \tag{3}$$

where α_i is called the "plasticity coefficient" of the synapse. This formulation neatly expresses the basic assumption that, in the conjunction hypothesis of learning, a synapse changes *in proportion to the correlation of the input and output signals* (Kohonen 1984).

McNaughton et al. (1978) emphasise that, in the rat fascia dentata, postsynaptic discharge is no prerequisite for long-term potentiation to appear, which is in contrast to Baranyi and Fehér's (1981) results. This type of synaptic plasticity might therefore be assigned to category (b). Even in this homosynaptic case, however, *cooperativity* of many input fibres is required, insofar as a minimal number of such fibres have to be activated synchronously in order to generate long-term potentiation (McNaughton et al. 1978). Therefore, it is doubtable whether the term "homosynaptic" can be used in a strict sense.

In recent years, the "heterosynaptic" cross-influence exerted by parallel inputs, or the *cooperative interactions between parallel inputs*, on plastic changes has attracted much attention and opened an exciting new field of research. For example, Baranyi and Fehér (1981) found, for neurones of the cat motor cortex, that time-locked repetitive stimulation of two convergent excitatory pathways produced a long-lasting facilitation of the excitatory postsynaptic potentials (EPSPs) evoked by one of them. The important point in these findings was that, for the effect to appear, it was necessary that the other pathway elicited a postsynaptic spike, thus requiring interaction of the two presynaptic as well as the postsynaptic signals. This would then be an example of category (c). It is noted in passing here that except for heterosynaptic facilitation, long-lasting heterosynaptic depression has also been described (Lynch et al. 1977). Theoretically, such a possibility of heterosynaptic cooperativity was considered by Kohonen (1984). If two presynaptic fibres converge on a postsynaptic neurone at a sufficiently close distance from each other, they may permanently facilitate each other, for instance through conformational changes of the membrane proteins. Such changes could be induced by correlated inputs. In analogy to Eq. (3), we could formulate such a situation by the two equations:

$$d\mu_i/dt = \alpha_i \xi_i \xi_j, \tag{4a}$$
$$d\mu_j/dt = \alpha_j \xi_i \xi_j, \tag{4b}$$

where the indices i and j refer to the two inputs. This formulation suggests that the synapses from the two inputs to the postsynaptic cell are facilitated *in proportion to the correlation between the input signals*. Such a hypothesis, including possible membrane mechanisms, has recently been outlined more concretely by Finkel and Edelman (1985) to account for Edelman's (1979) group selection theory. Interestingly, the authors combined a short-term synaptic modulation (as required by von der Malsburg, Sect. 1.3.2), taking place in the postsynaptic neurone, with a long-term presynaptic change. The former was supposed to depend on correlated activity in presynaptic inputs, and one possible molecular mechanism was also proposed and is illustrated in Fig. 6. Since this postsynaptic mechanism would depend on the synaptic depolarisation effected by the correlated inputs, the two (or more) inputs should be situated close to each other in the dendritic tree of the postsynaptic cell in order to interact optimally. Perhaps this requirement is one reason for the fact that this possibility of presynaptic interaction has so far seldom been considered theoretically and rarely found experimentally.

Some experimental evidence for heterosynaptic cooperativity has recently been obtained. For example, McNaughton et al. (1978) reported that, above a certain threshold, long-term potentiation of perforant path synapses in rat fascia dentata increased with the number of stimulated presynaptic fibres (see above), and was stronger with concurrent conditioning stimulation of the

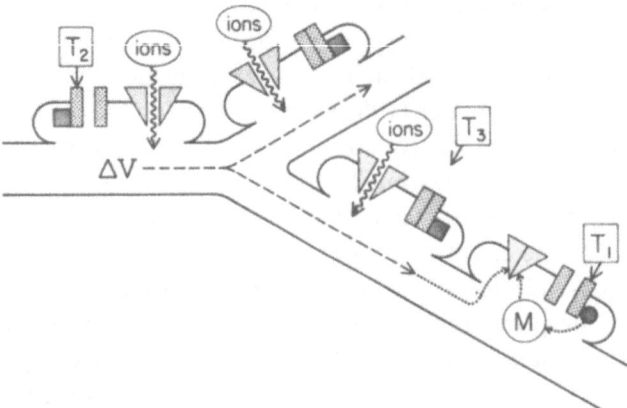

Fig. 6. Operation of one possible mechanism for the postsynaptic rule. Schematic of four synapses on a bifurcating dendritic arbor. *Shaded rectangles* represent open or closed receptor-operated channels (ROCs); *shaded triangles* represent voltage-sensitive channels (VSCs). Transmitter (T_1) binding to the lower right synapse activates the receptor-associated structure (*shaded square* = inactivated; *shaded circle* = activated) and begins a biochemical cascade producing modifying substance, *M*. *M* modifies local VSCs in the appropriate voltage-dependent state. Binding of transmitter, T_2, at the leftmost synapse leads to a local postsynaptic potential (PSP) (ΔV), which is electrotonically conducted to other synapses (*dashed line*), changing the state of their VSCs. The voltage change reaches the lower right synapse (*dotted line*) at a time when the concentration of modifying substance is high, leading to increased modification of the VSCs and a change in the channel population distribution. The other synapses are not modified because the relative timing of local transmitter binding (e.g. T_3 has not yet bound) and heterosynaptic inputs do not meet the permissive conditions. (With permission from Finkel and Edelman 1985; their Fig. 1)

medial and lateral perforant paths than with separate stimulation of either pathway alone. Lee (1983) reported that a greater degree of long-term potentiation resulted when adjacent afferents to the hippocampal stratum radiatum were co-actively stimulated, as compared to that observed with activation of a single pathway alone. This *cooperativity* did not depend on an increased postsynaptic discharge during conditioning, as evidenced by the absence of greater long-term potentiation following orthodromic and antidromic co-activation of the postsynaptic CA1 pyramidal cells [category (d)]. Lee (1983) proposed that this cooperative influence occurs in the vicinity of the synapse potentiated. Also, Levy and co-workers (Levy and Steward 1983; Levy et al. 1983) showed that responses in the rat dentate gyrus elicited by stimulation of the sparse contralateral projection from the entorhinal cortex could be potentiated or depressed on a long-term scale when conditioned by strong stimulation of the corresponding potent ipsilateral pathway. For potentiation to occur, the crossed pathway had to be activated during a critical temporal window (with respect to the conditioning ipsilateral stimulation). On the other hand, depression occurred if the crossed pathway was inactive or was activated at a time outside the critical time window. The critical time window coincided with or slightly preceded (up to about 20 ms) the conditioning stimulation. Whether potentiation or depression of the crossed projection occurred was therefore determined by its own state of activity (Levy and Steward 1983). The ipsilateral system must be strongly activated to exert its conditioning effect, which implies that activation of individual spine synapses is not sufficient. Levy and Steward (1983) suggest that there is some "retrograde" interaction between a process initiated within the main dendritic shaft and individual spines (see Fig. 6). This dendritic process would begin at a threshold depolarisation of a sufficient number of spines. Note that such a process would then require synchronous activation of many converging synapses, though in the conditioning pathway. Also note again the possible importance of cell architecture, since the synaptic interaction would have to take place between synapses on the same dendritic tree (defined by a stem shaft), or branches thereof. Levy and Steward (1983) also discuss some mechanisms possibly involved in this kind of potentiation, the details of which cannot be reproduced here (see also Voronin 1983; Eccles 1983). One particular point deserves our interest, however. Behavioural trace conditioning can be optimally effective for intervals between the conditioned and unconditioned stimuli up to a few hundred ms or even longer. But at these long intervals, paired conditioning in the experiments of Levy and co-workers produced depression at the synaptic level. If the synaptic type of conditioning were to underly the behavioural type, a delay circuitry would be required (such as a reverberating loop) that would allow a response to circulate around the limbic system several times. This loop would preserve the activity elicited by a conditioned stimulus during trace conditioning until the conditioned and unconditioned stimuli could temporally converge upon synapses and cells involved in associative potentiation. (It is noted in passing that there is also some evidence for heterosynaptic depression; see N. McNaughton and Miller 1986.)

As noted above, the precise molecular mechanisms underlying long-term potentiation (LTP) are not yet known. Models would have to account for two features: the *input specificity* (long-term potentiation occurs for those inputs that are active during the "print" phase), and *cooperativity* (i.e. the influence exerted by parallel inputs; see above). After models for each single feature had been proposed, Wigström and Gustafsson (1985) recently proposed a combination of the two mechanisms. A single volley in a presynaptic pathway is suggested to release transmitter(s) which combine with two types of postsynaptic receptor. One of the activated receptors rapidly opens ionic channels which predominantly pass sodium and potassium ions entailing an EPSP. The other transmitter–receptor interaction has a more delayed and prolonged effect that appears conditional on a sufficient depolarisation of the subsynaptic membrane. A significant part of the current through these ion channels is carried by calcium which then induces processes leading to LTP. Wigström and Gustafsson briefly reviewed experimental evidence that is in accordance with the two transmitter–receptor processes of different time-course and channel specificity, and mention that these processes may also occur in *spinal α-motoneurones*. The authors then add some interesting remarks as to the possible control of the potentiation process in nerve cells. As outlined above, LTP is supposed to be critically dependent on the local potential of the subsynaptic membrane. This potential is determined by both the excitatory and inhibitory inputs to that region. Wigström and Gustafsson (1985) consider two different cases. Firstly, the active excitatory input to a cell could be distributed rather evenly over the dendritic surface. In this case, the temporary (< 100 ms) *removal of inhibitory input* to a specific region would allow a local potentiation of other specific inputs. Secondly, wide electrotonic spread (which depends on cell architecture) could prevent an effective isolation of a subsynaptic membrane as envisaged in the previous case. In the latter case, the local potential would largely be governed by the overall depolarisation of the neurone, whereby the conditions for potentiation would resemble those in the Hebb model. Wigström and Gustafsson (1985) emphasise that the predominance of one or the other situation will not depend on the mechanism for LTP induction itself, but will be determined by properties of the cells involved, such as the electrical isolation of the subsynaptic membrane and the location of the inhibitory synapses. These ideas again stress the role of precise cell anatomy, and, perhaps more importantly, the possible role of inhibitory mechanisms in this particular context. Excitatory and inhibitory inputs to many types of cells (for instance, spinal α-motoneurones) are not in general uniformly distributed over the cell surface so that, in other contexts, the notion of "strategic sites" of inhibitory inputs was coined long ago.

As with models of cerebral cortex, models of cerebellar function (Marr 1969; Albus 1971; Eccles 1977) were based on the modifiability of synapses from parallel fibres to Purkinje cells (or even stellate and basket cells: Albus), in order to devise schemes of cerebellar learning. An important difference between Marr's and Albus' theories is that the former assumed the above synapses to increase in strength with concomitant activation of climbing fibres, whereas the latter postulated the reverse. There is now accumulating experimental evidence for Albus' alternative (Ito et al. 1982; brief review: Andersen 1982; but see Racine et al. 1986). Ito and Kano (1982) demonstrated that concomitant repetitive (low-frequency) stimulation of climbing fibre and parallel fibre inputs to extracellularly recorded Purkinje cells (the climbing fibres being activated by stimulating the inferior olive 4–12 ms prior to parallel fibre activation) decreased the synaptic efficacy of parallel fibre input for long periods (at least 1 h; note that this is a heterosynaptic influence). It is noteworthy that, although a spike in a climbing fibre always elicits a spike in its Purkinje cell, the synaptic modification is *not* dependent upon the appearance of output spikes. Rather the long synaptic depolarisation (lasting several hundreds of milliseconds), which is evoked by climbing fibres in Purkinje cells, appears to be the determinant (Ito et al. 1982). This long depolarisation can be reduced or suppressed by inhibitory input to the Purkinje cell from basket and stellate cells, whereby the depressant effect of climbing fibre stimulation on parallel fibre synaptic efficacy can be decreased or abolished (Ekerot and Kano 1985). These authors proposed that the long depolarisation evoked by climbing fibre stimulation increases the calcium concentration in different dendritic branches contacted by the climbing fibre, the mechanism of action of the heterosynaptic influence on parallel fibre synapses thus being compatible with the model of Wigström and Gustafsson (1985) outlined above. It follows from this mechanism based on the long depolarisation that the climbing fibre activation may precede the parallel fibre activation by up to several hundred milliseconds in order to be able to exert its depressant effect, but the strongest effect is still exerted at short intervals (20 ms or so: Ekerot and Kano 1985). Ekerot and Kano conclude that, consequent on their findings, the Marr-Albus hypothesis on cerebellar learning has to be revised. This hypothesis held that *all the parallel fibre synapses* active in temporal contiguity with a climbing fibre impulse would be equally modified. Their finding that inhibition can depress the heterosynaptic plasticity effect of conjunctive climbing and parallel fibre stimulation suggests a more sophisticated mode of cerebellar learning. In this mode, the local organisation of excitatory and inhibitory synaptic inputs to the Purkinje cell dendritic tree will be of paramount importance, enabling the parallel fibre synapses to be modified differentially. That is, the strongest depression of the efficacy of parallel fibre synapses will occur in those parts of the dendrites subject to the strongest net excitation. It is important to note again the possible significance of the spatial organisation of excitatory input shaped by that of inhibitory input (for the importance of the spatial distribution of excitatory input to Purkinje cells see also A. Pellionisz 1979). So far, it has not yet been established that these synaptic modifications last long enough to account for motor learning. Another possible function of climbing fibres arises from their short-term conditioning effects on Purkinje cell responses to parallel fibre inputs. Ebner et al. (1983) and Bloedel et al. (1983) put forward the hypothesis that climbing fibres act to increase the responsiveness of Purkinje cells to mossy fibre inputs for some 100 ms or so after a climbing fibre spike. Interestingly, correlations between sets of climbing fibres activated by natural inputs may be of particular importance in this respect. These correlations may be furthered by electrotonic interactions between those inferior olivary neurones (Llinás et al. 1974; Sotelo et al. 1974), which project to circumscribed and functionally linked cerebellar cortical areas (Bloedel et al. 1983). These short-term functions of climbing fibres do not exclude their possible long-term effects mentioned above (Bloedel et al. 1983).

It may be noted in passing that Braitenberg (1967, 1977) proposed a model in which the cerebellar cortex is envisaged to work as a biological (stopwatch) clock in the millisecond range, and in which again synchronisations between climbing fibres due to electrotonic interactions in the inferior olive (see above) play a major role. However, as noted by Pellionisz (1986), this idea is burdened with the "problem of the demonstrably too short span of such timing" (p. 255). Indeed, in the cat Purkinje cells aligned along parallel fibres and thus getting common input are apparently not correlated over distances larger than ca. 100 μm (Ebner and Bloedel 1981 b).

Whereas the results on synaptic modifications reviewed above are not so much related to specific physiological functions, synchronised patterns of neural activity are also considered to play an essential role in the ontogenetic formation of topographic and functional maps in the central nervous system. Two examples are mentioned here.

In mammals, long-term modification of excitatory synaptic transmission from retinal afferents to cortical cells probably depends on electrical activity of the related cells in such a way that postsynaptic discharge is essential for the modification process (Singer 1983). In an early sensitive period of ontogeny, this modification may help to stabilise specific retinocortical connexions involved in orientation specificity and matching of projections from binocular receptive fields to common cortical target neurones. Singer and co-workers established "rules", which closely resemble those postulated by Hebb (1949). They can be summarised as follows. (1) The strength of excitatory synaptic transmission from corticopetal afferents to cortical target neurones increases if these inputs are active in temporal contiguity with the postsynaptic cells. (2) This strength decreases if the postsynaptic cell, but not the presynaptic fibre is active. (3) Irrespective of the amount of activity in the presynaptic fibres, plastic changes of synaptic efficacy do not occur when the postsynaptic cell is inactive (Singer 1983).

Singer (1983) argues teleologically to explain the necessity of the above modificatory processes. Animals with binocular vision need to develop cortical nerve cells with two corresponding receptive fields, one in each eye. The retinal loci, with which each such cell must be connected via afferent pathways must be precisely matched regarding their location on the retina. The problem faced by the organism is that the appropriate connexions in the mature system cannot be anticipated with any great precision, because retinal correspondence depends on the size of the eyes, the position of the eyes in their orbits and the interocular distance in the grown animal, and these parameters depend on epigenetic influences. This problem could be solved by some sort of functional labelling of the pathways originating from corresponding retinal loci. These afferents should convey identical activity patterns when the animal is fixating a target with both eyes. If these common activity patterns were used to induce Hebbian modifications in the connexions of the afferents with the common target cells, synaptic plasticity could be employed to selectively stabilise those connexions in the appropriate way. Hence, correlated activity in the two afferent pathways would be ideally suited to optimise the correspondence of binocular connexions (see heterosynaptic plasticity). Now, Singer (1983) argues that, if processes of synaptic modification were used for such purposes, selection of appropriate afferent–target connexions according to function could be successful only when the Hebbian processes are "gated". Selection should then occur only when the kitten is actually fixating a target with both eyes and not in all the many other instances in which the eyes are not properly aligned. "In the latter conditions, Hebbian modifications would lead to competition between the afferents from the two eyes and cause disruption rather than optimisation of binocular connectivity" (Singer 1983; p. 93). In other words, and this is readily grasped intuitively, for statistical reasons (large numbers of incoming fibres, high incidence of states of misalignment) a nonselective Hebbian mechanism would favour disorganisation rather than the development of orderly connexions.

After briefly reviewing the experimental evidence for the effect of nonspecific modulatory systems which increase cortical excitability and facilitate synaptic modification ("gating"), Singer (1983) presents a model of the gating mechanism. A critical prerequisite for the adaptive changes to occur appears to be a strong enough depolarisation of the dendritic membranes of the cortical target cells (see Fig. 7). This is reminiscent of models briefly reviewed above. Singer (1983) hypothesises that, again, the occurrence of an adaptive change could be the influx of Ca^{2+} ions through activated, voltage-dependent Ca^{2+} channels. He lists four aspects which render this hypothesis attractive. Firstly, it presents a basis for heterosynaptic plasticity and could explain the fact that many different non–retinal projections to striate cortex interfere with the Hebbian modifications. Secondly, free Ca^{2+} ions in the cytoplasm trigger a variety of biochemical processes. Thirdly, the membranes of developing neurones contain many Ca^{2+} channels. Fourthly, a heterosynaptic control of long-term synaptic plasticity (in excitatory synapses) has recently been demonstrated in the cerebellar cortex. Singer (1983) here refers to the work of Ito et al. (1982) mentioned above, and adds that activation of the climbing fibres produces a strong depolarisation of Purkinje cell dendrites which is sufficient to reach the threshold of dendritic Ca^{2+} channels. Note that this mechanism, which is based on a strong postsynaptic depolarisation, could also account for some of the results reported by Levy and Steward (see above). However, whereas the synaptic modifications investigated and discussed by Singer (1983) occur during a brief sensitive period in early ontogeny, those studied by Levy and Steward and others can also be elicited in

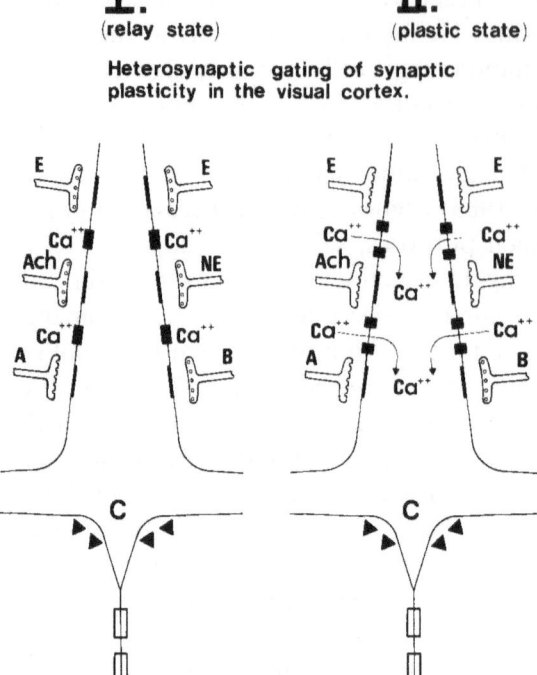

I. (relay state)   **II.** (plastic state)

Heterosynaptic gating of synaptic plasticity in the visual cortex.

Fig. 7. Heterosynaptic gating of synaptic plasticity in the kitten striate cortex. The specific synaptic connexion from an input (*A*) to a cortical cell (*C*) is hypothetised to operate in two states: a relay state (*I: left*) and a plastic state (*II: right*). For plastic changes to occur at the particular synapse, it is a necessary, but insufficient condition that both the presynaptic terminal and the postsynaptic cell be active (*I*). An additional condition is that neuromodulatory influences facilitate the signal transmission, such that a threshold is reached for the influx of Ca^{2+} (*II*). These influences are hypothetised to be cholinergic (*ACh*) and noradrenergic (*NE*) inputs, which have to work in conjunction to be effective; *E* unspecific excitatory synapses. (Courtesy of W. Singer)

mature organisms. The early ontogenetic synaptic modification can be modulated by various input systems, which could be necessary to define a plastic state. Among the modulating input systems are probably noradrenergic and cholinergic inputs (Bear and Singer 1986; see Fig. 7). A model of a self-organising neural network leading to ordered projections to the visual cortex was recently presented by Singer and co-workers (Frohn et al. 1987).

Fawcett and O'Leary (1985) recently emphasised the importance of electrical nerve cell activity for establishing topographical mappings, concentrating on retinotectal, retinocollicular or retinocortical projections but adding some interesting remarks on neuromuscular mappings. The processes underlying the establishment of the former ordered projections during a critical (sensitive) period are different in amphibians and fish than in reptiles, birds and mammals. But it appears that the basic mechanisms might be essentially the same. Fawcett and O'Leary propose that two such mechanisms are at work, one using chemical labelling of the projecting axons and their target cells, the other involving special patterns of electrical activity in the projecting axons. The first mechanism sets up a rough and imprecise topographic mapping, the second refines upon the former (see von der Malsburg; Sect. 1.3.2). This refinement would to a large extent rely on spatially ordered patterns of electrical activity in retinal ganglion cells; that is, Fawcett and O'Leary (1985) assume that neighbouring retinal ganglion cells would exhibit synchronous firing (due to their common inputs), whilst distant ganglion cells would not, and hence they would also not cooperate in establishing the appropriate topographical mapping. More concretely, the authors then consider two possible models of the way in which the impulse synchronicity might exert its action. The first is based on the action of the nerve growth factor (NGF), and may be explained as follows. Consider the retinotectal projection (Fawcett and O'Leary 1985; their Fig. 3). Two (or more) neighbouring ganglion cells discharging in a synchronised manner should terminate close to each other on each of their target cells in the tectum. Due to their synchronous firing they produce a large depolarisation in the postsynaptic cell which in response releases a large amount of nerve growth factor. Any fibre originating from another retinal locus and erroneously connecting to the same target cell would not fire synchronously with the former axons and would not cause such a large depolarisation whereby less trophic factor would be released. The trophic factor released is taken up preferentially by the synchronously firing axons, whereby their terminals are strengthened and increased in number, and the parent neurones survive. The erroneously projecting cell, however, will receive less trophic factor, retract its terminals and eventually die. Note that this model relies heavily not only on synchronous firing, but again on defined architectural relations between axon terminals and postsynaptic cells.

The second model makes less specific assumptions. The trophic factor would be released nonspecifically at the terminals of all the optic fibre and have the sole function of a ganglion cell survival factor. The competition for synaptic space on a tectal target cell would be governed by other factors, such that fibres with erroneous terminations would compete unsuccessfully for synaptic space, thus establish fewer synapses and take up less trophic factor. This would ultimately lead to the death of their retinal ganglion cells. The factors involved in the competition are not specified.

Fawcett and O'Leary (1985) then address the question whether similar mechanisms as proposed for the retinotectal (or equivalent) projections might be at work in other parts of the nervous system where topographical mapping takes place. They point out that so far there is little evidence bearing on this issue, but that electrical activity might be involved in the refinement of topographical connectivity patterns in many parts of the nervous system.

As examples, they treat two systems, the projection from the dorsal lateral geniculate nucleus to the visual cortex, and the connexions of α- and β-moto-neurones to their skeletal muscle fibres. The latter system is particularly interesting for things to come (Chaps. 2 and 3). At an early ontogenetic stage, when the skeleto-motoneurones establish their initial connexions with the muscle fibres, there is a phase of neuronal death, during which roughly half of the motoneurones die. Subsequently, there is a period of "pruning" of their terminal arbors. Electrical activity has been suggested to be involved in both of these latter processes. Fawcett and O'Leary (1985) argue that, despite there being clear differences between the control of these refining processes in the neuromuscular system and in the retinofugal projections, there may also be important similarities. But they stress that "the real analogy to the processes we have been discussing seems to be events which happen within a single muscle" (p. 205). The initial innervation of a muscle by a motor axon covers a much larger area than in the adult; and this crude mapping is only later refined to the adult pattern both by motoneuron death and by the elimination of polyneuronal innervation.

Thus, the establishment of topographical mappings appears to depend upon particular patterns of neuronal activity, possibly upon synchronisation between the firings of parallel elements converging on a postsynaptic cell. However, the precise structure of these firing patterns under somewhat natural conditions has still to be elucidated. The experimental patterns used so far seem to be rather unnatural because many parallel channels were artificially synchronised.

Whereas the above long-term modifications of synaptic efficacy can account for the enduring establishment of connexions, they cannot explain the fast buildup and destruction of functional cell assemblies as proposed by von der Malsburg (Sect. 1.3.2). However, as briefly mentioned above, Finkel and Edelman (1985) have already proposed a molecular mechanism for short-term synaptic modulation. Whether these short-term processes have a relation to the facilitation described at the neuromuscular junction (see above) and at rat dentate synapses by McNaughton (1978) remains to be established.

1.3.4 Summary

In the preceding sections we have reviewed a number of brain models based explicitly on correlations or, more specifically, on synchronisation between discharge patterns of neurones. The empirical content of these models, that is, the extent to which they are supported by experimental data, varies. Abeles' notion of "synfire chains" that led to a far-reaching concept of brain function originated from multi-unit recordings from cortical neurones and the application of the three-cell correlation technique. This method established the existence of correlated firings of different cortical cells, the time intervals of the correlated discharges extending over hundreds of milli-seconds. Explanation of such long intervals necessitated the assumption of very long chains of parallel sets of neurones which could be traversed by synchronised waves of firings ("synfire chains"). An essential feature of Abeles' theory is that synchronised firing of sets of neurones embodies ano-ther *code* above and beyond the commonly accepted code of mean firing rates (see App. G). This code makes use of correlated *fluctuations* of discharges of nerve cells around their respective (more or less fast-varying) time averages. Synchronised firings could be a means of establishing neural assemblies building up and dissolving on a fast time-scale. These assemblies, always being composed temporarily of many, though changing cell groups, display a huge versatility and, hence, capacity of information processing. An impor-tant contribution of von der Malsburg appears to be the idea of "synaptic modulation" based upon the manner and degree of correlation between neuronal firings which are assumed to change synaptic strength on a fast time-scale.

Synaptic strength has been shown to be modifiable at all levels of the nervous system, from the neuromuscular junction to central synapses in cortical and cerebellar structures. But these plastic processes vary in time-course and underlying molecular mechanism. The importance of particular firing patterns, synchronised or otherwise correlated, has also recently been hypothesised and, at least partially, supported. The synaptic changes caused by these firing patterns are usually long-term "plastic" rather than short-term "modulatory" changes as required by von der Malsburg's model. Synaptic plasticity may play a role in the ontogenetic "refinement" of topographic mappings (Fawcett and O'Leary 1985), the acquisition of skilled motor behaviour (learning), and the deposition of "engrams" (memory). The quali-ty of short-term "synaptic modulation" and its role in complex information processing by neural assemblies have still to be demonstrated, but this modu-lation is an interesting possibility to be considered also at other levels of the nervous system. Facilitation is certainly a phenomenon encouraging the search for short-term processes akin to modulation. Particular interest would receive heterosynaptic short-term processes which could be sensitive to syn-chronous or otherwise correlated input patterns (see Finkel and Edelman 1985).

The way in which synaptic strength is modified by neuronal activity is not yet clearly established. The Hebbian scheme of synaptic efficacy being chan-ged by concomitant activity of the pre- *and* postsynaptic cells has gained some experimental support, although the exact mechanism remains obscure, particularly with respect to the fine structure of pre- and postsynaptic activity that is of importance (what is "concomitant" or "simultaneous" activity?). Strong enough depolarisation of the postsynaptic cell appears to play a role.

The cooperativity of synchronised parallel presynaptic inputs could then be explained by the strongly depolarising effects they exert. If such cooperation could occur between *specific* presynaptic inputs, then the local anatomy of their connexions to particular dendritic structures of the postsynaptic cell would be essential. The "gating mechanism" of Singer (1983) produced by unspecific inputs would probably not require such a fine-structured anatomy. A very interesting possibility would be *cooperativity of synchronised inputs* that is *not dependent upon the postsynaptic discharge* as suggested by Lee (1983). This would possibly again require specific anatomical arrangements of the cooperating synapses on the postsynaptic membrane surface.

Quite generally, however, it is difficult to connect synaptic plasticity with higher-order brain functions, as outlined by Lynch and Baudry (1984) for LTP and memory. They argue that, firstly, the proposed cellular mechanism must produce functionally meaningful and extremely persistent neural changes adequate to account for the behavioural manifestations of memory and that, secondly, it must be amenable to experimental manipulation. These two requirements of explanatory power and testability are not easy to fulfil, for which Lynch and Baudry (1984) discuss some reasons. So far, the relation between synaptic plasticity and learning can be demonstrated only for simple behavioural paradigms in relatively simple organisms, such as Aplysia (Rayport and Kandel 1986).

1.4 Brainstem

The phylogenetically old brainstem might be regarded incapable of developing as intricate data processing possibilities as cortical structures. However, as the examples in the following sections show, this potency is quite elaborate. Two examples will be considered, one related to auditory perception, and the other to different states of the functional organisation of the reticular formation.

1.4.1 A Spatio-Temporal Cross-Correlation Model for Acoustic Pitch Detection

The model of Loeb et al. (1983) is perhaps the most concrete and specific hypothesis at present available to explain complex perceptive processes. However, it is supposed to work at subcortical levels, viz. in the brainstem. This model deserves closer attention. Loeb and co-workers first consider classical models of acoustic pitch detection, namely the so-called "place-pitch" and "periodicity-pitch" theories, and discard them as being unable to account for pitch detection in the frequency range of 500–5000 Hz. For this

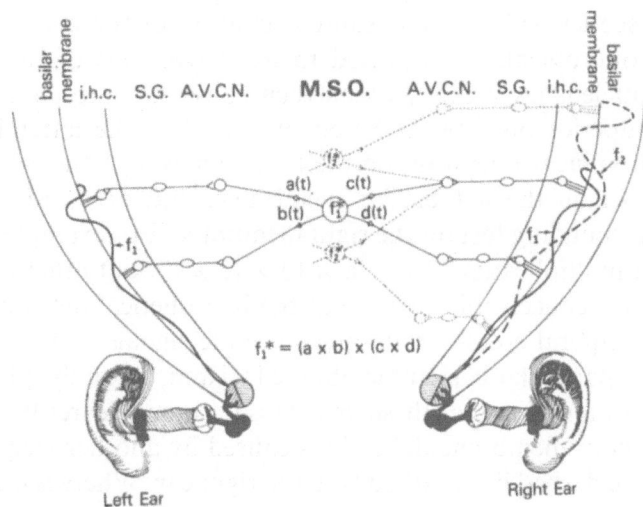

Fig. 8. An ascending pathway which might subserve pitch perception by detecting synchronicity among spatially distributed auditory nerve fibres. The *line* marked f_1 indicates the basilar membrane deflections caused by that frequency of sinusoidal stimulation presented in phase to both ears. This is detected by the inner hair cells (*i.h.c.*), conveyed as neural impulses in the spiral ganglion (*S.G.*) cells of the auditory nerve to a simple relay in the anteroventral cochlear nucleus (*A.V.C.N.*), and binaurally to the medial superior olive (*M.S.O.*). There a hypothetical detector cell, f_1^*, becomes maximally activated by the nonlinear summation of four synchronously phase-locked inputs: $a(t)$, $b(t)$, $c(t)$, and $d(t)$. *Dashed* basilar membrane *curve* f_2 indicates the deflections caused by another frequency which would be detected by the two *MSO* detector cells labelled f_2^*. Note that activity in a given auditory fibre and AVCN cell can contribute to the sensation of more than one pitch. Conversely, the detection of a given pitch can be performed using activity originating from different places on the basilar membrane. Note also that as long as none of the *MSO* input functions are allowed to have a value of zero, the detector cells will function for monaural as well as binaural acoustic stimuli. (With permission from Loeb et al. 1983; their Fig. 5)

band, they present an alternative hypothesis in which "the detection of synchronicity between two phase-locked signals derived from sources spaced a finite distance apart on the basilar membrane can be used to extract spectral information from the spatiotemporal pattern of basilar membrane motion" (p. 149). Figure 8 (their Fig. 5) illustrates a "prototypical form of the model". It is a schematic diagram of some pertinent structures of the auditory system. The cochleas of both ears are unrolled. The lines marked f_1 indicate the instantaneous basilar membrane deflections which are caused, at a particular instant, by that frequency of sinusoidal stimulation in phase to both ears. It is here assumed that two small groups of inner hair cells (i.h.c.) separated on the basilar membrane by a whole wavelength of deflection are excited at the two (supra-threshold) maxima of (unilateral) basilar membrane deflection (see left side). The ensuing excitation of the spiral ganglion (S.G.) cells is tonotopically transmitted to respective cells of the anteroventral cochlear nucleus (A.V.C.N.) which then converge onto common cells of the medial superior olive (M.S.O.) denoted by f_1^*. The latter M.S.O. cells perform a nonlinear operation on the four signals a (t) and b (t), coming from distinct loci on the left basilar membrane, and c (t) and d (t), coming from corresponding loci on the right membrane. For example, the operation could be a multiplication; $f_1^* = (a \times b) \times (c \times d)$, but other nonlinear operations are also conceivable. Note that the hypothetical detector cells f_1^* act as spatio-temporal cross-correlators, since the signals they receive are, firstly, related to particular basilar membrane loci and, secondly, phase-locked to particular motion phases at these loci. This becomes clearer by looking at the instantaneous membrane deflections caused by another frequency as represented by the dashed line marked f_2 in the right ear. Whereas one (the upper) of the two hair cell groups tuned to frequency f_1 would also respond to f_2, the lower group would not, because at this site f_2 is out of phase with f_1. However, to detect f_2 there could be other "partners" for the upper group as represented by dashed lines. The signals arising from these loci would then converge on specifically tuned M.S.O. cells denoted f_2^*. "Note that activity in a given auditory nerve fiber and AVCN cell can contribute to the sensation of more than one pitch. Conversely, the detection of a given pitch can be performed using activity originating from different places on the basilar membrane" (p. 155). It is also not necessary to have the pitch-related hair cells spaced a full deflection cycle apart, a fraction of 0.3–0.4 of a cycle would do. Moreover, the same model could account for localisation of sound sources by using phase differences of the signals generated in both ears.

Some further features stressed by Loeb et al. (1983) deserve attention. (1) The hypothetical mechanism for the nonlinear, time-sensitive interaction of the inputs, which might occur distally in the dendritic tree, renders possible multiple, independent cross-correlation processes within a given MSO cell. The fact that MSO neurones have a unique dendritic architecture may account for their high temporal sensitivity. (2) The critical interactions between small increments of time and space in the proposed model would have interesting implications for the normal development of such a system. According to Loeb et al. (1983), it is improbable that such a precise connectivity could be completely specified genetically without the impact of experience. They then refer to Hebbian synaptic modification, where synchronicity of presynaptic and postsynaptic signals play a critical role in the development of self-organising perceptual systems. It would have been natural to extend these ideas to correlations among presynaptic events.

It is noted in passing here that Torre and Poggio (1978) proposed a neural implementation of directional selectivity in visual motion detection, which was based on nonlinear spatio-temporal interactions of parallel synaptic inputs to local dendritic areas of a neurone. Marr (1982) later suggested another possibility.

1.4.2 Activity Patterns in the Reticular Formation

The reticular formation of the lower brainstem is an intricate meshwork of neurones with many inputs and outputs from and to various structures. Most of the neurones receive *multimodal* inputs from cortical, cerebellar, cardiovascular, respiratory and somato-sensory afferents. This has caused much trouble in interpreting the functions of the reticular formation.

Schulz et al. (1985) have intensively studied the statistical interactions of neurones in the canine lower-brainstem reticular formation, probably including the nucleus reticularis gigantocellularis and the nucleus reticularis caudalis. This region contains the ascending reticular activating system as described by Moruzzi and Magoun with large and medium-sized neurones possessing long ascending and descending axons. Langhorst et al. (1983) and Schulz et al. (1983) reported that the neurones recorded in this region are under the influence of cardiovascular, respiratory and somato-sensory systems. The above authors reported results on 88 cells recorded for periods from 30 to 250 min. and 35 double and 6 triple recordings were obtained with a single electrode ("neighbouring" neurones). In two cases 5 cells, and in another two cases 4 cells were recorded with two electrodes inserted on both sides of the brainstem RF ("distant" neurones). In all, 73 pairs of neuronal discharge patterns were investigated. The authors found five types of interrelationship which were correlated with different functional states of the CNS so that different types of correlation could appear with a change in this state.

Type 1: Flat cross-correlograms. Flat cross-covariance histograms (CCHs: conventional cross-correlograms where the mean firing rate has been subtracted from the bin contents) were found at some time in 53 of 73 recordings, 14 were flat throughout. The remaining 20 pairs exhibited some type of correlation throughout the recordings. The flat correlograms were found when the cells fired relatively regularly (with symmetrical interspike interval histograms) at rates above 16 pps.

Type 2: Cross-correlograms with short-latency peaks indicating strong coupling between the discharges of two neighbouring neurones. An example of this type is shown in Fig. 9 A, on two different time scales. The discharge patterns of the two neurones were fairly irregular. The upper CCH has a peak to the right of the ordinate at zero time, indicating a strong coupling of discharge of the two cells. Such peaks occurred to the right or the left of the ordinate and were restricted to within ± 5 ms of the time origin. Occasionally the position of the peak switched from one side of the time origin to the other in the course of a long recording. Strong couplings of this sort were found in

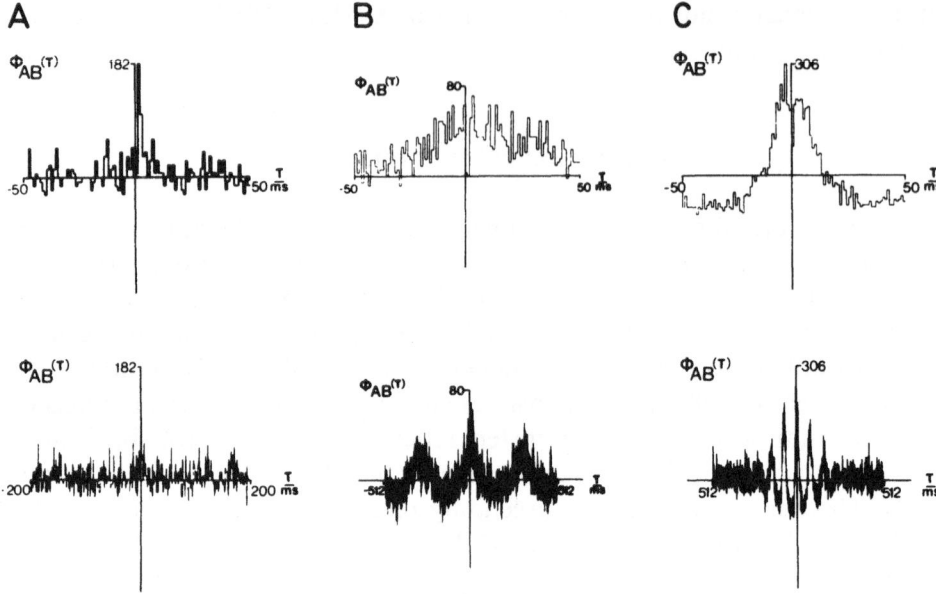

Fig. 9 A–C. Three types of cross-correlograms found for reticular neurones. Respective *upper* and *lower* cross-correlograms in each column are on different time scales, otherwise the same. *Left column* (**A**) Example of a neurone pair simultaneously recorded with short-term synchronisation of their discharges. Both cells had interval distributions of exponential shape. The cross-correlograms (with mean discharge rate subtracted) exhibit a strong peak just to the right of zero lag. Mean discharge rates for neurone **A**: 9.5 pps, for neurone **B**: 8.7 pps; *Middle column* (**B**) Cross-correlograms of a neurone pair discharging rhythmically coupled. The period in the *lower* cross-correlogram is about 300 ms. In the *upper* cross-correlogram no time shift between neurone discharges is seen. Discharge rates of neurone **A**: 8.1 pps, of neurone **B**: 5.4 pps; *Right column* (**C**) High-frequency oscillations (ca. 13.5 Hz) in symmetrical cross-correlograms for two neurones simultaneously recorded with one electrode. Mean discharge rates for neurone **A**: 11.7 pps, for neurone **B**: 10.0 pps. (With permission from Schulz et al. 1985; right columns of their Figs. 7, 8, and 11)

25 neighbouring cell pairs, never in pairs recorded from two different electrodes situated at a distance of at least 2 mm on both sides of the brainstem. They concurred with relatively regular or irregular discharge patterns of the single neurones. Schulz et al. (1985) argue that these strongly coupled discharges originate predominantly from common inputs, because the discharge of neighbouring neurones could be similarly influenced by afferents from somato-sensory systems. It is of particular interest to note the spatial restriction of this kind of correlation, which is reminiscent of a similar restriction in cortical networks (Sect. 1.3.1).

Type 3: Rhythmic cross-correlograms. An example of this type is shown in Fig. 9 B. A symmetrical broad peak around zero delay is repeated in the CCH on both sides, though with lower amplitude. The rhythm determined from such repetitions is between 2 and 5 Hz. Such CCHs were found in 27 pairs. The discharge rates of the cells were in the range of 0.3–7 pps. The spikes often occurred in bursts separated by 200–500 ms intervals. The bursts of several cells occurred in a rhythmically correlated fashion, and showed no phase shifts for neighbouring cells. The rhythm in the neuronal discharge was in turn correlated with low-frequency, high-amplitude EEG waves (theta waves), although the former lagged behind the latter. Schulz et al. proposed "that rhythmically coupled discharges are induced by inflows to the neuronal network and are not generated within the neuronal network of the reticular formation . . ." (1985; p. 55). They then suggested that the probable source of the rhythms which concur with the theta rhythm in the EEG may be a common oscillator in the hippocampus.

Type 4: Combination of rhythmic cross-correlograms with short-term synchrony. The previous two types could be combined, this combined CCH being observed in 8 pairs. The neurones correlated in this way fired at rates from 6 to 11 pps. Again high-amplitude EEG waves in the theta band occurred. Schulz et al. (1985) observed that the pattern of combined strong and rhythmic couplings could be transformed to exclusively rhythmic couplings by any experimental interventions which decreased the mean discharge rate. Experimental increase of pressoreceptor input always induced such discharge forms when the mean discharge rates decreased.

Type 5: Cross-correlograms exhibiting high-frequency oscillations. 28 cell pairs showed repetitive CCH peaks rhythmically recurring at high frequency (Fig. 9C). For neighbouring neurones the CCH maxima of the high-frequency oscillations occurred symmetrical to the ordinate. Schulz et al. (1985) observed similar high-frequency oscillations for neurones recorded with two electrodes from different sites, but the CCH maxima were then arranged asymmetrical to the ordinate. In the functional state associated with high-frequency oscillations, the correlated neurones fired at similar mean rates in a range of 8–16 pps and with small fluctuations in interspike intervals. Schulz et al. (1985) suggested that the coordinated behaviour of the cells in the state of high-frequency oscillation might result from activity of pressoreceptor afferents.

The interpretation that Schulz et al. (1985) give of their results is of particular interest in the present context. They first remind us that, in 1958, Moruzzi propounded a concept of the relation between mean discharge rate of reticular neurones and the functional organisation of the brainstem. Moruzzi distinguished three activity levels in the network. The lowest level represents the resting activity in a drowsy animal and is maintained by the integration of various afferents. The neuronal discharge patterns are then rhythmically correlated at low mean rates. This state probably corresponds to the "tonus central" discussed by Bremer and Terzuolo. The second level is characterised by the recruitment and influence of new afferents from cortical and peripheral sources impinging on the reticular neurones whose discharge rates are thereby raised. Moruzzi postulated that, at this stage, the reticular system is organised in functional modules. Schulz et al. (1985) feel that their results support Moruzzi's hypothesis. The combination of somato-sensory afferents converging on neighbouring reticular neurones induces the organisation of the network into subpopulations. These subpopulations are postulated to evoke differentiated activity patterns in various peripheral and cortical effector systems. The third level of activity is characterised by high discharge rates of the reticular neurones caused, for instance, by the activation of nociceptor or chemoreceptor afferents. Under these circumstances, the reticular network loses its differentiated organisation and again acts as an entity to evoke an emergency reaction.

On this view, the three activity levels are associated with three different forms of functional organisation of the reticular neuronal network, and on account of the different activity patterns, the reticular system exerts different influences on its effector systems in these three states. Schulz et al. (1985) suggest that, in conditions of low activity levels, the reticular network acts as an "energiser" that maintains the resting activity of the nervous system by its rhythmic activity patterns. At the intermediate activity levels, the network exerts different influences on different effector systems. The reticular formation then has the largest regulatory potency based on the most differentiated functional organisation. This organisational state is again quitted when a large number of neurones fire at high rates.

Schulz et al. (1985) emphasise that their results show that reticular neurones are capable of well-coordinated and discernible discharge patterns. Neurones which are functionally coupled by synchronised discharge may be considered to act as subpopulations in the network. Each subpopulation is built up and hence functionally defined by the number and the type of afferents acting on the constituent neurones and by the intrinsic activity pattern within the network. The configuration of subpopulations will change when the composition of effective afferents changes.

A particularly interesting aspect of the speculative considerations of Schulz et al. (1985) is their attempt to relate their concept of reticular subpopulations to the columnar processing units of the cortex defined by Mountcastle (1979). A reticular subpopulation is connected by axon collaterals to only a limited number of other subpopulations, but can be combined with them in a variety of patterns. Hence, its contributions to different neural populations can vary in time and can be differently emphasised. In general, this concept is reminiscent of Abeles' "multifunction" of synfire chains (Sect. 1.3.1) and uses the same basic ideas. Schulz et al. (1985) imagine that the activities of different populations have different functions within the entire neural network, such that the population which receives the most "essential" information is determining the function of this overall system. It is the functional state of and not simply the anatomical connectivity within, the network which determines the flow of information and the pathways followed through such a system with its manifold feedback mechanisms. If the relative predominance of these pathways can be changed, this could correspond to changes of the functional properties of the system. Then, specific functions, for instance cardiovascular and respiratory regulation, need not be localised in circumscribed centres, but are performed by differently interconnected plurifunctional subpopulations. This again, as with Abeles' synfire chains (Sect. 1.3.1), provides for a certain immunity to damage of parts of the network. "The dynamic organization of reticular neurones in subpopulations and the coordination of the activity of these populations is thought to be the mechanism which enables coordinated regulation of respiratory, cardiovascular and somatomotor systems for the adaptation of the organism to different functions" (Schulz et al. 1985; p. 59).

This concept as outlined above has some features in common with the ideas of Abeles (1982) and von der Malsburg (1981, 1985). Firstly, it stresses the importance of "functional units" consisting of groups of neighbouring neurones which are correlated (synchronised), mostly by common inputs; different functional units are able to interact, however, through the many connexions internal to the reticular formation. Secondly, though not pointed out explicitly, these "functional units" must be flexible, loose, rapidly changing associations of neurones since they are generated by changing inputs from the external (and internal) body environment. Thirdly, the functional cell associations are relevant for operational states in which the organism is awake and executes discriminative tasks requiring attention. Since attention normally scans the environment, it is immediately clear that the "functional units" must co-vary with it. What is not addressed in the concept of Schulz et al. (1985) are processes of synaptic modulation according to von der Malsburg (see also Crick 1984). Such processes might help to temporarily stabilise reticular subpopulations.

Whether the "functional units" of correlated reticular neurones are indeed as local as postulated by Schulz et al. (1985), remains to be established. They have computed conventional cross-correlograms for pairs of neurones, and Abeles' results with three-cell correlations demonstrate that the former method fails to detect more complex patterns of cell interactions possibly involving more cells than just the locally adjacent ones recordable with one electrode. Thus these cell interactions could encompass more extended networks. Nonetheless, Abeles (1982) using a different, "population statistics", approach, also found that cell populations with statistically homogeneous properties (i.e. common synchronising inputs) cover a restricted area of the cerebral cortex (e.g. 3×3 mm, although such an estimate obviously depends on experimental and anatomical conditions).

It is worth emphasising once again that reticular neurones receive converging multimodal inputs from presso-, chemo- pulmonary mechanoreceptors and somato-sensory receptors as well as from cortical and cerebellar afferents. The mean discharge rate of a reticular cell would therefore tell nothing specific about any of the above modalities. This would be consistent with the hypothesis of the nonspecificity of the reticular formation. However, a neural code represented by the correlated firing of groups of neurones could maintain specificity in the face of multimodal inputs to each single constituent cell (see Abeles 1982; Sect. 1.3.1). Whether this is a working principle of the cerebral cortex as well as of the brainstem awaits experimental verification which probably will come only in some distant future, if at all. (Some remarks on the different codes embodied in mean rates and correlated firing patterns are made in App. G).

Similar, though not totally identical, results were reported by MacGregor and Lewis (1977). They found that the tendency for a midbrain neurone to show correlated firing with other midbrain neurones appeared to increase with the degree of rhythmicity apparent in the neurone. Also, broad-peak cross-correlograms showed that adjacent cells (for example, those recorded with a single electrode) tended to exhibit in-phase broad rhythmic correlations, whilst neurones separated by hundreds of microns (recorded with separate electrodes) tended to show out-of-phase broad rhythmic correlations. The authors then add some speculations on the underlying causes of these correlations. They suggest that local regions in midbrain tissue could sometimes exhibit synchronised firing upon which any other afferent activation is superimposed. In contrast, distant regions would be out of phase with each other. This could come about either by the propagation of a slow wave through the midbrain or by the existence of disjunct subsystems of neurones, each subsystem tending to fire synchronously within itself, but not necessarily with other systems. Whilst this holds generally, there are some exceptions in that some local neurone pairs are out of phase and some distant neurone pairs are in phase. "One might speculate that the hypothetical wave advanced in a not-totally homogeneous fashion extending isolated branches of synchronizing influence into as yet uninvaded regions or, alternatively, that separate functionally synchronized subsystems might interpenetrate anatomically" (p. 270).

1.4.3 Summary

In the preceding sections, two different examples were given for complex signal processing at the brainstem level. The first example is a model proposed by Loeb et al. (1983) to account for acoustic pitch detection. This model is based, in a very specific manner, on correlations between spiral ganglion cells and the evaluation of the spatio-temporal patterns of these correlations by brainstem nuclei. The nonlinear evaluation operation is thought to be performed by medial superior olive cells which act as spatio-temporal cross-correlators. Three features of this model are of particular interest. Firstly, a given auditory nerve fibre can contribute to detection of different pitches, i.e. its function is only defined by its temporary integration into an entire ensemble of other neurones, the point of interest being that this is highly typical of neural assemblies (Sect. 1.3). Secondly, cell architecture may again play an important role in evaluating cross-correlated inputs. Thirdly, the ontogenetic development of the appropriate synaptic connexions is thought to follow lines of synaptic plasticity similar to those discussed for retinotopic projections (Sect. 1.3.3).

The second example is concerned with a seemingly quite different topic, namely the different functional states of the lower-brainstem reticular formation. Schulz and co-workers (1985) have found four different types of cross-correlated firing patterns of neighbouring and distant cell pairs: short-term synchrony, rhythmic correlations, combined short-term synchrony and rhythmic correlation, and high-frequency oscillations (in the CCH). Short-term synchrony occurred only between neighbouring cell pairs. The four patterns of correlation were associated with different mean firing rates and with three different functional states of the reticular formation which are characterised by the ways in which neurones cooperate and interact, and the types of input to the reticular formation. The middle state is the most differentiated one in that neurones appear to be grouped into subpopulations which are defined by particular input patterns and ensuing short-term synchronisation of reticular neurones. Schulz et al. interpret their results in terms of Mountcastle's module concept in which subpopulations of neurones in an otherwise distributed system are dynamically formed and dissolved. Synaptic plasticity is not discussed in this context, this probably being due to the conventional view that synaptic plasticity involves long-term processes and not short-term changes. It would be interesting to combine the previous authors' view with that of von der Malsburg's concept of synaptic modulation (Sect. 1.3.2) which is assumed to establish and stabilise neuronal assemblies on a short-term time base. In any case, the view of Schulz and collaborators is a step towards solving the problem of the multimodality of most reticular neurones, which might lead one to mistakenly envisage the reticular network as an unspecific system incapable of differentiated regulatory actions.

2   Correlations Between Spinal Neurones

> " 'Mass communication is defined as the transfer, among
> groups, of information that a single individual could not pass to
> another'. It is not such a bad image, the brain as an ant colony!"
>
> D.R. Hofstadter[5]

As shown in the preceding sections, correlations between firing patterns
of different neurones in supraspinal structures have led to quite unique
models of brain function, which specifically rely on such patterns and which
interpret them as a code distinct from the commonly accepted mean firing
rate code. Perhaps surprisingly, something equivalent has not happened for
the spinal cord level although it will now be shown that similar correlation
patterns also occur at this level. Indeed there is some reluctance on the part
of many workers to try and set up such hypotheses. Instead, correlations
between firings of neurones of the same kind are used as methodological tools
to investigate synaptic connectivity, and indeed very successfully so. A prerequisite for building new theoretical frameworks based on neuronal correlations may be the development of the *module or neural assembly concept* which
– as discussed in the Sect. 1.4 – has already "diffused" down at least to the
brainstem. If this conceptual prerequisite is in fact essential, the time may be
ripe for new views of spinal cord function, all the more so as Mountcastle
(1979) and Szentagothai (1983) have already tried to apply it to the spinal
cord, albeit from different perspectives. However, the modules are not always
clearly defined even at the cortical level (e.g. Creutzfeldt 1983), and they are
even less so at the spinal level. The functional modules have yet to be
identified. Before entering functional discussions, however, a description of
the correlation patterns at spinal cord level is due. In the following, the
anatomy of the spinal cord is not treated extensively except in special contexts. The reader is referred to the recent monograph of Brown (1981). Since

---

[5] Hofstadter DR (1980) Gödel, Escher, Bach: An eternal golden braid. Vintage Books, New
York, p 350. A more complete quotation would run as follows: "In his book *The Insect Societies,*
E.O. Wilson makes a similar point about how messages propagate around inside ant colonies:
'(Mass communication) is defined ...' "

the material to be presented now goes into much detail, it may be advisable to read the summaries first.

The brainstem is known to exert significant actions on spinal cord neural systems. For example, vestibulo-, rubro- and reticulospinal systems influence segmental motoneuronal and interneuronal systems (e.g. Lundberg 1979; Baldissera et al. 1981). The brainstem also mediates some of the influences exerted by autonomic systems on spinal motor systems (Meyer-Lohmann 1974, Schulte et al. 1959a; Schulte et al. 1959b). Finally, the respiratory drive to the intercostal motor output derives from brainstem structures. Some of these influences cause correlations (among other effects) between discharge patterns of spinal neurones. Also segmental inputs as well as segmental interneurones cause such correlations. This section will deal with details of these correlations, their causes and effects. The organisational scheme followed in presenting the material is to treat the elements of the so-called spinal stretch reflex arc, starting with motoneurones, progressing to muscle receptors and closing the loop back to motoneurones, with some remarks added on interneuronal systems which affect signal transmission through this loop.

2.1 Correlations Between α-Motoneurone Discharges

α-Motoneurones, the "final common path" to skeletal muscle (except for β-motoneurones, Sect. 2.2) have been much investigated, in man and animal, for correlations between their discharge patterns, because such statistical time relations would bear upon the amplitude distribution of physiological and pathological types of tremor. (Dietz et al. 1976, have demonstrated a direct correlation between the degree of synchrony between two motor units and the amplitude of physiological tremor.) The results have been controversial (e.g. Gelfand et al. 1963; Person and Kudina 1968; Milner-Brown et al. 1975; Kranz and Baumgartner 1974; Mori 1973, 1975; Shiavi and Negin 1975; Dietz et al. 1976; Datta and Stephens 1980; Datta et al. 1985; Sears and Stagg 1976; Kirkwood et al. 1982a, b; Connell et al. 1986; Davey et al. 1986; Ivarsson et al. 1986). There may be several reasons for this controversy. One is that the term "synchrony" was not always clearly defined and that, in fact, it was often used in different senses (see Freund 1983). Another is that, in many cases, tendencies towards synchronous (or other time-related) firing of two α-motoneurones may not have been detected for statistical reasons, e.g. when the recording length was too short (see Sears and Stagg 1976). A third is that discharge correlations among motoneurones may appear and disappear according to changing experimental circumstances.

The types of correlation encountered are briefly reviewed here, essentially following Kirkwood et al. (1982b), who have intensively studied the correlations between respiratory α-motoneurones in cats. Thereafter the causes of these correlations are discussed.

2.1.1 Types of Correlation Between α-Motoneurone Discharges

1) *Short-term synchronisation* occurs between α-motoneurones (Milner-Brown et al. 1975; Dietz et al. 1976; Datta and Stephens 1980; Datta et al. 1985; Sears and Stagg 1976; Kirkwood et al. 1982 a, b; Connell et al. 1986; Davey et al. 1986; Ivarsson et al. 1986) in much the same way as between cerebro-cortical neurones (Abeles 1982; Sect. 1.3.1), cerebellar Purkinje cells (Ebner and Bloedel 1981 b), retinal ganglion cells (Mastronarde 1983 a), reticular neurones (Schulz et al. 1985; their type 2; Sect. 1.4.2). The discharge probability of one α-motoneurone is increased above average in the temporal vicinity (within a few milliseconds, therefore "short-term") of the spike of another α-motoneurone. This is illustrated in Fig. 10 A and B for cat respiratory (thoracic) α-motoneurones. The narrow peak in the cross-correlogram may be superimposed on broader bases corresponding to "medium" or "long-term" ("broad-peak") synchronisation as shown in Fig. 10 C and D. (This combination also occurs in cross-correlograms for retinal ganglion cells; Mastronarde 1983 b). An interesting feature of the short-term synchrony between thoracic respiratory α-motoneurones is its characteristic spatial distribution. Its strength declines over a few segments so that no short-term synchrony is detectable for α-motoneurones separated by more than three or four segments (Kirkwood et al. 1982 a).

2) Kirkwood and co-workers (see above) also described *broad-peak synchronisation* between discharges of respiratory α-motoneurones, the peaks extending to roughly ± 20 ms on both sides of the reference zero time. An example is given in Fig. 10 C and D. This type of correlation appears to be enhanced after ipsilateral hemisection of the thoracic spinal cord above the locus of recording (Kirkwood et al. 1984). As short-term synchronisation, it shows a spatial distribution although it is still detectable at a segmental separation of five segments (Kirkwood et al. 1982 a). (Similar broad peaks have been described for retinal ganglion cells: Mastronarde 1983 b; cerebro-cortical neurones: Abeles 1982; cerebellar Purkinje cells: Ebner and Bloedel 1981 b; respiratory medullary neurones: Hilaire et al. 1984).

3) *Long-term rhythmic synchronisation* is present if two neurones are modulated by the same low-frequency rhythm, such as respiratory (Sears and Stagg 1976; Kirkwood et al. 1982 b), locomotor or delta-theta EEG rhythms (Schulz et al. 1985; their type 3). It can also be an artefact if two independent but fairly regularly discharging neurones are recorded and cross-correlated over too short a period of time (see, e.g. Clark et al. 1981).

4) *Cross-correlograms exhibiting high-frequency oscillation* (h.f.o) (at frequencies between 60 and 120 Hz) were found for inspiratory (external intercostal) α-motoneurones (Kirkwood et al. 1982 b). This type of correlation is illustrated in Fig. 10 E and F. Its strength, as that for short-term synchrony, declines with spatial distance of the motoneurones although at a slower rate

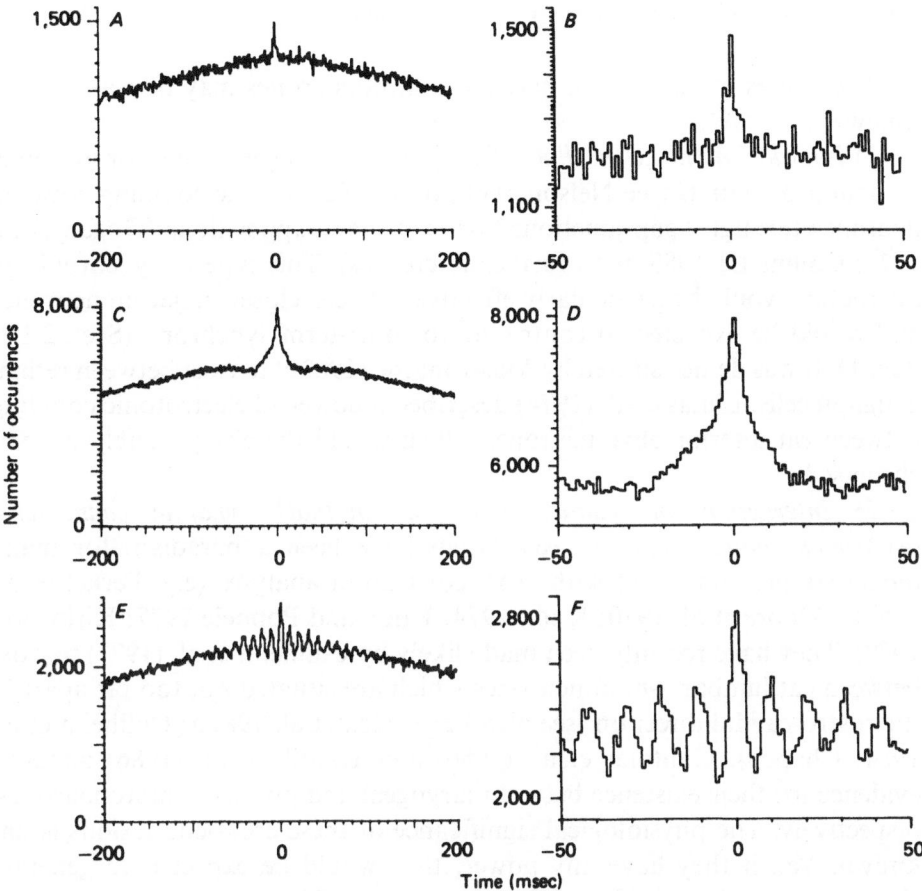

Fig. 10 A–F. Cross-correlation histograms showing three different forms of α-motoneurone synchronisation, in each case between the efferent discharges of T6 and T7. Each histogram in *left hand* column, **A, C, E**, from a different animal: **A** moderately deeply anaesthetised, end-tidal CO_2 5.5%; **C, E** lightly anaesthetised, end-tidal CO_2 4.0 and 6.0 % respectively. **B, D, F**, the same histograms with different time and amplitude scales. (With permission from Kirkwood et al. 1982b; their Fig. 1)

(Kirkwood et al. 1982a). It corresponds to the type 5 correlation between reticular neurones described by Schulz et al. (1985; Sect. 1.4.2; see also Hilaire et al. 1984), albeit showing a different frequency. This type is also found in phrenic motoneurones (M. I. Cohen 1979).

5) A *special type of long-term synchronisation* is found in low-frequency, particularly pathological types of tremor in man, in which α-motoneurone discharges may be "synchronised" (better: correlated) over prolonged time-spans of some tens of milliseconds (see Freund and Dietz 1978; Freund 1983).

These various kinds of correlation may be superimposed on each other depending upon the state of the animal preparation (see Kirkwood et al. 1982b, 1984) or of the human subject/patient (see Freund and Dietz 1978).

2.1.2. Causes of α-Motoneurone Correlations

The causes of correlation between α-motoneurones may be classified as follows:

1) *Interaction via extracellular field potentials* as generated, for example, by action potentials (see Nelson 1966) or via electrotonic coupling between neurones through "gap-junctions" or membrane appositions (Gogan et al. 1977; Collins III 1983, with further references). This type of synchronising interaction would be particularly effective between closely adjacent neurones and would be expected to contribute to short-term synchrony (Sect. 2.1.1, item 1). It was demonstrated by Mastronarde (1983c) to exist between retinal ganglion cells. Llinas et al. (1974) described a powerful electrotonic coupling between cat inferior olive neurones which would thereby probably be synchronised.

2) *Interaction via (unidirectional or mutual) synaptic connexions.* (a) *Monosynaptic excitatory connexions* have been a paradigm for many model studies concerned with cross-correlation analyses (e.g. Perkel et al. 1967b; Moore et al. 1970; Knox 1974; Knox and Poppele 1977; Kirkwood 1979). They have recently been made likely by Cullheim et al. (1977) to exist between cat lumbar α-motoneurones which are situated not too far apart in the rostro-caudal direction (see also Lagerbäck et al. 1981a; Cullheim et al. 1984; Chap. 3). Gauthier et al. (1980) and Khatib et al. (1986) adduced evidence for their existence between laryngeal and phrenic α-motoneurones, respectively. The physiological significance of these cross-connexions is unknown. Yet, if they have any power, they would be expected to generate short-term synchrony (Sect. 2.1.1, item 1). (b) *Disynaptic cross-connexions* between spinal α-motoneurones exist via Renshaw cells. These inhibitory interactions also exhibit a certain spatial pattern (see Windhorst 1979a; Kirkwood et al. 1981; Sect. 4.3.1). However, their role in synchronising or desynchronising α-motoneurone discharges has been a matter of controversy (e.g. Gelfand et al. 1963; Adam et al. 1978; Kirkwood et al. 1982b; Sect. 2.5). If they contribute at all to α-motoneurone synchronisation, then probably to the broad-peak (Sect. 2.1.1, item 2) or rhythmic type (Sect. 2.1.1, item 4; cf. also Sect. 2.9.2.1). It is worth mentioning that cortical pyramidal cells are also interconnected intracortically by the above two types of interconnexion: a short-range excitatory one (via recurrent axon collaterals) and a longer-range inhibitory one (via inhibitory stellate cells) (see Creutzfeldt 1983).

3) *Common input* by branching of a presynaptic input fibre and divergence of its branches to many α-motoneurones is certainly one of the most important sources of synchronisation. It is probably responsible for short-term synchronisation (Sect. 2.1.1, item 1) if the input fibre exerts relatively brief synaptic actions in the postsynaptic cells (of the order of a few milliseconds). For example, consider the well-studied case of monosynaptic connex-

ions from Ia fibres onto homonymous α-motoneurones (for a review see Henneman and Mendell 1981). Each Ia fibre branches to contact nearly all homonymous α-motoneurones (see below). An action potential running orthodromically up a Ia fibre can be assumed to arrive almost simultaneously (with very slight latency differences) at most of the Ia fibre synaptic terminals and, hence, after a brief synaptic delay, to cause nearly synchronous EPSPs in the α-motoneurones contacted. The rise times and durations (half-width) of these EPSPs are relatively short (of the order of 1 and 4–5 ms, respectively; see Fleshman et al. 1981). Knox (1974) and Knox and Poppele (1977) presented a neurone model demonstrating that spikes in the postsynaptic cell are predominantly elicited during the rising phase of the EPSP in response to the positive time derivative of the EPSP. Although Kirkwood and Sears (1978), in a modified model, showed that, in addition to the EPSP derivative, the EPSP itself also contributes to spike initiation (see also Kirkwood and Sears 1982b; Fetz and Gustafsson 1983; Gustafsson and McCrea 1984), the time span over which this can occur is still very brief (a few milliseconds); and since this span is roughly the same in all α-motoneurones, they tend to produce action potentials nearly synchronously. This mechanism applies not only to Ia fibre input, of course, but to any common input system to motoneurones. Indeed, Kirkwood et al. (1982a) presume that the short-term synchronisation they found between respiratory α-motoneurones derives from bulbospinal neurones providing the common respiratory drive to the α-motoneurones. The same common input mechanism was also suggested by Schulz et al. (1985) to explain the short-term synchronisations between medullary neurones (Sect. 1.4.2) and by Mastronarde (1983a) to account for the same correlations between retinal ganglion cells. Common input also causes long-term correlations if the postsynaptic neurones produce longer-lasting responses (e.g. several impulses) to each input event (cf. Mastronarde 1983b) or if it mediates common slow rhythmical modulations of the excitability of the postsynaptic neurones. These excitability changes can be excitatory and inhibitory (Moore et al. 1970). Indeed, correlations between neurones of the same type are often used as an indirect means of studying the input organisation to the cells. An excellent example for the potential potency of these methods (in conjunction with other techniques) is the work of Mastronarde (1983a, b).

An interesting observation on synchronised α-motoneurone discharge in man was made by Datta and Stephens (1980) and Datta et al. (1985). They found that the degree of short-term synchronisation is higher between pairs of high-threshold motoneurones than between pairs consisting of one high-threshold and one low-threshold motoneurone, and assumed that this difference was due to the joint arrival of EPSPs onto high-threshold motoneurones from common presynaptic neurones not shared with low-threshold motoneurones. This interpretation would be consistent with evidence from animal and human experiments for the existence of afferent pathways which selectively excite high-threshold motoneurones and can alter the normal recruitment order (low-threshold before high-threshold units).

4) *Correlated common input:* The synchronising actions of common inputs are modified if the various diverging *inputs are also mutually correlated.* For example, the h.f.o. synchronisation between intercostal α-motoneurones (Sect. 2.1.1, item 4) is probably due to the same kind of synchronisation between discharges of bulbospinal neurones (see Kirkwood et al. 1982b). These authors also attribute the "broad-peak synchronisation" (Sect. 2.1.1, item 2) to the synchronised firings of as yet unidentified interneurones in the spinal cord and discuss the possible origins of the synchronisation within the interneurone pool. In their view, there are at least three possibilities, which are not mutually exclusive. Firstly, the common presynaptic input by branched axons producing short-term synchronisation of α-motoneurones could also exert its effects on pools of interneurones. "If it is sufficiently strong and if there are many interconnexions, then the multisynaptic chains involved could produce a gradual broadening of the narrow peaks which are appropriate to a monosynaptic common input system" (p. 132). Note that multisynaptic chains are equivalent to synfire chains (Sect. 1.3.1), but that a problem with this sort of signal transmission through many successive synaptic connexions is also clearly stated by the above quotation. Secondly, the interneurones involved in synchronised firing could have patterned discharges, e.g. they could fire in bursts. Thirdly, this synchronisation could also be evoked by an external stimulus, such as the cardiac pulse which can synchronise intercostal muscle spindles (Kirkwood and Sears 1982a).

Kirkwood et al. (1982b) conclude their discussion with some interesting functional considerations. They suggest that the pathways involved in such synchronisation as considered here could also be operative in a normal animal, since synchronisation among intercostal α-motoneurones may be required for such tasks as vocalisation or the defensive reflexes of the respiratory tract. Although the strongest broad-peak synchronisation they observed was an abnormal feature insofar as it resulted from acute lesions it "may well represent the time scale or mode of activation of pathways contributing to normal intercostal movements, which are part of a spectrum ranging from ballistic movement, via respiratory to those which contribute to posture as a whole" (p. 133). On this view, therefore, synchronisations are not merely undesired epiphenomena resulting from common inputs and leading, in the worst case of synchronised α-motoneurone activity, to increased tremor amplitudes (Dietz et al. 1976). This subject will be taken up in Chap. 3.

Rudomín and co-workers (Rudomín et al. 1969; Rudomín and Madrid 1972; Rudomín et al. 1975; Rudomín 1980; see also Rall and Hunt 1956) consider synchronisation in a particular interneuronal pool as an important mechanism of signal transmission through some spinal pathways. They maintain that lumbar α-motoneurone excitability is correlated by synchronised activity among interneurones controlling presynaptic inhibition on terminals from common input fibres to the motoneurones (Sect. 2.9.2.2). The sources of synchronisation within these interneurone pools could be the same as those discussed above by Kirkwood and co-workers.

2.2 Correlations Between β-Motoneurone Discharges

β-Motoneurones innervating extrafusal as well as intrafusal muscle fibres have not yet been deliberately investigated with respect to correlations between their discharge patterns. There is no reason to believe, however, that they should not, at least in principle, follow the patterns described above for α-motoneurones, although their input patterns could be different from those of α-motoneurones (see Hulliger 1984). Indeed, many of the results reviewed above and assigned to α-motoneurones may unknowingly have been obtained on β-motoneurones.

2.3 Correlations Between γ-Motoneurone Discharges

2.3.1 Types of Correlation Between γ-Motoneurone Discharges

γ-Motoneurones of cat semitendinosus, gastrocnemius and soleus muscles have recently been shown to be synchronised to various degrees depending upon the kind of preparation (decerebrate with intact spinal cord or spinalised) and upon various peripheral stimuli (Murthy and Yoon 1979; Ellaway et al. 1982b; Ellaway and Murthy 1982, 1983, 1985a, b; Davey and Ellaway 1984; Connell et al. 1986). These correlations were frequently observed in spinal cats and consisted of the *short-term type of synchronisation* with a peak in the cross-correlogram centred at zero time and falling off to control values within 5 to 10 ms if the two units had about the same conduction velocities. Examples are shown in Fig. 11 (left and right column). In certain cases, there were shifts of the peak of the cross-correlogram which could be accounted for by a difference in axonal conduction latency of the spikes down to the recording site (Ellaway and Murthy 1985a). In a small number of spinal cats, a *stronger, nearly synchronous firing* within 1–2 ms was observed, which could occur superimposed upon the broader base described previously (Ellaway and Murthy 1985a; cf. Connell et al. 1986). Other asymmetrical types of correlation with either peaks or troughs in the cross-correlograms were also found. An example of a *cross-correlogram displaying a trough* is shown in Fig. 11 (middle column). This type of correlation has so far been found in only one animal by Ellaway and co-workers.

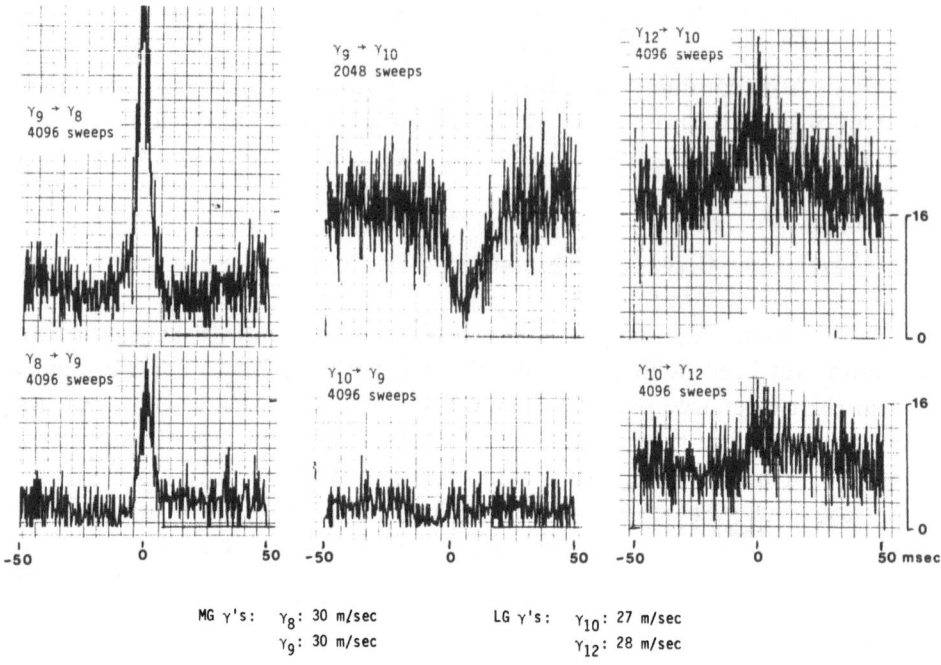

MG γ's: γ$_8$: 30 m/sec   LG γ's: γ$_{10}$: 27 m/sec
           γ$_9$: 30 m/sec   γ$_{12}$: 28 m/sec

Fig. 11. Cross-correlograms of the background discharges of γ-motoneurones in the decerebrated spinal cat. *Left column* correlograms between medial gastrocnemius units; *Right column* correlograms between lateral gastrocnemius units; *Centre* correlograms between a medial and a lateral gastrocnemius unit. (Courtesy of P.H. Ellaway)

Nonetheless, the dips were consistently present between a number of γ-motoneurone pairs in which one unit was from the medial gastrocnemius and the other from the lateral gastrocnemius muscle, and never between pairs with both units from either the medial gastrocnemius or the lateral gastrocnemius which always had the expected peak of synchrony (P. Ellaway, pers. comm.).

Synchrony of firing was a common feature in spinal cats (170 pairs out of 188 pairs in 30 cats), few correlations (6 out of 47 pairs) appeared in decerebrate cats (Ellaway and Murthy 1985a). The authors concluded that, in intact, anaesthetised cats, the activity of the brain can suppress the tendency towards synchronous discharge. The site of origin of this suppression probably lies between the colliculi and the first cervical vertebra (Ellaway and Murthy 1985a). The degree of synchronisation, i.e. the height of the short-term peak relative to the mean firing probability, appeared to be inversely correlated with the discharge frequency of the correlated cells (Ellaway and Murthy 1983, 1985a). However, Davey and Ellaway (1984) now found synchrony to be dependent more upon the *regularity of firing* (i.e. variability of interspike interval) rather than on the rate of the two discharge trains, such that the more irregular the discharges the greater the synchrony. This observation agrees with the finding that regularity of discharge decreases as frequency falls. However, both efferents must show marked irregularity before substantial synchrony occurs. Note that these findings conform with those of Schulz et al. (1985; Sect. 1.4.2) on reticular neurones with respect to the interrelations between mean firing rate, regularity of discharge and tendency towards synchrony. The parallelism between the degree of synchronisation (between cells) and the variability of discharge (of individual cells) would be explicable if it arose from some common rhythmic input (strong enough to modulate postsynaptic firing), which acts on both efferents (hence their synchrony) and which interferes with the 'spontaneous' (or 'preexisting') firing rhythm of each target cell (hence the variability, caused perhaps by some beating of rhythms). The reduction of synchrony with increasing firing rate indicates that the synchronising input becomes less influential, as other (nonsynchronising) inputs gain influence. This could be due to both presynaptic (increasing presynaptic inhibition?) and/or postsynaptic (saturation?) mechanisms.

2.3.2 Causes of γ-Motoneurone Correlations

The most likely cause of γ-motoneurone synchronisation appears to be *shared presynaptic input*. Alternatives, such as synchronisation between discharges of independent presynaptic fibres (Sect. 2.1.2, item 4) or interactions of favourably sited neurones through electric field effects or electrotonic synaptic coupling (Sect. 2.1.2, item 1), appear less likely (Ellaway and Murthy 1985a). Connell et al. (1986) argue that the preponderance of narrow peaks in the correlograms, or of an accentuated central region to the peak with half-widths in the range of 3–4 ms, suggests that a substantial part of the synchronised discharges results from activity in common presynaptic fibres (Knox and Poppele 1977; Kirkwood 1979). The fact that some correlogram peaks were wider could indicate the contribution of some synchrony between different presynaptic axons. However, if synchrony in the presynaptic input neurones played a role, then probably synchrony between the interneurones involved in the flexion reflex rather than between the primary afferents. In a preliminary study, Davey and Ellaway (1985) found no indication of a tight synchrony between primary afferents of either hair or pressure receptors excited by stimulation of the skin of the foot.

Ellaway and Murthy (1985b) presented evidence that the afferents which cause synchronised background discharges of γ-efferents to triceps surae are restricted to the ipsilateral segment. Natural stimulation in the contralateral sural field of innervation did not cause additional synchronisation, whilst stimulation in the ipsilateral field did. Mild pressure applied to the ipsilateral plantar cushion in the spinal cat did not contribute an extra component of synchronised firing, although it excited the γ-motoneurones. This indicates that excitation and enhancement of synchronous firing are two different effects, which need not be coupled, and that this is probably due to different properties of the pathways involved. Thus, one pathway may augment both the general excitability of certain cells and their tendency towards synchronous discharge, whilst another may have only the first effect. Homonymous or heteronymous synergistic muscle afferent input did not contribute to γ-efferent synchrony.

It is of interest to know whether the synchronising segmental input is distributed to γ-motoneurones in a spatially organised (topographical) manner. This might be possible since, although γ-efferents branch to innervate different spindles, the distribution of these efferents to large hindlimb muscles is confined to the muscle region innervated by a single nerve fascicle. Therefore, the synchrony of discharge of γ-motoneurones distributing their axons to different nerve fascicles was investigated. Ellaway and Murthy (1985b) did not find a topographical pattern. Pairs of neurones projecting to different fascicles, even to the different heads (gastrocnemius medialis and lateralis) of the muscle, were not less synchronised than pairs projecting to one fascicle. However, this homogeneity of distribution of synchrony is not generally found between gastrocnemius and soleus. Eight homonymous pairs of γ-motoneurones (five gastrocnemius and three soleus) exhibited synchrony of their background discharge, but this was not evident for any of the six gastrocnemius-soleus pairs studied. However, synchrony appeared in these six pairs when the heel was stimulated by light stroking or gripping. It may therefore be concluded that the sural input giving rise to additional synchrony is shared between γ-efferents to the two muscles, but the afferent input which simply synchronises the background discharges is not (Ellaway and Murthy 1985b). That is, γ-motoneurones to synergists may have both shared and independent afferent connexions. This may also apply to muscles not generally acting as synergists, such as the semitendinosus and gastrocnemius. The question therefore is whether such reflex pathways are under phasic gain control, and synchrony would appear only under particular conditions (Ellaway and Murthy 1985b).

2.3.3 Correlations Between α- and γ-Motoneurone Discharges

Connell et al. (1986) studied, in spinalised cats, the synchronised firings of α- and γ-motoneurones coactivated during the flexion reflex. Short-term synchrony was present both for pairs of α-motoneurones and for pairs of γ-motoneurones. For γ-efferents, the tendency towards synchronous firing was stronger during reflex excitation than during "background" firing. Additional synchrony was introduced in particular when the skin of the heel was stimulated. Unlike the synchronised firing of pairs of motoneurones of the same type, synchrony between α- and γ-efferents was either absent or less common and weaker. Connell et al. (1986) proposed therefore, that the reflex pathways to the two types of motoneurone may largely be segregated. In other words, in the flexion reflex the afferent input appears to be conveyed to the α- and γ-motoneurones through largely nonoverlapping sets of interneurones. In this context, the authors discuss the possible nature of the skeleto-motor neurones involved and suggest that the *fusimotor* action of β-motoneurones would require them to receive the same synaptic inputs as the γ-motoneurones of the same function (static or dynamic). Such shared inputs could then, in the spinal animal, result in synchrony of discharge. Those α-motoneurones showing short-term synchrony with γ-fusimotoneurones could therefore be β-motoneurones. On the other hand, these β-motoneurones, due to their skeleto-motor function, would share inputs involved in the direct activation of skeletal muscle fibres and may therefore be expected to show, in addition, synchrony with the purely skeleto-motor α-motoneurones.

2.4 Summary

α-Motoneurones may be correlated in a variety of patterns. This has been most thoroughly demonstrated for the thoracic respiratory α-motoneurones of cats breathing spontaneously. Short-term synchrony, broad-term synchrony, rhythmic correlation and high-frequency modulation as well as combinations thereof were found depending upon experimental conditions, e.g. state of animal preparation (depth of anaesthesia, pCO_2 level etc.) and lesions performed. These patterns have also been described for neurone pairs at supraspinal levels. The causes of these various correlation patterns are different, ranging from interactions between α-motoneurones themselves to different presynaptic inputs which may in turn be correlated. Since at least three of the above patterns (short-term synchrony, broad-peak synchrony and high-frequency oscillation) show a spatial distribution in that their strength declines with segmental separation of the motoneurones, the various synchronising inputs must also be assumed to be spatially organised, albeit in different ways for the different types of correlation. It is worth recalling that such spatial patterns were also found for the correlations of cortical neurones (Sect. 1.3.1) and of nerve cells of the lower reticular formation (Sect. 1.4.2).

Ellaway and co-workers described two major types of correlation between γ-motoneurone discharge: (a) short-term synchrony (as between α-motoneurones; Sect. 2.1.1, item 1), the highly synchronous discharges being an extreme case, and (b) inverse correlations (found in only one experiment so far). Other types such as broad-peak, rhythmical and high-frequency oscillation synchronisation have not as yet been found, they might exist in other preparations. For example, spinalisation would preclude the occurrence of high-frequency oscillation. Occasionally, the short-term synchrony could be superimposed upon broader bases. As to the origin of the short-term synchronisation of γ-efferents, the most likely cause appears to be the action of *shared presynaptic input*. This presynaptic input probably originates in sural nerve afferents from the heel region. It is not compartmentalised to γ-efferents within a muscle.

The importance of γ-motoneurone synchronisation appears to be obvious. Ellaway and Murthy (1985a) discuss this issue in relation to tremor. It is known that any tendency for skeleto-motoneurones to become synchronised will create tremor and is therefore disadvantageous. Since in spinal cats the discharge of γ-efferents is almost always synchronised to some degree, a synchronising influence in the muscle spindle loop would be fed back to α-motoneurones. This could establish synchronised reverberating loops. "Thus, the intrinsic interest in short-term synchrony between γ-motoneurones lies in its contribution to both normal and pathological tremor states" (Ellaway and Murthy 1985a; p. 228). This topic will be taken up in Chap. 3, where the role of correlations between parallel neuronal elements will be given a somewhat more differentiated interpretation. Be this as it may, the role of motoneurone correlations may also be seen from a different angle.

Firstly, it is not certain that, under more natural conditions, correlations (particularly tendencies towards synchronous firing) between γ-motoneurone discharge patterns are suppressed as far as possible and, if present, display no topographical or otherwise ordered pattern. The faculty of supraspinal structures to depress γ-motoneurone correlation was demonstrated by Ellaway and Murthy (1985a, b) in somewhat artificial preparations whose central nervous system was probably in a tonic and rather undifferentiated state. The emergence of synchronised firing among γ-motoneurones upon stimulation of cutaneous inputs suggests that more differentiated input patterns might enhance synchrony. This could also occur with inputs descending to the spinal cord, perhaps in conjunction with segmental inputs, during the performance of more natural motor acts. Under these conditions, the γ-motoneurones might receive a differentiated input preprocessed by interneuronal systems, and might also produce a more topographically ordered correlation pattern.

Secondly, during early ontogeny, correlations between motoneurones, including γ-motoneurones, might play a role in establishing topographical mappings as suggested by Fawcett and O'Leary (Sect. 1.3.4). Such mappings exist for α-motoneurones (Chap. 2), but probably also for γ-motoneurones (Huhle 1985). However, the degree to which the fine grain of neuronal firing patterns is of importance in this regard remains to be elucidated.

2.5 Correlations Between Muscle Afferent Discharges

As pointed out by Ellaway and Murthy (previous section), an "intrinsic interest" in synchrony of γ-motoneurone firing arises from this synchrony being a potential "synchronizing influence in the muscle spindle loop feeding back to motoneurones". More generally, this also applies to synchrony between α-motoneurone discharge patterns and its influence on spindle as on Golgi tendon organ afferents.

α- and β-motoneurones each innervate a number of skeletal (extrafusal) muscle fibres which are usually distributed over a restricted "territory" of a particular muscle. Each of these territories contains a certain number of muscle spindles and Golgi tendon organs whose discharge patterns are then altered and possibly correlated (among each other) by contraction of the respective motor unit (Fig. 12 A). Hence, there is a *divergence* of the effects of each motor unit on anything from a few to a substantial number of muscle spindle and Golgi tendon organ afferents. On the other hand, since territories of different motor units overlap, there is a *convergence* of effects on each muscle spindle or Golgi tendon organ from varying numbers of motor units (Fig. 12 B). Similar divergence–convergence distribution patterns hold for the fusimotor innervation of muscle spindles by β-motoneurones (Fig. 12 C) and γ-motoneurones (Fig. 12 D). These connexions thus establish a multi-input, multi-output system of high complexity that can in principle be represented by one or more matrices collecting the parameters which characterise signal flow through the system. The question then is whether these matrices have a nonrandom internal structure, i.e. whether the manifold signal flow lines exhibit an inhomogeneous but ordered distribution of gains or other parameters (cf. Windhorst 1978 b).

Essentially the same divergence–convergence structure is found in the feedback connexions of muscle spindle and Golgi tendon organ afferents onto motoneurones. By virtue of their extensive divergence, muscle afferents may exert both mono- and polysynaptic influences on the discharge patterns of many motoneurones and interneurones and, hence, cause correlations between their firings. Since the peripheral (muscular) and central (spinal) divergence structures are coupled, tendencies towards synchrony of firing of parallel neuronal elements could spread and reinforce each other until the whole multichannel loop oscillates. (This may also occur in other networks of similar structure which are found ubiquitously in the central nervous system.) This problem is dealt with in more detail in the next chapter. The question of concern here is which types of correlation are found between the firing patterns of these afferents, how they originate and how they modify the correlations between motoneurones and other neurones.

Correlations between muscle afferents have not yet been studied very extensively. For various reasons, *Ia fibres from primary muscle spindle endings*

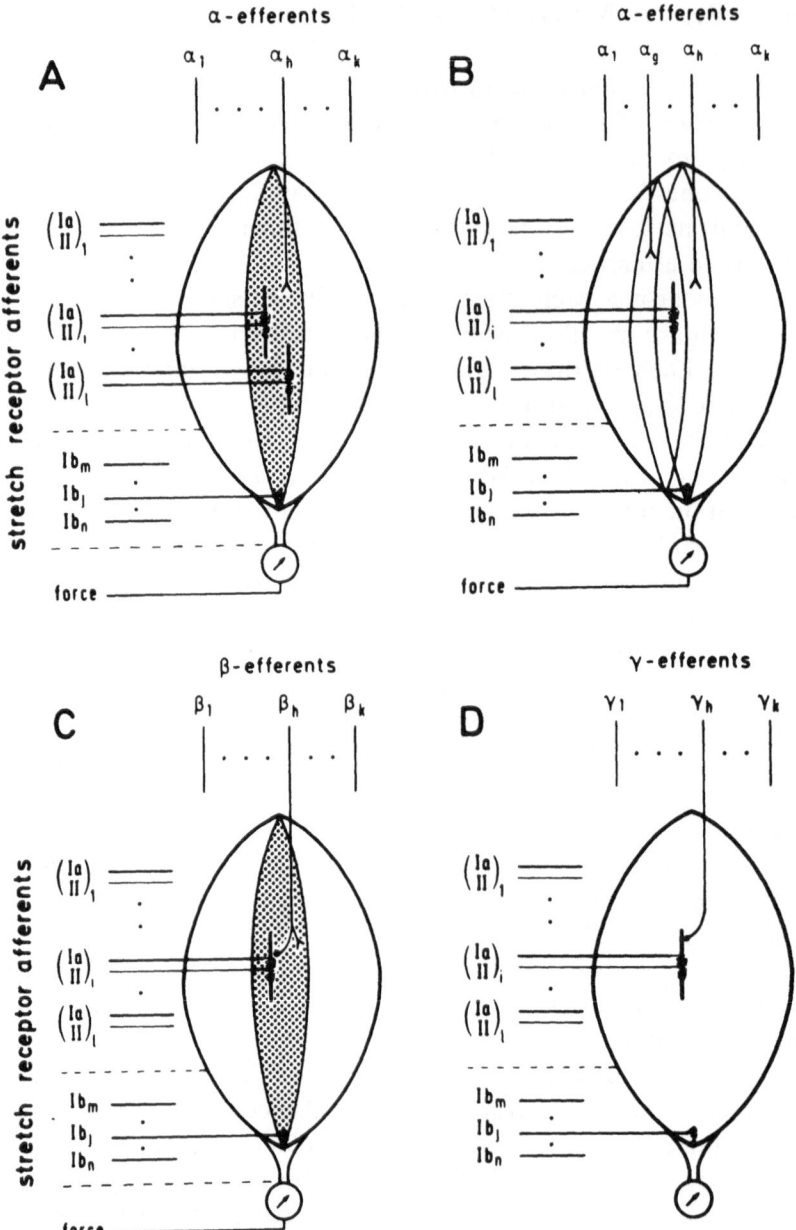

Fig. 12 A–D. Parallel structure of signal flow through skeletal muscle from motor efferents to muscle spindle and Golgi tendon organ afferents. **A** Divergent action of an α-efferent on several (here two) muscle spindles and Golgi tendon organs (only one shown for simplicity); **B** Convergent action of several (here two) α-efferents on a spindle and Golgi tendon organ; **C** Dual action of β-motoneurone axons on spindles via extrafusal and intrafusal effects; **D** Innervation of spindles by γ-efferents. The fusimotor effects on spindles sketched in parts **C** and **D** also show a convergence-divergence structure as do the skeletomotor effects sketched in parts **A** and **B**. Fusimotor effects in parts **C** and **D** represent a layer of signal transmission parallel to the skeletomotor effects in parts **A** through **C**

have found most interest (Windhorst 1977; Windhorst and Meyer-Lohmann 1977; Windhorst 1978a; Schwestka et al. 1981; Windhorst and Schwestka 1982; Osborn and Binder 1987; Kirkwood and Sears 1982a). This is due partially to their high sensitivity to small mechanical perturbations reaching them from intra- as well as extrafusal sources, and partially to their relatively strong monosynaptic effects on α-motoneurones. (In part, the interest in spindle afferent correlations also originates from their relevance to interpreting the results of spike-triggered averaging used to establish the central connectivity of Ia fibres; see, e.g. Kirkwood and Sears 1982a; Hamm et al. 1985b).

Since muscle afferents do not interact synaptically at their site of origin (in the muscle), correlations between their discharge patterns would be expected to arise from common input sources. Thus, correlations between Ia fibres, between group II fibres from secondary muscle spindle endings and between the former and the latter could result from

1) common fusimotor (β- and/or γ-) innervation of different spindles from branches of the same input fibres (Windhorst 1977; Fig. 12C and D);

2) synchronisation of two or more fusimotor (β- and/or γ-) fibres which innervate different spindles (Murthy and Yoon 1979);

3) any combination of the mechanisms listed in 1 and 2;

4) internal dynamic muscle length changes which originate from skeleto-motor (α- and/or β-) activity, that is from the unfused contractions of motor units during muscle contraction, and which may commonly affect the firing of two (or more) spindles (Fig. 12A and C);

5) other sources such a blood pressure pulsations (Ellaway and Furness 1977; Kirkwood and Sears 1982a) and mechanical artefacts (e.g. Kirkwood and Sears 1982a), which may influence the firings of different spindles;

6) external (perhaps imposed) length alterations, in particular short-lasting disturbances, arising from abrupt changes of the mechanical load.

These manifold sources often render interpretation of empirically observed correlations between Ia fibres difficult.

The present account will concentrate on the correlations introduced by extrafusal activity into discharges of spindle afferents originating in the same skeletal muscle (item 4).

2.5.1 Correlations Between Ia Afferent Discharges in Decerebrate Cats

Correlations between firing patterns of different Ia afferents were described by Windhorst (1977, 1978a) and Windhorst and Meyer-Lohmann (1977) in decerebrate cats. These were attributed to spontaneous fluctuations of extrafusal muscle contraction (and possibly also of concurrent γ-action) which, due to unfused contractions of the active motor units, produced

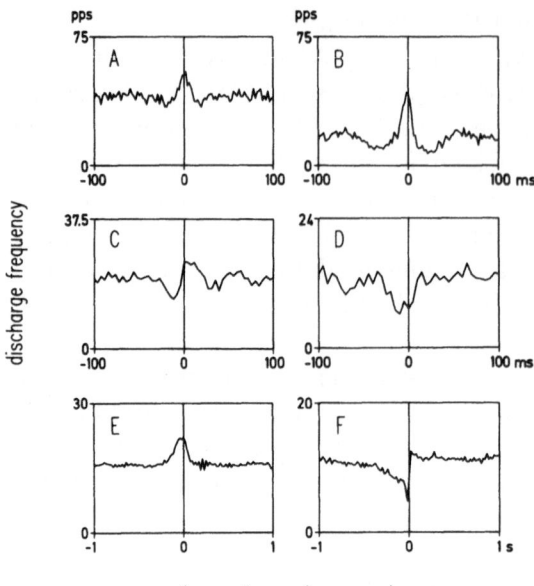

time τ from reference pulse

Fig. 13 A–F. Examples of cross-correlations between muscle afferents. **A** and **B** Positive corre-alations between Ia afferents from spontaneously contracting triceps surae muscles in decere-brate cats (two experiments); **A** data length 102 s, bin width 2 ms, 5039 ref. events; **B** data length 129 s, bin width 2 ms, 2447 ref. events. **C** Correlation between a Ia fibre from a primary spindle ending (ref.) and a group II fibre from a secondary ending (corr.) originating both in the cat medial gastrocnemius muscle; extrafusal activity was generated by stochastic stimulation of three motor units; data length 200 s, bin width 4 ms, 6045 ref. events; **D** Inverse (negative) correlation between two Ia afferents, one (ref.) originating in the proximal cat semitendinosus muscle compartment, the other (corr.) in the distal compartment; stochastic stimulation of three motor units; data length 87 s, bin width 5 ms, 1392 ref. events; **E** and **F** Two examples of correlations between cat soleus Ia afferents with longer time courses (note the changed time scale: *abscissa*); **E** stochastic stimulation of two motor units; data length 230 s, bin width 20 ms, 2990 ref. events; **F** stochastic stimulation of three motor units; data length 305 s, bin width 20 ms, 3829 ref. events. The last four cross-correlograms (**C–F**) were obtained from experiments performed in cooperation with U. Niemann and C. Schwarz

stochastic internal muscle length changes which affected groups of muscle spindles. Examples of these correlations are presented in Fig. 13 A and B, the muscle of origin of the Ia fibres being the triceps surae. It can be seen from these examples that the peaks in the correlograms are usually broader than the "short-term" peaks found, for instance, in motoneurones (Figs. 10 A, B, and 11, left column). They thus resemble more closely "broad-peak" correla-tions, which were also found by Abeles (1982) for pairs of cerebro-cortical neurones and by Ebner and Bloedel (1981b) for pairs of cerebellar Purkinje cells. The reason for the peaks being broad for spindle afferent correlations is easy to grasp. Whereas the presynaptic input events causing synchrony in

α-motoneurones have a brief ("sharp") time course (Sect. 2.1.2, item 3), the common extrafusal inputs to spindles have a slow time course (motor unit contractions).

Windhorst and co-workers tried to pin down the internal muscle length changes eliciting the correlated Ia discharges by averaging muscle tension fluctuations with respect to discharges of the correlated (and of noncorrelated) Ia fibres. They suggested that so-called "positive" correlations (with peaks near zero time in the correlograms as in Fig. 13 A and B) occurred most often between afferent fibres whose spindles were situated in some vicinity within the muscle and were thus influenced by similar internal length changes. "Negative" correlations between afferents with troughs near the time origin (Windhorst 1978 a) were seen occasionally for spindles oriented in series to each other (with respect to extrafusal muscle fibre course). Remember that the negative type of correlation was also found between discharges of retinal ganglion cells (Mastronarde 1983 a), cerebro-cortical cells (Abeles 1982; see Fig. 2 B), cerebellar Purkinje cells (Ebner and Bloedel 1981 b) and γ-motoneurones (see Fig. 11, middle column).

Osborn and Binder (1987) have recently re-investigated some early results of Windhorst and co-workers including other muscle afferents. In decerebrate or decapitate cats (treated with L-Dopa and nialamide), 49 pairs of afferents (6 Ia–Ia combinations; 7 group II–II; 8 Ia–Ib; 10 Ia–II; 12 Ib–II) were tested for correlations, 14 (28%) showing significant temporal correlations. The authors concluded that, in general, the proximity of muscle receptors (tested by gently probing the muscle) was "a poor predictor of the degree of correlation". They also described 5 (out of 12) Ib–II afferent pairs, which displayed negative correlations, and suggested that they might be due to the "dyad" receptor arrangement described by Marchand et al. (1971). But these correlated pairs were not always located close together in the muscle. Osborn and Binder (1987) concluded that although two receptors may appear to be in close vicinity, they may not be responsive to the same set of motor units. They are probably right, but their findings do not contradict Windhorst's findings as much as it might appear at first glimpse. Firstly, Osborn and Binder (1987) investigated all kinds of stretch receptor afferents, not only pairs of Ia afferents. Secondly, proximity per se is indeed no exact predictor of the kind and strength of correlation, even for Ia-Ia pairs, one reason being that probing the muscle from the surface cannot establish the exact anatomical relation between receptors. Depth is not well estimated, and, moreover, the anatomical attachment of spindles and Golgi tendon organs to particular sets of motor units (Botterman et al. 1978) may be more important than mere distance between receptors (see Binder and Osborn 1985). Indeed, even muscle spindles (or Golgi tendon organs) situated at opposite ends of muscle fibres (classified as "distant" receptors by Osborn and Binder 1987) should be expected to be correlated by contraction of these fibres. These receptors could then be considered as functionally close to each other. This does not preclude the possibility that neighbouring receptors are more likely to be correlated than remote ones. At least Binder and Osborn (1985) found that tendon organs located in the same region of tibialis posterior are much more likely to be excited by the same motor units than are tendon organs in widely separated regions of the muscle. This may apply more generally also to spindles and other muscles. Nonetheless, it may seem appropriate to replace the (essentially two-dimensional) "spatial proximity" used to define association of receptors by a "functional proximity" defined in terms of motor units having a common influence on the discharge patterns of the respective receptors. These two definitions of "proximity" may differ to a greater or lesser extent depending, firstly, upon muscle structure, and, secondly, upon the exact fixation of muscle spindles to extrafusal muscle fibres, these factors being little known at present. (Hamm et al. 1985 b used a synchronisation index to demonstrate weak tendencies towards synchronous discharge of the two types of spindle afferent and Golgi tendon organ afferents with other muscle afferents in *passive* medial gastrocnemius muscle.)

"Negative" correlations need not only originate from spindle-Golgi tendon organ associations as "dyads" (review by Botterman et al. 1978; see also Richmond and Abrahams 1975 b; Richmond and Bakker 1982), which are necessarily in close spatial proximity, but they may also occur between receptors (also spindle pairs) separated by some distance (see Schwestka et al. 1981; see also next section). The functional in-series arrangement of spindle receptors supposed to lead to "negative" correlations is particularly well defined in muscles composed of separate in-series compartments, such as the semitendinosus and certain neck muscles (cf. Fig. 36). Examples for such negative correlations are displayed in Figs. 13 D and 15 (left and right columns, lower row). Negative correlations may also be supposed to occur for pairs of Ib fibres from Golgi tendon organs, because motor units directly coupled by one or two muscle fibres to one Golgi tendon organ might excite this receptor on contraction and "unload" a neighbouring tendon organ (cf. Binder and Osborn 1985).

One might argue that there is little sense in studying "intermodal" correlations such as those between spindle and Golgi tendon organ afferents. However, this argument overlooks the possible convergence of these afferents onto spinal neurones. For instance, Ib and Ia afferents commonly excite laminae V–VI interneurones mediating "nonreciprocal inhibition" (Sect. 2.9.1.1). Correlations in discharge of the converging fibres could therefore be of significance for the effects exerted on such interneurones. In this context, "sensory partitioning", that is the predominant sensitivity of both types of receptor to motor unit activity in their vicinity, is probably an important determinant. It should also be mentioned that *cutaneous receptors* may be quite sensitive to muscle activity occurring beneath their receptive fields (Kuipers et al. 1986), and again the related afferents converge on the above interneurones as on other spinal neurones receiving Ia and/or Ib input.

2.5.2 Correlations Between Spindle Afferent Discharges Time-Related to Motor Unit Twitches

The findings described for spindle afferents from triceps surae muscles of decerebrate cats were essentially confirmed in experiments in which several motor units of cat hindlimb muscles were activated by electrically stimulating their axons (Schwestka et al. 1981; Windhorst and Schwestka 1982; Fig. 13 C–F, and Figs. 14 and 15). This experimental approach has the advantage that the input patterns, i.e. the motor unit activation patterns, are known and controllable. The cross-correlations between afferent discharges as shown in Fig. 13 were calculated without explicit regard to the inputs (motor unit contractions). However, the temporal relation of such correlations to the generating inputs is of importance for the effects exerted by correlated discharges at the postsynaptic (spinal or even higher) level. Here are therefore illustrated typical examples of cross-correlations between spindle afferent discharges time-related to motor unit contractions. The method of calculating such event-related cross-correlations is described in Appendix C.

In the experiment underlying Fig. 14, two motor units of the cat medial gastrocnemius muscle were stimulated with independent random patterns at mean rates of $\lambda_1 = 7.8$ pps and $\lambda_2 = 7.7$ pps. The activities of three muscle spindle afferents were recorded simultaneously, of which two (Ia fibres) were selected for the subsequent analysis. These two afferents are denoted by A and B and had mean discharge rates of 39.4 pps and 23.8 pps, respectively. In Fig. 14, the upper correlogram in the right column was computed for the two input patterns and demonstrates their independence. The lower right correlogram is that between the afferents' discharge patterns and shows a small broad-peak positive correlation. The average action of the two motor units (denoted MU 1 and MU 2; see top) on each of the two afferents A and B is represented by the peristimulus-time histograms (PSTHs) superimposed in the lower row (left two columns). MU 1 had a relatively strong effect on afferent A (thick line) in that it produced an initial rate reduction during tension rise (due to spindle unloading) followed by a "relaxion discharge" and a further oscillation in rate; its effect on afferent B (thin line) was weaker. The effects of MU 2 on the two afferents were the other way round. The middle row in the left two columns shows the arithmetic means between the lower PSTHs (see below); and the upper row shows the cross-correlation coefficients related to the times of activation of either motor unit.

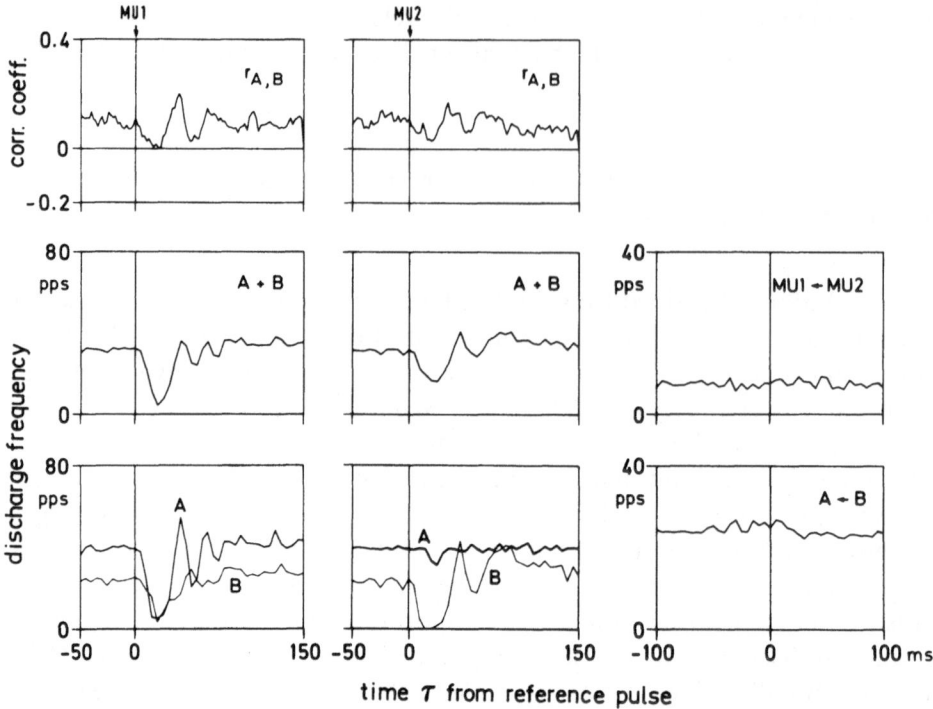

Fig. 14. Interactions between two motor units (denoted *MU 1* and *MU 2*) and two Ia afferents (denoted *A* and *B*) in the medial gastrocnemius muscle. The two motor units were stimulated randomly at mean rates of $\lambda_1 = 7.8$ pps and $\lambda_2 = 7.7$ pps. The afferents had mean discharge rates of 39.4 pps (*A*) and 23.8 pps (*B*). The two cross-correlograms in the *right column* show the statistical independence of the two input patterns (*top*) and a weak tendency of the two afferents to discharge in a broadly synchronised manner (*bottom*). The lower panels in the *left two columns* display conventional PSTHs showing the average effects of the two motor units (one column for each motor unit) on the two afferents (superimposed). The *middle row* displays arithmetic means of the PSTHs below; the *upper row* shows stimulus-related cross-correlation coefficients. The latter were calculated for a pulse width $\Delta t = 18$ ms. Bin width for the correlograms, 5 ms. Number of reference events: *left column (MU 1)* 2430; *middle column (MU 2)* 2413; *right column, upper plot* 2430; *right column, lower plot* 11846. (Windhorst, Koehler and Schwarz, unpublished)

One might argue that the stimulus-related cross-correlation (upper row) adds hardly any information to that contained in the PSTHs. Within a postsynaptic neurone contacted by the two correlated afferents, the firing probability might be presumed to be adequately predicted by the summed impulses of the two (or more) presynaptic fibres. Therefore, the stimulus-related averaged summed pulse density could be a sufficient measure. In the case of the example illustrated in Fig. 14, such a sum is represented by the sums of the PSTHs or, in this case, by their arithmetic means (sums divided by 2), as shown in the curves in the middle row. At first glimpse, the time-courses of these curves indeed appear similar to the time-courses of the

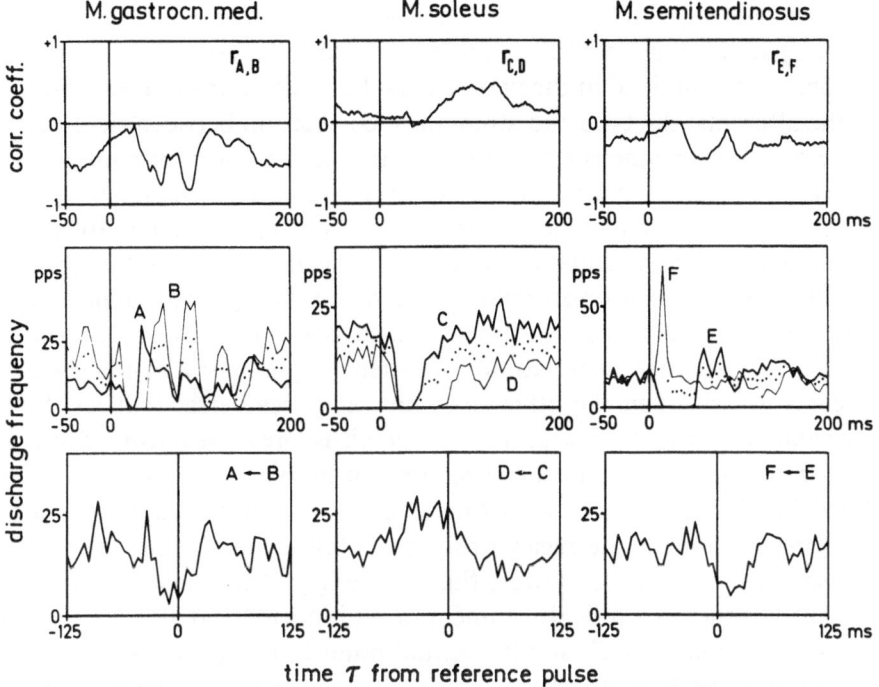

Fig. 15. Stimulus-related correlation functions (*upper row*), PSTHs (*middle row*) and CCHs (*lower row*) for three pairs of spindle afferents from three hindlimb muscles (*columns*). Afferent *A* and afferents *C* through *F* were Ia fibres; afferent *B* had a conduction velocity of slightly over 72 m/s and was also classified as Ia, although it might perhaps better be referred to as "intermediate". In each case a single motor unit was stimulated at about 8.5 pps (medial gastrocnemius: semiregular; soleus: periodic; semitendinosus: stochastic stimulation). Δt for medial gastrocnemius and semitendinosus examples: 30 ms, for soleus example: 50 ms. Bin widths for the correlograms, 5 ms. Number of reference events between 486 and 834. (With permission from Windhorst, Koehler and Schwarz 1987; their Fig. 4)

stimulus-related cross-correlations. The above argument nonetheless does not hold for the following reasons. Each single PSTH estimates the *mean pulse density* (as a function of τ around reference events) of one output averaged with respect to the temporal sequence of reference events. It thus "selfishly" reflects the time-averaged relation of each single output to the particular input, without regard to other parallel outputs. Thus, two PSTHs for two outputs (or the sum of both) tell us nothing about how and to what degree the two output signals co-vary, i.e. change their pulse placing from one stimulus occurrence to the next in a correlated manner. The stimulus-related correlation function tells us exactly what PSTHs cannot: It estimates *the degree to which the firings of two outputs are correlated in subsequent "realisations" (stimulus occurrences)*, quite apart from the mean pulse densities expressed by the two PSTHs. It thus determines the degree of co-variation of the firing density fluctuations around their respective means (see Eq. (C1) in App. C) and thus adds information on the firing patterns of the two afferents beyond that contained in the two PSTHs. In addition to the mean firing rates of the afferents, the type and strength of correlation between their discharge patterns co-determine the firing probability and pattern of commonly contacted postsynaptic neurones. For example, if at a particular time after an event the two afferents tend either to fire or not to fire in a coupled fashion (yielding a positive correlation coefficient), the firing probability of the postsynaptic cell is disproportionately increased during occurrence of those events eliciting the common afferent discharges and not during the others. In contrast, a negative correlation coefficient expressing a tendency for one afferent to fire when the other does not, and vice versa, reduces the firing probability of the postsynaptic cell, which is prevented from firing in response to the random occurrences of nearly synchronous afferent spikes (Sect. 2.8.2). The event-related time-course of such correlations is a factor that, in addition to the mean firing probabilities of the two afferents (expressed in the PSTHs), changes the excitability of the postsynaptic cells.

These rather theoretical arguments are supported and visualised by experimental observations. First, an essential point to recognise in Fig. 14 is that, on closer inspection, the time courses of the stimulus-related correlations (upper row) do not correspond very well to those of the PSTHs (lower row) or their sums (middle row), particularly in the left column. (Of course, this does not in general preclude the possibility of similar time courses of PSTHs and stimulus-related cross-correlations, as for example in Fig. 15, middle column.)

The difference of information contained in PSTHs and stimulus-related cross-correlations is even more obvious in the examples displayed in Fig. 15 (left and right columns). Two examples from the medial gastrocnemius (left column) and the semitendinosus (right column) illustrate negative types of correlogram (lower row), i.e. the pulse density of one afferent is depressed around the (reference) pulses of the other. The stimulus-related cross-correlations (upper row) have roughly the same shape with negative coefficients almost everywhere. This implies that the occurrence of a pulse in one afferent train tended to concur with the absence of a pulse (at about the same τ) in the other, and vice versa. The strength of this tendency varies with τ and is greatest at two values of τ which fortuitously nearly coincide for the two examples (medial gastrocnemius and semitendinosus). This type of negative correlation indeed amounts to an inhibitory effect, in the statistical sense, with respect to the firing probability of the postsynaptic neurone commonly contacted by the two afferents (see Sect. 2.8.2), quite in contrast to what summed PSTHs (dots in upper plots) would suggest. As seen in the stimulus-related correlation functions (upper row, left and right plots), this effect has a definite time course bound to the motor unit contraction.

The source of the negative correlations is most easily understood for the semitendinosus (cf. Figs. 13 D and 15, right column). This muscle is composed of two compartments in series, separated by a tendinous inscription (cf. Sect. 3.3.2, and Fig. 36 A). The stimulated motor unit and spindle E in Fig. 15 were located in the distal compartment, whereas spindle F was situated proximally and was therefore pulled upon by contractions of the distal motor unit, thereby producing the "early" discharge in the PSTH (right column, middle plot, thin curve). The near coincidence of this early firing with the spindle pause in spindle E was probably one, though not the only, cause of the negative correlation which shows up in the CCH (lower left plot). At least the time course of the correlation function $r_{E,F}$ is not totally accounted for only by this relation of early discharge and spindle pause.

Early discharges may however also occur in muscles not separated by tendinous inscriptions into serial compartments (Fig. 15, left column). This spindle discharge feature was extensively studied during the 1950's and 1960's (Granit 1970; Matthews 1972). It could result from three causes (1) Ephaptic excitation of spindle sensory endings by motor fibre action potentials arriving in the muscle; this cause is unlikely to be of any importance when single motor units are activated. (2) Initial mechanical stretch of the spindle if it is situated near the origin or insertion of the muscle fibres being activated (Robrecht 1971). F. Richmond wonders "if b_2c spindles, which can be deeply embedded in the musculotendinous junction, could be a particularly common source for early discharges. However, the in-series effects may occur not only at the musculotendinous junction but even in the muscle mass. Many long muscles are composed of short, in-series fibers, whose recruitment patterns might cause an early discharge of spindles" (Letter of 6 August 1984). This is confirmed by anatomical data on the cat splenius muscle (Richmond et al. 1985), the cat sartorius, tenuissimus and semitendinosus muscles (Loeb et al. 1987). (3) Skeleto-fusimotor β-innervation (Emonet-Dénand et al. 1975; Jami et al. 1982) could simultaneously activate intrafusal and extrafusal muscle fibres. Especially static fusimotor fibres (whether γ or β) can "drive" Ia fibres at high frequencies (up to about 60–80 Hz, Boyd 1981) and may, under favourable circumstances, be expected to generate 1 : 1 spike translation even at lower rates. Static β-axons innervate (extrafusal) muscle units of the fast type (FR and FF according to Burke 1981) (Jami et al. 1982; Murthy 1983). Yet the negative correlation for the medial gastrocnemius muscle (Fig. 15, left column) was apparently not primarily due to the involvement of early discharges, but to the different timing of later spike occurrences in the two spindle fibres (see PSTHs middle plot).

Within a total of about 200 stimulus-related correlation functions calculated for pairs of afferents from the three muscles (medial gastrocnemius, semitendinosus and soleus) without regard to the strength of effects exerted by the stimulated motor unit on the afferents, about one third presented random fluctuations of the coefficients around zero (without characteristic pattern), one third showed predominantly positive coefficients (as $r_{C,D}$ in Fig. 15), one sixth exhibited predominantly negative coefficients (as $r_{A,B}$ and $r_{E,F}$ in Fig. 15), and one sixth of the coefficients varied between positive and negative values. Thus the predominantly negative correlation functions occurred half as often as the predominantly positive correlation functions. Also, predominantly negative correlations seem to be rarer in soleus than in the other two muscles.

It must be stated that all these fine features of afferent discharge patterns originating from muscle receptors have not yet been sufficiently studied experimentally. This applies particularly to "intermodal" interactions of various types of muscle afferents, the study of Osborn and Binder (1987) and of Hamm et al. (1985b) being the only exceptions at present. It is also important to emphasise that all the previous considerations have so far excluded muscle afferents of group III and IV, some of which are sensitive to mechanical stimuli (e.g. Schmidt et al. 1981; Ellaway et al. 1982a) and exert oligo- and polysynaptic effects on motoneurones, for which correlations between firing patterns of these afferents might be very important. Furthermore, various kinds of afferents from nonmuscular receptors, such as joint, cutaneous and Pacini afferents, whose discharges might also be modulated by unfused motor unit contractions, could be of importance, particularly since these afferents converge on various spinal interneuronal systems, so that their effects might depend on "intermodal" correlations as well. Needless to say, no relevant studies on such correlations are available so far.

2.6 The Effect of Fusimotor Input on Spindle Afferent Correlations

Fusimotor (γ- and β-) innervation may have complicated effects on muscle afferent correlations, which are difficult to assess. On the one hand, β- and γ-fibres, by diverging to several spindles, provide common input to the latter and can thus be expected to correlate muscle afferent discharges (see above), the more so the more the input fibres are synchronised themselves. However, on the other hand, since many fusimotor fibres usually converge on one spindle, an *uncorrelated* input on these fibres could also de-correlate Ia fibre discharges correlated by other sources. This mechanism works by the injection of uncorrelated noise into the encoders of different spindle afferents whereby spike initiation is randomised. Note that, in this role, the fusimotor system would function in a diffuse, undifferentiated and unspecific manner. (A similar role was attributed by Adam et al. 1978 to the Renshaw cell network which was supposed to be able to de-correlate α-motoneurone discharges; see also Gelfand et al. 1963, and Sect. 3.4.1). Indeed, Inbar et al. (1979) showed that in decerebrate cats with intact ventral roots (hence high fusimotor output) the correlations between discharges of Ia fibres (induced by external inputs muscle stretches and 60-Hz line current) were much weaker than with cut ventral roots (abolished fusimotor output to the muscle under study). In a way, these authors were lucky in using decerebrate cats because Ellaway and co-workers (Sect. 2.3.1) found little synchronisation between γ-fibres in those preparations. Also, the fusimotor activity may be unnaturally high and undifferentiated in decerebrate cats. As argued above (Sect. 2.4), the patterns of correlation between fusimotor fibres might show sharper and physiologically more meaningful profiles under more natural conditions than found in decerebrate or even spinal cats. Thus, this case is still open and may not be settled in the form of a clear alternative as presented above. The interaction of the various input systems impinging upon spindles (and Golgi tendon organs) may thus create complicated but physiologically significant correlation patterns in the muscle afferent feedback to the spinal cord.

2.7 Summary

Muscle spindles and Golgi tendon organs are multiple-input systems in that they are subjected to skeletomotor (α), fusimotor (γ), skeletofusimotor (β) and other (mechanical) inputs. Since each single input may modulate the discharge of smaller or larger numbers of afferents, the latter's firings may be correlated by the mechanism of common input. Some studies are available on correlations between spindle and tendon organ afferents resulting particularly from skeletomotor activity originating either spontaneously in decerebrate (or decapitate) cats or by artificial stimulation of motor units. The patterns of correlation found range from tendencies towards synchronous discharge (although usually of the broad-peak type) to inverse correlations. Intermediate forms also occur. These patterns are not unique, since they were also described for neurones at supraspinal and spinal levels. If they result mainly from extrafusal activity, the different patterns of afferent correlation are probably associated, at least to some extent, with different spatial arrangements of the receptors within the muscle.

If correlations between afferent firing patterns are studied in temporal relation to motor unit activation, they can be shown to vary in sign and degree in the course of motor unit contraction. That is, both positive and negative twitch-related correlations, whose degree varies in the course of the twitch, are found. In conjunction with mean firing rates, this variability would co-determine the discharge probability of any postsynaptic neurone commonly contacted by the correlated afferent fibres. That is, positive correlations would enhance, and negative correlations reduce, the probability of postsynaptic discharge. Whilst this argument may appear plausible, the significance of such features remains to be established experimentally (but cf. the computer simulations described in Sect. 2.8.1). This becomes immediately obvious if one considers that the size of the central effects will strongly depend on the degree of divergence (e.g. of spindle projections) and on the extent of compartmentalisation. In addition, it is likely to depend on the time constants of postsynaptic potentials elicited in the receiving neurones, as well as on numerous other factors (as discussed in Sect. 2.8.2).

Fusimotor activity may be imagined to have contrasting effects on the correlated activity of muscle spindle afferents. On the one hand, if parallel fusimotor fibres exhibit uncorrelated firing patterns themselves, they could de-correlate muscle spindle afferents, since each one of the latter probably receives input from a different set of fusimotor fibres. On the other hand, fusimotor fibres, by diverging to different spindles, provide them with common input and could thereby correlate their firing patterns. Correlations between discharge patterns of fusimotor fibres could enhance this effect. The effect of fusimotor input on spindle afferent correlations is therefore complicated and not easily assessed. Under natural conditions of muscle contraction, the set of spindles in each muscle may be assumed to integrate the different patterns of inputs originating from the sets of active α-, β- and γ-motoneurones and to generate an afferent firing pattern with an intricate, though not necessarily diffuse, shape. These shapes very probably depend upon the task to be performed (cf. Sects. 4.11, 4.12).

2.8 Effects of Correlations Between Presynaptic Fibres on Spinal Neurones

What effects particular patterns of correlation between muscle afferents have on spinal postsynaptic neurones cannot be predicted in general because they also depend upon the pattern of convergence on those neurones. For example, compare two cases. An α-motoneurone of a cat hindlimb muscle is known to receive monosynaptic input from nearly 100% of its homonymous Ia afferents, the average unit EPSP amplitude being of the order of 100 µV. Dorsal spinocerebellar tract (DSCT) cells receive monosynaptic input from a much smaller sample of Ia fibres (10–18 in the case of the gastrocnemius muscle; Kröller and Grüsser 1982), the EPSP amplitude being much larger (Tracey and Walmsley 1984, with further references). Thus, in the latter case, the importance of discharge fluctuations in afferents for postsynaptic firing should be greater than in the former case. Ia inhibitory interneurones mediating reciprocal inhibition might occupy an intermediate position.

More generally, the effects of correlations between presynaptic (input) terminals on postsynaptic discharge properties depend upon a number of factors which have not yet been studied very extensively in the present context, neither theoretically nor experimentally. Some of the reasons for this deficiency will soon become apparent. If information can be carried and processed in terms of the particular code which is defined by correlations between neuronal firings as postulated by several brain models (Sects. 1.3.1 and 1.3.2), then this processing certainly is one relying on fine features of discharge patterns. The evaluation of these fine features by neuronal structures may again be expected to depend on fine features. What will be discussed in quite some detail in this section is therefore the various factors of structure and function of neuronal connexions that may influence the transmission and evaluation of correlations in presynaptic fibres. The connexions between muscle stretch receptor afferents and motoneurones provide a good example and basis for this discussion particularly because the monosynaptic Ia fibre-α-motoneurone synapse is the synapse studied best. Equivalent conditions, after adequate modifications, may however prevail all over the central nervous system. In this sense, the material presented in this section is in many respects paradigmatic.

2.8.1 An Introductory Model

To get a first impression of the effects to be expected, it is worth introducing and briefly reviewing a notable theoretical investigation.

An important computer simulation study related to the above problem has been presented by Segundo et al. (1968). These authors modelled a postsynaptic neurone whose characteristics with respect to membrane potential and threshold behaviour after a spike could be varied. Simulated were 64 presynaptic terminals, the statistical properties of the spike trains carried by them being variable over a wide range. The presynaptic discharge patterns ("forms") could be relatively regular or irregular (Poisson), independent of each other or correlated (synchronised) to different degrees. The main conclusions run as follows. (1) If the postsynaptic cell is influenced by few and strong presynaptic fibres, the temporal fine structure of the output spike train depends on the presynaptic form, i.e. is influenced by the statistical structure of the presynaptic spike trains. (2) With completely independent input spike trains on the presynaptic fibres, as the number of these inputs increases and each becomes weak, the individual presynaptic form becomes less important, and the postsynaptic cell generates the same output regardless of the detailed structures (regular or irregular) of the corresponding input trains. (3) The individual presynaptic form is important again when sets of presynaptic fibres are correlated, e.g. when they show tendencies towards synchronised firing.

Segundo et al. (1968) discuss some possible functional implications among which the most important ones are the following. The authors suggest a functional classification of presynaptic fibre terminals based upon their influence on the postsynaptic output. Strong terminals eliciting large postsynaptic potentials are dominant and exert a powerful influence by way of the statistical structure of the presynaptic input spike trains. In contrast, weak terminals eliciting small postsynaptic potentials have an influence that is contingent upon the degree of interdependence. If their discharge patterns are weakly or not at all correlated (and their mean discharge rates are not very low), these terminals may act exclusively by way of their mean rates and provide a smooth variation of the postsynaptic membrane potential and firing rate. If, on the other hand, there is a tendency towards synchronous firing, the set of weak but correlated terminals can assume a relatively dominant role and impose a precise relation between the timing of presynaptic events (synchronous inputs) and postsynaptic discharge. As might be expected, intermediate forms between these two cases may also occur.

Based on these considerations, Segundo et al. (1968) conclude that the degree of correlation between weak presynaptic input fibres must be very significant in natural operation where shifts in presynaptic correlation may be expected to concur with shifts in the form of presynaptic discharge patterns. This last idea is supported by the results of Schulz et al. (1985) described in Sect. 1.4.2. In this respect, the *degree of input correlation* has a significance of its own, independent of any other statistical parameters of the input discharge patterns (such as regularity of firing). It can lead to remarkable output modifications, irrespective of other input features. (This might argue for their role as a separate code.) The degree of input correlation is important also indirectly since it determines whether the numerous weak presynaptic inputs simply act as a bias related to their mean discharge rates or, alternatively, dominate the input as a complex function of several presynaptic statistical properties. Segundo et al. (1968) argue that their observations may shed some light on the possible functional roles of the small postsynaptic potentials which account for most of the input to neurones in centres with complex connectivity, such as the spinal interneuronal pools or the brainstem reticular nuclei. They imagine that presynaptic inputs may reflect different neural states (e.g. sensory conditions) by way of different discharge forms and different degrees of correlation. The role of weak inputs could therefore "alternate operationally between that of establishing a uniform background by way of the mean rates when uncorrelated, and that of eliciting rapid adjustments by way of their precise timing when they become more synchronous" (p. 169).

These considerations, general and speculative as they may be, may provide the "ideological" background for the things to come.

where y is the normalised membrane potential V ($y = V/\sigma$) and is assumed > 0. This probability P can thus be changed by a number of factors (see Abeles 1982; p. 22):

a) P increases by reduction of T (depolarisation of average membrane potential or lowering of threshold), and vice versa;

b) P increases by increase in σ, and vice versa.

The variable σ, in turn depends upon several factors.

α) Membrane conductivity (increase reduces σ, and vice versa);

β) Synaptic input patterns.

The relevant factors related to the last item become clear by simple modelling (Abeles 1982). Assume that

1) there are N excitatory synapses on the neurone;

2) each presynaptic terminal, on arrival of a spike, produces an EPSP of the form $A.\exp(-t/\varepsilon)$ (see Fig. 16C) where the amplitude A is presumed constant for all synapses;

3) all the presynaptic fibres fire at average rates of λ in an *uncorrelated* manner;

4) synaptic potentials sum linearly.

Then, the total variance, σ^2, of the transmembrane potential is given by:

$$\sigma^2 = N.\lambda.A^2.\varepsilon/2 \tag{6a}$$

$$\sigma = A.\sqrt{N.\lambda.\varepsilon/2} \tag{6b}$$

Thus, there are four factors determining σ, if inputs are considered uncorrelated. However, σ is also increased when presynaptic fibres tend to fire synchronously (see below) because large depolarising potential transients occur more often than randomly. Conversely, σ is decreased when presynaptic terminals are correlated negatively or inversely. Based on these (and more detailed) assumptions and calculations, Abeles concluded that synchronous activation of input fibres may be about ten times as effective in eliciting postsynaptic spikes as asynchronous input activity (see his Chap. 7). (This statement should be qualified by adding that the "efficiency factor" depends on the strength and temporal exactness of input correlation, see below). He therefore proposed that *cortical neurones* may act as *"coincidence detectors"*. This is a very strong hypothesis. However, matters may be much more complicated in firing neurones (see below).

B) α-Motoneurones firing repetitively. In continuously firing α-motoneurones the influence of synaptic noise on discharge probability is modulated by the average time-course of the post-spike membrane potential. Calvin and Stevens (1968) have investigated this situation. They proposed a simple model of the spike-generating mechanism which essentially consisted of a linear (depolarising) rise of membrane potential after a preceding spike (see also Calvin and Schwindt 1972; Schwindt and Calvin 1972; Schwindt 1973; Fetz and Gustafsson 1983). Superimposed on this slope was synaptic noise which determined the exact timing of the next spike when the average depolarising potential shift approached threshold. Thus, membrane potential fluctuations, including large depolarising transients caused by synchronised presynaptic inputs gain increasingly more influence upon spike initiation with time passed since the preceding spike. Exceptions may occur due to the fact that some motoneurones at times show "delayed depolarisation" after a spike, whereby a second action potential may follow closely upon the first one, thus producing "double discharges". Hence, during this period of delayed depolarisation, the structure of synaptic noise is also of importance. After a double discharge, the next spike is usually delayed by more than a mean interspike interval, thus also delaying the influence of synaptic noise. However, "re-excitation" of α-motoneurones need not be related to delayed depolarisation (Gogan et al. 1984).

The depth and duration of afterhyperpolarisations differ in the different motoneurone types (see Burke 1981), as do Ia EPSP amplitudes (see below). Thus the influence of input correlations should vary with motoneurone type. This is also the reason why totally different conditions may prevail in other spinal neurones such as Renshaw cells or Ia inhibitory interneurones.

2.8.2.2 Significance of the Temporal Structure of Input Correlations

We shall now start discussing specific features predominantly of the monosynaptic Ia fibre-motoneurone connexion.

As reported above (Sect. 2.5), the correlations between any two Ia fibres are not only of the synchronous type. To estimate the effects of different correlation patterns and strengths on motoneurone firing probability, Windhorst (1979c) also used a simple model that bore some similarity to that of Abeles (1982).

However, instead of considering all N synaptic inputs to the neurone, only *two* excitatory ones were taken into account. Thus the contribution of the i-th subset of $N_i = 2$ terminals to postsynaptic membrane potential fluctuations was simulated. This restriction was introduced in order to study the effects of *recorded Ia fibre discharge patterns*, of which sometimes only two were available. Moreover, the situation is easier to grasp, because characterising the correlation patterns of three parallel inputs (instead of two) would need three CCHs (instead of one), and would hence complicate matters greatly. The two input trains were passed through linear filters approximating the motoneurone membrane characteristics. The emerging "membrane potential" fluctuations were added linearly, and amplitude densities were estimated from them (as done for real motoneurones by Calvin and Stevens 1968; see above). These densities were compared under two conditions: naturally correlated Ia fibre discharges and artificially de-correlated patterns, the latter being obtained by shifting one train with respect to the other by about 1 s and reiterating the procedure. (Note that this de-correlation preserved the time structure of each single spike train.) In detail, action potentials were converted by electronic means into standard rectangular pulses of 0.4 ms width. The train of these pulses was then passed through a cascade of first-order linear filters with time constants $\varepsilon_1 = 0.66$ ms and $\varepsilon_2 = 8$ ms. The time course of the response of this combined filter to single impulses resembled that of an EPSP recorded by Burke (1967; EPSP labelled 1707 AII in his Fig. 11). Thus, this assumption was more realistic than the equivalent one of Abeles (see above). Nonetheless, other assumptions were the same, such as the equality of amplitude and time-course of the EPSPs generated by the two presynaptic fibres, and linear summation of EPSPs.

Consider an example. If the Ia spike trains underlying the cross-correlogram of Fig. 13 A are used as (superimposed) inputs to the filter, "membrane" potential fluctuations with a density as shown in Fig. 17 A (lower graph: curve labelled T) result. Potential amplitudes on the abscissa are given in arbitrary units. The density labelled T is not Gaussian, as also seen from the respective cumulative probability distribution in B (left curve) which lacks the typical S-shape. The non-Gaussian shape results from the specific EPSP shapes and the low number of spike trains involved. This amplitude density would hence reflect the contribution of the two (moderately synchronised) inputs to the overall potential fluctuations. De-correlation of the input spike trains (by shifting them 1 s in time with respect to each other) results in a density as given by curve C. The difference T-C is displayed in the upper graph (note the change in ordinate scale!) and shows that potential amplitudes have been redistributed in that, with partially synchronised inputs, more low and high and fewer medium-sized amplitudes occur than with uncorrelated inputs. This is typical for "positive" input correlations. Note that this redistribution effect would increase the standard deviation σ (see above). In other words, if again an arbitrary threshold is assumed, for instance that symbolised by a vertical dashed line at an abscissa value of 120 (in A, lower graph), the probability of crossing this threshold would be higher with synchronised than with uncorrelated inputs. This is made more explicit in B where the cumulative probability distributions for amplitudes above arbitrarily assumed threshold values are plotted for test (T) and control (C) cases and various thresholds. The final values (amplitudes at 200), which are the probabilities of crossing the given threshold, are always greater for T than for C curves, and the ratios of these values $(T/C|_{\infty})$ obviously increase with increasing threshold (see Windhorst 1979 c). This means that partial synchronisation of inputs is relatively more effective for a greater distance between average membrane potential and threshold.

Since it has been argued that the time derivative of EPSPs may also be of importance in spike generation (Knox 1974; Kirkwood and Sears 1978,

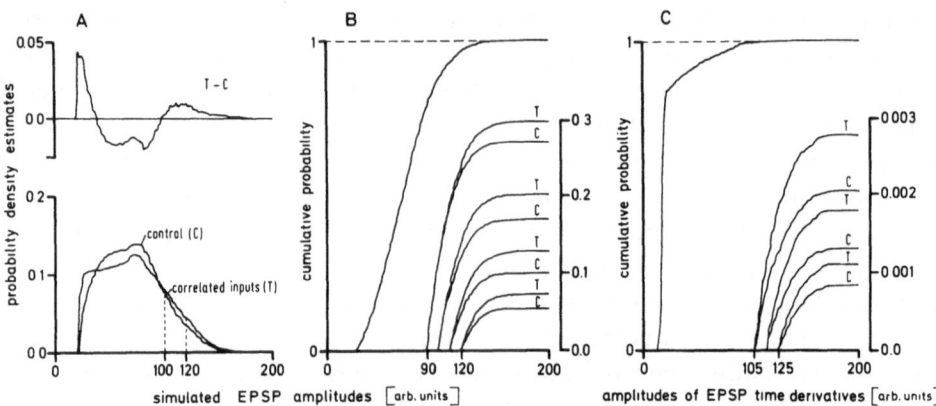

Fig. 17 A–C. Amplitude distributions of simulated postsynaptic potential fluctuations for correlated and uncorrelated Ia inputs. Action potentials were electronically converted into rectangular pulses of width $\Delta t = 0.4$ ms, which were then passed through a second-order low-pass with time constants $\varepsilon_1 = 0.66$ ms and $\varepsilon_2 = 8.0$ ms. "Postsynaptic potential fluctuations" produced in this way by two Ia fibre spike sequences were superimposed, and the amplitude density of the resulting waveforms was determined using an averager (sampling rate 5 kHz, recording period 2 min corresponding to 600,000 samples). The cross-correlogram for the two Ia fibres considered in this example is shown in Fig. 13 A. A *lower graph*: Amplitude densities for correlated (T) and de-correlated (C) inputs; *upper graph*: Difference $T-C$ amplified twice; **B** Cumulative probability distributions (integrated densities); the *left curve* (to which the *left ordinate* is assigned) is the integral of the density T in A determined over the entire amplitude range; the *right curves* (to which the *right ordinate* is assigned) are integrals of the densities in A (and labelled correspondingly), but with integration starting at arbitrary lower amplitude values from 90 to 120 (which values could be taken as deviations of membrane potential from its mean in μV, but this is not essential); **C** Same arrangement as in **B** but the underlying potential fluctuations were electronically differentiated with respect to time before being submitted to the same analysis as in B

1982b; Fetz and Gustafsson 1983; Gustafsson and McCrea 1984), the same analysis as outlined above was repeated after differentiating the "membrane potential" fluctuations electronically. Figure 17C shows cumulative probability distributions for a total-range control (C: left curve) and "supra-threshold" probability for test (T) and control (C) cases. Again, for the chosen range of arbitrary thresholds, the T curves exceed the C curves, implying that synchronisation in the inputs may enhance the velocity of depolarising transients and hence increase the probability of spike initiation, as expected. However, the absolute probabilities are low in this range and might not play a significant role. Also, the quantitative relation in which absolute amplitude and time derivative combine to initiate spikes is still a matter of controversy, and it may vary from neurone to neurone, and it need not be the same under different conditions (see above references).

This type of analysis has also been carried out with "negatively" correlated inputs as occasionally found between Ia fibres from spontaneously contracting triceps surae muscles in decerebrate cats. The amplitude densities obtained in these cases are the opposite of those described above for positive input correlations, i.e. large depolarising potential transients occur less frequently. Statistically, therefore, negative correlations between inputs are "inhibitory" in the sense that they reduce the postsynaptic firing probability as compared to noncorrelated inputs.

An attempt was also made to quantify these various effects of "positive" and "negative" input correlations as a function of their strengths and arbitrary thresholds (Windhorst 1979c). The evaluation proceeded as follows. Cross-correlation histograms (CCHs) such as those displayed in Fig. 13 A and B, were sampled at 20 equidistant instants between -24 and $+24$ ms. These amplitude values were used to calculate arithmetic means and standard deviations, from which a coefficient of variation (CoV_t) was determined as a measure of the relative depth of CCH modulation. This measure contains a random component resulting from the stochastic fluctuations of even flat CCHs. This part (CoV_{ran}) was estimated empirically by shifting in time (by ca. 1 s) one afferent spike train with respect to the other before again computing CCHs (which were flat: de-correlated), and proceeding as before. The CCH modulation component truly due to the correlation ($CoV[cor]$) was then extracted from the overall CoV_t using the formula:

$$(CoV[cor])^2 = (CoV_t)^2 - (CoV_{ran})^2$$

For the input spike trains exhibiting the particular correlation, the ratios of cumulative probability of exceeding threshold before (T) and after de-correlation (C) were then determined as described above (Fig. 17). These ratios were plotted – with the arbitrary threshold value as an additional parameter – against the $CoV[cor]$ values as shown in Fig. 18. The data were obtained from three representative experiments (different symbols) which yielded multiple measurements of CCHs under different conditions (CCHs from two of these experiments are shown in Fig. 13 A and B). The following points are of interest:

1) On the average, the relative change of suprathreshold probability (the above ratio) increases with the strength of positive correlations ($CoV[cor]$). Conversely, this holds also for negatively correlated spindle afferents whose CoV's are plotted as negative values on the abscissa. In this case, the cumulative probability of threshold crossing is reduced, i.e. the negative correlations produce an inhibitory effect in the statistical sense.

2) The higher the threshold, the greater the relative change of suprathreshold probability.

3) For high thresholds, physiologically occurring strengths of correlation between Ia fibres can increase the cumulative probability of threshold crossing by factors up to $2.5-3$.

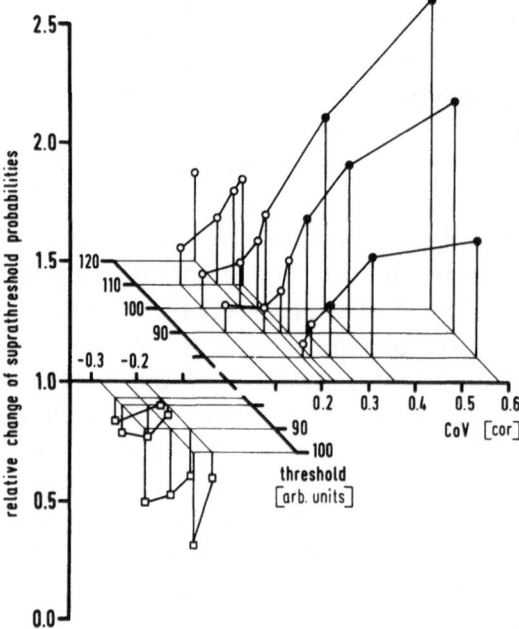

Fig. 18. Changes in the probability of crossing threshold induced by correlated Ia input fibres. The ratios of (threshold-dependent) cumulative (firing) probabilities for the correlated case to the de-correlated case (see previous figure) are plotted against the correlation-dependent modulation of the CCHs, as expressed by a coefficient of variation ($CoV[cor]$) of the CCHs. The arbitrary threshold values serve as an additional parameter. Results from three experiments (different symbols) on decerebrate cats. The different values for each experiment (symbol) were obtained with the triceps surae muscle stretched to different lengths

Essentially similar results were obtained when a faster EPSP was used for the simulation (EPSP 1618 K from Burke 1967, simulated by time constants of $\varepsilon_1 = 0.145$ ms, and $\varepsilon_2 = 2.4$ ms).

It has become clear above that the time pattern of the input correlations is of great importance for their effects on postsynaptic excitability. This can be refined by intuitive reasoning. Take a positive input correlation. Its effect of producing large depolarising potential transients will of course depend upon the strength of correlation which may be quantified by some measure of the excess of the correlogram peak (at $\tau = 0$) above the mean level (see Kirkwood et al. 1982a). However, this is evidently not the only parameter of importance. Another is the relation between time-course of the correlogram and time course of the EPSPs generated by the correlated input fibres. For example, if the EPSPs are both brief, the input spikes must be accurately synchronised in order to produce neatly superimposed EPSPs and thus to attain large depolarising potential transients, whereas this requirement is less rigid with longer EPSPs. Thus the effect of certain input correlation patterns obviously depends also on postsynaptic filtering characteristics – as already indicated by the fact that the time constant ε appears in Eqs. (6).

2.8.2.3 Distribution of EPSP Amplitudes and Time-Courses

A) Factors Influencing EPSP Amplitudes and Time-courses. The assumption of equal EPSP amplitudes and time-courses is not warranted for Ia fibre-α-motoneurone connexions. Instead, these parameters vary widely (for review see Henneman and Mendell 1981; also: Fleshman et al. 1981; Harrison and Taylor 1981).

a) The first factor co-determining EPSP amplitudes is the *type of α-motoneurone* contacted (for classification of motoneurones see, e.g. Burke 1981). Ia EPSPs on average decrease in size from S-type over F(int)- and FR-type to FF-type motoneurones (at least in cat medial gastrocnemius α-motoneurones: Fleshman et al. 1981; Harrison and Taylor 1981). (Incidentally, this does not hold for synapses from group II muscle spindle afferents on homonymous α-motoneurones; Munson et al. 1982.) According to Fleshman et al. (1981), this dependency of Ia EPSP amplitudes on motoneurone types results not only from differences in motoneurone input resistance but predominantly from other type-specific motoneurone properties. Among the latter are total membrane surface area (cell size, determining motoneurone capacity), specific passive membrane properties (capacity and conductance), cell geometry and voltage-sensitive conductances, which, under various conditions, may influence the recruitment order of the motoneurones (for discussion and further references see Pinter et al. 1983). The usual recruitment of motor units at various levels of steady (and increasing) muscle contraction follows the above sequence (from S- to FF-type motoneurones). Thus, at each level of steady muscle contraction, there are always motoneurones firing repetitively (e.g. S-type motoneurones), others whose membrane potentials are depolarised close to threshold (e.g. FR-type motoneurones) and still others whose membrane potentials are far from threshold (e.g. FF-type motoneurones). (Pinter et al. suggest that "the absolute voltage threshold and threshold depolarization are not fixed but may vary depending on depolarization caused by ... background synaptic current"; 1983, p. 649). This implies that any synchronised firing between two (or more) Ia fibres would have different effects on the different types of motoneurone, such that the state of activity, distance from mean membrane potential to threshold and EPSP amplitude all play roles as outlined above. Moreover, the strength and form of the correlations between Ia fibres may also depend on the level of extrafusal activity (Windhorst and Meyer-Lohmann 1977), and hence recruitment. It is interesting to note that recruitment order and EPSP amplitudes are matched in such a way that, in silent (not yet recruited) motoneurones, distance T of mean membrane potential from threshold is inversely correlated with EPSP amplitude. Hence, since EPSP amplitude is directly related to σ [see Eq. (6b)] in Abeles' model, the ratio T/σ, determining post-synaptic excitability, becomes higher (excitability becomes lower) from S-type to FF-type motoneurones. In the same sequence, the synchronised inputs become relatively, though not absolutely, more important for spike generation (see above). Noted in passing, Harrison and Taylor (1981) have argued that the above recruitment order in conjunction with the ordered strength of Ia fibre-motoneurone connexions explains the increase of stretch reflex gain with increasing muscle contraction.

Not only EPSP amplitude, but also the probability of functional connexions between muscle afferents and α-motoneurones depends on motoneurone size as recently shown by Clamann et al. (1985). In pooled data, larger motoneurones received contacts from a higher proportion of group II spindle afferents than did smaller cells. (This is somewhat at variance with results of Munson et al. 1982; but the influence of axonal conduction velocity of the motoneurone was not seen in all experiments of Clamann et al. 1985.) A 100% connectivity was only reached between Ia afferents and motoneurones when both were large. The authors present the following probabilistic explanation. Because of the greater surface area of a large motoneurone, the chance of any terminal arborisation occurring within its domain should be greater. If the dendritic trees expand within about the same tissue volumes, but their complexity increases with conduction velocity, the density of dendrites within this volume increases with the total cell surface area or size. Hence, any axon collateral arbor developing within the receptive volume of a large motoneurone should have a higher probability of making contact with it.

b) The second factor influencing EPSP size is the *conduction velocity* of the Ia and group II spindle afferents. There exists a positive correlation between the two parameters (Mendell and Henneman 1971; Kirkwood and Sears 1982a; Clamann et al. 1985). Lüscher et al. (1979, 1980) have extended these results by showing that fast-conducting afferent fibres give off more terminals to a *population* of motoneurones than do slowly conducting ones, this probably being due to the distribution of more terminals to each single motoneurone as well as to a larger number of motoneurones. This applies to both Ia and group II muscle afferents though with different slopes, the dependency being stronger for Ia than for group II fibres (Lüscher et al. 1979). Also, Munson et al. (1982) showed that fast spindle group II fibres project to more homonymous (medial gastrocnemius) α-motoneurones than do slow ones. Clamann et al. (1985), who recently confirmed the earlier findings of Lüscher et al. by recordings from individual α-motoneurones, give the following explanation. Apparently, larger fibres have a greater capacity for forming functional connexions. It has long been known that larger fibres in the mammalian nervous system split into more branches and terminals than do small fibres. Apparently, there is no conspicuous exception to this rule. The correlation of the amplitude of single-fibre EPSPs with the diameter of the stem presynaptic fibres suggests that this diameter is finally reflected in the size of the axon collateral arbor that it forms on individual motoneurones. This follows the same chance principle as the projection frequency (previous paragraph). The large importance of chance in such a process would then provide an explanation for the random way in which the influence of fibre size is expressed in the experimental results of Clamann et al. (1985). Windhorst (1978 b) employed this chance principle to explain topographical relations in monosynaptic connexions between Ia afferents and α-motoneurones (see below).

Since conduction velocity is directly correlated with the dynamic sensitivity of the afferent to velocity of stretch (Matthews 1963; see Discussion in Lüscher et al. 1979), the propensity of two rapidly conducting afferents to generate correlated discharge patterns in response to common intra- or extrafusal mechanical events may be higher than that of two slowly conducting afferents. This applies at least to the conventional division of the muscle spindle afferents into the two classes of Ia and group II afferents. Indeed Binder and Stuart (1980 a) showed that Ia fibres are more sensitive to single motor unit contractions than group II fibres are (see Sect. 3.2.3). Lüscher et al. (1979) present interesting speculations in this respect which are worth being briefly reviewed. They base their arguments on, firstly, evidence of different locations of the synapses from group II and Ia afferents on the motoneurone somato-dendritic membrane, and, secondly, a suggestion by Rall (1959). Rall's theoretical studies of neurones suggest that the more dendritically located synapses, which generate slower EPSPs at the soma, would be more appropriate for the maintenance and regulation of the general excitability of the motoneurone pool. They would be well suited to process input from group II spindle afferents with their predominantly static sensitivity. In contrast, the more proximally located input would be better suited to cause rapid reflex discharge. This input could predominantly derive from spindle group Ia afferents with their exquisite sensitivity to small rapid changes in muscle length. The tonic synaptic bias delivered via the dendrites by input from secondary endings could thus be envisaged to modify the effectiveness of the input from Ia fibres. It is worth comparing these suggestions with the more general ideas of Segundo et al. (1968) discussed above (Sect. 2.8.1).

Another interesting finding and some further correlations may complement the view outlined in the previous paragraph. Maier (1981) in the pigeon and Eldred et al. (1974) in mammals found a positive correlation between the number of intrafusal muscle fibres per spindle and the speed of contraction of the parent muscle. It is likely that more intrafusal fibres would require a larger afferent axon because this would have to give off more branches (according to the same principle as that prevailing at the afferent's central end; see above explanation by Clamann et al. 1985). Indeed, in mammals complex spindles have a parent afferent fibre of larger diameter than simple spindles do (Adal and Barker 1962). If all these correlations hold generally, the spindle afferents from a fast muscle would be more effective centrally than those of a slow muscle. Contraction speed of motor units co-determine their effects on spindles (Binder and Stuart 1980a).

Different sites of synaptic terminals within the somato-dendritic system, all other relevant factors being constant, determine the time-course of EPSPs but hardly their amplitudes (Mendell and Weiner 1976; Iansek and Redman 1973). Proximal Ia synapses thus do not produce significantly larger EPSPs, but rather faster ones, which, due to a higher rate of rise, might more easily elicit postsynaptic spikes than distally generated EPSPs (Lüscher et al. 1979). On the other hand, this would require more precise timing of synchronous spikes on Ia fibres impinging proximally than on those making contacts distally if synchronisation of inputs is to play a role in spike initiation (see above). Also remember that the range and form of the amplitude densities, an example of which is shown in Fig. 17A, and their standard deviation σ depend on both parameters, amplitude and time-course of EPSPs [see Eqs. (6)]. The above idea is weakened, however, by the finding of Mendell and Henneman (1971) that a single Ia fibre can elicit EPSPs of very different time course in different motoneurones indicating that the synapses are distributed more widely than presupposed. (This distribution might reflect another order, for example a topographical one; see next paragraph.) Also, Lüscher et al. (1979), using postsynaptic population potentials (PSPPs) as estimators of the effects of single muscle afferents on populations of motoneurones, found only a weak negative correlation between conduction velocity and PSPP rise time.

c) The third important factor co-determining EPSP amplitude (but not time-course) is *"partitioning of Ia EPSPs"*. This term implies that EPSPs evoked in an α-motoneurone by Ia afferents originating from different parts of the homonymous muscle differ in average amplitude (Hamm et al. 1985a). This effect may be determined by two main factors: "topographical organisation" ("location specificity") and species specificity. Topography or location specifity holds if Ia afferents from different parts of a muscle enter the spinal cord at different levels and thereby exert different actions on homonymous α-motoneurones located at different levels. Species specificity obtains if Ia afferents exert stronger monosynaptic actions on motoneurones projecting back to their muscle region of origin than on motoneurones projecting to other regions, even if these motoneurones are located close to each other in the spinal cord; the Ia afferents may then be envisaged as "seeking out" their "own" motoneurones. Both factors may contribute to partitioning of EPSPs to different degrees.

One prerequisite for topographical connexions as defined above is a weighted distribution of the monosynaptic effects of Ia afferents entering the spinal cord at different rostro-caudal levels on α-motoneurones located at different spinal levels. This requirement has been shown to be fulfilled. Lüscher et al. (1980) have used their new method of recording PSPPs to demonstrate spatial order in the Ia fibre-motoneurone connexions. These results have recently been confirmed by Clamann et al. (1985) by recording from individual α-motoneurones. The monosynaptic excitatory effect of a Ia fibre onto homonymous (and heteronymous) α-motoneurones is distributed not homogeneously but in a weighted fashion. Recall that medial gastrocnemius (and other) motoneurone pools are organised in columns extending rostro-caudally in the ventral horn of the (cat) spinal cord (Romanes 1951; Burke et al. 1977). According to Lüscher et al. (1980), the number of synaptic terminals given off by a Ia fibre to its homonymous motoneurones depends on the fibre's rostro-caudal entry point into the spinal cord. The distribution is asymmetric in the rostro-caudal direction, extending further rostrally than caudally, probably due to the way a Ia fibre branches and distributes its principal collaterals after entering the cord (Brown and Fyffe 1978; Ishizuka et al. 1979). Such a weighted distribution of Ia excitatory effects had previously been postulated by Windhorst (1978 b, 1979 a), in order to account for focussed signal transmission through the spinal cord (and the stretch reflex system as a whole). The mechanisms proposed by Windhorst (1978 b) to explain the weighted Ia strength were different from those suggested by Lüscher et al. (1980) who argue "... that the density of terminals given off by the afferent fiber is greatest around its entry point and becomes sparser in more remote primary collaterals" (p. 980). The latter authors are probably right (see also Lucas et al. 1984). This "thinning out" of terminal arborisations with distance between afferent entry point and motoneurone location also leads to a reduction of the probability of the afferent connecting to a motoneurone as shown by Clamann et al. (1985), who provide the following explanation for this "distance effect". A Ia fibre running longitudinally up or down the spinal cord, projects a series of primary collaterals ventrally to make connexions with motoneurones. As the total number of collaterals given off by a Ia fibre increases with distance from the entry point, the diameter of the Ia fibre and its collaterals should decrease progressively. Since the diameter of a stem axon or any of its collaterals is reflected in the extent of the terminal arbors (see above), a series of progressively thinner collaterals should have successively smaller terminal fields, whose probability and capacity to form connexions should decrease accordingly. The distance effect would then simply be a corollary of the fibre size effect.

The second requirement for topographical partitioning is that Ia afferents originating from different muscle regions show a propensity to enter the spinal cord at different levels. This requirement is fulfilled in some muscles but not in others (Sects. 3.3.1 and 3.3.2). Note, however, that even if topography is absent, species specificity can still establish partitioning of EPSPs. A brief overview of the existing data base is presented in Sects. 3.3.1 and 3.3.2.

Topographical and species specific factors can thus be expected to play some role in determining the strength of monosynaptic connexions from Ia afferents to homonymous α-motoneurones and, hence, the possible effect of synchronised inputs. However, this type of organisation differs widely from muscle to muscle in an apparently haphazard manner. In whichever muscle EPSP partitioning exists, any particular α-motoneurone receives different information from different parts of the muscle. The most influence on that motoneurone will be exerted by Ia afferents originating from the muscle region to which it projects its motor axon. Since muscle spindles will be influenced most powerfully by motor unit activity in that muscle region, the ensuing correlations between their afferent discharge patterns may be assumed to be of particular importance for the "own" α-motoneurones. This suggestion was put forward by Windhorst (1978 b). It will be discussed again in relation to tremor in Sect. 3.3. It may also be suggested, in line with the proposal of Fawcett and O'Leary (1985; Sect. 1.3.3), that neural activity patterns carried on Ia afferents at early stages of ontogeny contribute to the establishment of the specific connectivity patterns discussed here.

d) A fourth, presynaptic factor is also worth mentioning. Amplitudes of EPSPs evoked monosynaptically in α-motoneurones by single Ia afferents are variable. The EPSP amplitudes may vary with the passage of time (see Henneman et al. 1984, with further references). If one Ia afferent sends more than one presynaptic terminal to an α-motoneurone, some of the synapses may be silent for some time and active during another period. Henneman et al. (1984) discuss this phenomenon. For silent synapses to become active in the authors' experiments, at least three possibilities could have played a role. (1) Henneman et al. (1984) observed a spontaneous decrease in the mean discharge rate of the afferents; this may have decreased the probability of conduction failure at the branch points in the terminal axon collaterals. (2) There may have been electrical interactions between α-motoneurones and primary afferents establishing a low-resistance pathway from the motoneurone to the presynaptic fibre. Thus, the small depolarisation observed in the motoneurones may have been carried into the very terminal branches of the axon collaterals of the afferent fibre. This depolarisation, if it occurred just distal to an axon branch point, could have increased the safety factor for impulse propagation through that point. (3) Activation of silent synapses might underlie the enhancement of synaptic transmission after acute spinal cord transection reported by Nelson et al. (1979).

As suggested in the first point (1), EPSP amplitude may depend on the frequency of arriving presynaptic impulses, as also shown by Honig et al. (1983). The second point (2) is interesting insofar as it opens the possibility that a depolarisation resulting from activation of synapses from one Ia afferent increases the probability of activation of synapses from another Ia afferent by a presynaptic mechanism. This "cooperative" interaction would require near simultaneous activations, and probably spatial proximity of synapses from the two afferents. It would be a cooperative mechanism which would enhance the probability of postsynaptic firing above and beyond that expected by the mere synchrony of afferent pulses (see above). The third point (3) refers to short- and long-term synaptic plasticity (Nelson and Mendell 1979; Nelson et al. 1979; Cope et al. 1980; see Sect. 3.5). Henneman et al. (1984) summarise these studies by emphasising "that the functional connectivity and transmission capacity of the Ia projection varies dynamically with the state of the spinal cord and central nervous system" (p. 160). This dynamic variability probably also applies to other synapses, and may thus be a basis for the relatively short-term formation and dissolution of functional cell assemblies at spinal cord level.

B) Factor Influencing Interaction Between Different EPSPs. One of the assumptions underlying the simple model calculations of Abeles and Windhorst (see above) was the linear summation of EPSPs generated by different presynaptic fibres. Now, it is well known that EPSPs, whether composite or individual, do not always sum linearly (e.g. Burke 1967; Kuno and Miyahara 1969; also B. L. McNaughton et al. 1981). Nonlinear summation would be expected to be particularly pronounced between EPSPs originating from closely adjacent synapses far out in the dendritic tree. Since Ia EPSPs measured in the soma do not show strong correlations between their amplitudes and time-courses (Mendell and Weiner 1976; Iansek and Redman 1973), those originating at distal dentritic sites must have appreciable amplitudes locally (Barrett and Crill 1974) due to the motoneurone cable properties (Redman 1973). At the peak of such a large EPSP, the driving potential for any other EPSP occurring simultaneously at the same site would be strongly reduced, so the two EPSPs would sum nonlinearly. Of course, the strength of this effect would depend on the relative location of the respective synapses within the complex architecture of the dendritic tree. Nonlinear summation would be stronger for synapses located more distally and for those sharing longer dendritic branches, through which the synaptic current must flow to reach the spike-initiating region. Such a mechanism for the processing of partially synchronised inputs was considered by Windhorst (1978b), but whether it plays a role is still uncertain. However, it is interesting to note that certain neck motoneurones apparently have input-specialised dendritic trees such that synaptic inputs originating from different sources are directed to different localised regions of the dendritic tree defined by respective stem dendrites (Rose 1982). The same appears to hold true for sacral α-motoneurones mediating a cutaneous reflex (Egger et al. 1980). There thus appears to be a "spatial segregation" of important inputs, which probably occurs on the dendritic tree and not at the soma. This implies that the site of integration of different inputs could be quite far away from the cell body (Rose 1982). This could give rise to complex spatial patterns of input processing with nonlinear interactions. Also remember the complex dendritic architecture of M.S.O. cells that was proposed by Loeb et al. (1983) to play a role in the spatio-temporal cross-correlation required in their model of acoustic pitch perception (Sect. 1.4.1).

The foregoing discussion has shown that the effect of correlations between muscle afferents on the discharge of α-motoneurones cannot be described in any simple way even when restricting attention to monosynaptic homonymous connexions. In fact, this matter has not yet been studied experimentally to a sufficient extent, this also being the reason for the almost totally theoretical discussion. It can be expected that input correlations influence motoneurone discharge in very different ways depending on a number of variable (indeed time-varying) temporal and spatial factors and circumstances. More generally, this applies to other spinal neurones as well. Indeed, the extensive discussion was intended to exemplify and stress the widely varying processing that input correlations must be assumed to be subjected to in different neurones with different properties.

2.8.3 Summary

The foregoing three sections were concerned with the importance of correlations between input channels for the transmission of signals from muscle spindle afferents to homonymous motoneurones and with the factors that influence them. A general computer simulation study by Segundo et al. (1968) indicates that the tendency towards synchrony of firing in presynaptic terminals exerts a significant effect upon the firing pattern of the postsynaptic cell. Input synchrony of physiologically observed strength among a larger number of weak presynaptic inputs is proposed to be able to raise the role of these inputs from a merely tonic background excitation to a more differentiated input.

Regarding the monosynaptic connexion from Ia (and spindle group II) afferents to α-motoneurones, the following factors may influence the significance of input correlations on postsynaptic excitability.

1) The state of motoneurone activity (rest or discharging) determines the time at which input correlations can become effective. More specifically, in a firing motoneurone, its refractory period and, more importantly, its postexcitatory hyperpolarisation would reduce its excitability for certain time periods after a discharge. This would preclude or at least reduce the effectiveness of larger depolarising transients, which possibly result from synchronised inputs, from reaching threshold.

2) The temporal structure of the correlations between muscle afferents ("positive", "negative", etc.) influences postsynaptic firing probability in such a way that, statistically, positive input correlations (tendencies towards synchronous firing) increase and negative correlations decrease postsynaptic firing probability as compared to that obtained with uncorrelated inputs. These changes in probability depend quantitatively on the sharpness, i.e. the time course of the input correlations.

3 A) Obviously, the degree of changes in the probability of postsynaptic discharge also depends upon the amplitudes and time-courses of the EPSPs elicited by the correlated inputs. These parameters are in turn dependent upon a number of factors. Firstly, the amplitudes of Ia EPSPs vary with α-motoneurone type, being largest in type S and smallest in type FF motoneurones (according to Burke's classification). Also, the probability of a functional connexion between a Ia afferent and a motoneurone is largest if both have fast conducting axons. Secondly, the probability of contact is higher and the amplitude of the EPSP produced by a Ia fibre greater if the afferent's conduction velocity is higher. Thirdly, besides "species specific" effects, topographic projection patterns of Ia afferents to homonymous α-motoneurones can be detected in the homonymous monosynaptic reflex organisation of certain muscles.

3 B) The spatial organisation of the correlated inputs within the dendritic tree of the postsynaptic cell is of importance insofar as closely spaced terminals far out on the dendritic tree may cause large depolarisations which, if they coincide, may sum nonlinearly thus reducing the efficacy of synchronised input firings.

Many of the aforementioned factors result from probabilistic principles according to which the connectivity between afferents and motoneurones is established. However, there are specific factors which may be superimposed upon the random background, such as topographical and "species specific" factors. Essentially the same applies to any part of the nervous system. The balance between the random and the specific "rules" may vary from muscle to muscle. Its development during ontogenesis is unclear but may be proposed to follow similar lines as those discussed by Fawcett and O'Leary (Sect. 1.3.3); that is, the establishment of specific connectivity patterns could depend on particular activity patterns of the connected elements. For this at least, the prerequisites are given. For example, synchronised discharges of Ia afferents, if appropriately located in the motoneuronal dendritic tree, might cause large enough depolarisations which are hypothesised to be an intermediate step in the development of long-lasting specific connexions. Early muscle activity generating appropriate efferent and afferent activity patterns is available. Moreover, on the shorter time-scale, it is very probable that all the above factors enable the system to perform highly sophisticated data processing tasks in which shorter-lasting alterations of synaptic efficacy ("synaptic plasticity" or potentiation, and "synaptic modulation" according to von der Malsburg; Sect. 1.3.2) presumably plays an important role. This matter should be investigated more intensively.

2.9 Interneuronal Circuits

Whereas the preceding account concentrated on monosynaptic connexions of muscle afferents to α-motoneurones in order to paradigmatically discuss the many factors modulating the effects of correlations between presynaptic fibres on postsynaptic neurones, the remarks in the following two sections will address interneuronal interfaces which complicate and modulate the signal transmission from afferents to motoneurones. Whereas monosynaptic connexions from Ia afferents to α-motoneurones are well known due to their easy experimental accessibility, the interneuronal systems are much less known and understood, but might turn out to be physiologically more significant than monosynaptic connexions. For example, Rymer and Hasan (1981) suggested "... that polysynaptic Ia projections may be a more important source of motoneuronal excitation than the more traditionally accepted monosynaptic Ia projections" (p. 111). However, and this is one of the keynotes of this monograph, it is not only the *amount* of excitation or inhibition that is of importance, but also the *pattern* and in particular the *degree of synchrony* of signal transmission, whose complexity and capacity is immensely augmented by interneuronal systems.

2.9.1 Signal Transmission to Motoneurones Through Interneuronal Systems

2.9.1.1 Oligo- and Polysynaptic Connexions to α-Motoneurones

It has always been generally agreed that the main connexions from group II spindle afferents to homonymous α-motoneurones were polysynaptic (review: Baldissera et al. 1981), so that the existence of monosynaptic connexions first had to be demonstrated explicitly. Also, it has long been accepted that autogenous inhibition of homonymous (and other) motoneurones from Ib afferents is mediated di- or trisynaptically (see below).

The evidence for polysynaptic excitation of homonymous (and heteronymous) α-motoneurones from Ia fibres has slowly accumulated (for a brief review of earlier work see Granit 1970; Matthews 1972; Hultborn and Wigström 1980; also Schomburg and Behrends 1978; Kirkwood and Sears 1982a; Powers and Binder 1985a, b). Jankowska et al. (1981b) investigated oligosynaptic excitatory effects from Ia fibres on various species of α-motoneurones and argued that these pathways involve other interneurones than the polysynaptic Ia effects on homonymous (and synergistic) motoneurones. The oligosynaptic Ia effects have the same distribution as the respective Ib effects whether excitatory or inhibitory (Fetz et al. 1979; Jankowska et al. 1981 a), so that "all the known reflex actions of Ib tendon organ afferents on ipsilateral motoneurones are also evoked from Ia muscle spindle afferents" (Jankowska et al. 1981 b; p. 423).

It is worth expanding a bit upon a particular interneuronal interface to α-motoneurones because – due to a special type of data recently gained on it – this system might be a paradigm. The classical interneuronal system mediating "autogenetic" or "synergistic" inhibition to α-motoneurones from Ib afferents has now been termed "nonreciprocal" inhibitory system (see Harrison and Jankowska 1985a) since about half of those interneurones influenced by Ib afferents of the triceps surae and plantaris muscles are also influenced by Ia afferents of the same origin (Jankowska and McCrea 1983; Harrison et al. 1983; Harrison and Jankowska 1985a, b). These interneurones, being part of the interneuronal pools of laminae V–VI, project collaterals to the upper lumbar segments where they inhibit cells of origin of the dorsal spinocerebellar tract (DSCT cells; see Harrison and Jankowska 1985a).

Harrison and Jankowska (1985a, b) have recently studied the patterns of convergence of different input systems on the lamina V–VI interneurones mediating nonreciprocal inhibition. These patterns are interesting because they may represent a more general principle of input organisation, which might apply to other spinal interneuronal systems as well. In general, this interneuronal system receives input from segmental afferent sources (Ia, Ib, group II, cutaneous, joint, flexor reflex afferents) and from descending systems (propriospinal, corticospinal, rubrospinal, dorsal reticulospinal and noradrenergic reticulospinal systems; see Fig. 12 in Harrison and Jankowska 1985a). But individual interneurones usually receive different patterns of converging input, exhibiting an individual profile with respect to their inputs. By a statistical analysis, Harrison and Jankowska (1985b) provided evidence that the different presynaptic fibres connect independently of each other within the population of interneurones, forming their contacts in a random way on the basis of their average projection frequencies to the interneurones as a group. Harrison and Jankowska (1985b) remark that the factors underlying the distribution of input to a given population of neurones, which are intermixed with neurones of different type, are as yet poorly understood. But these factors must include some that decide about *specific connexions to functionally different neurones* and some that are involved in *the random formation of synaptic contacts upon neurones of the same population*. For the latter type of connectivity, a probabilistic argument is advanced by Harrison and Jankowska (1985b), which is quite similar to that forwarded by Clamann et al. (1985; Sect. 2.8.2.3). The probability of formation of synaptic contacts from diverse afferents with interneurones in laminae V – VI should depend on the local density of synaptic terminals around the particular interneurone. For example, a greater proportion of these interneurones received input from the rubrospinal rather than from the corticospinal tract probably due to the larger overlap of the terminal region of the former tract with the location of the interneurones. Also, the terminal areas of Ib afferents are larger than those of Ia afferents, thus explaining that some interneurones receive Ib but not Ia input. The existence of a larger number of synaptic terminals per Ib collateral rather than per Ia collateral may work in the same direction and account for a generally stronger excitation of the interneurones by Ib afferents. Thus, certain topographical and other distribution patterns could here come into play directing the "randomness" in certain ways. For instance, remember that the distribution of the terminal arborisations of Ia fibres depends on the latter's entry height into the spinal cord (Sect. 2.8.2.3). Also, Ia and Ib fibres carry information on local muscle activity (Sect. 2.5). Spatial factors may then contribute to correlate the discharges of subpopulations of nonreciprocal Ia interneurones which are commonly contacted by a particular input system. This would also generate, at least to some extent, a spatial separation of one such subpopulation from others dominated by other input systems. Such a separation is supported by the fact that "these interneurones are distributed over a length of spinal cord at least twice as long as that of their target motor nuclei (i.e. over four or five segments or some thirty millimeters ...)" (Harrison and Jankowska 1985b; p. 411). On the whole, then, "it is not surprising that practically each one of the investigated neurones had a different pattern of input and that the 'fractionation' of input to these neurones is of the 'fine grain' type" (p. 413). It is interesting to compare these remarks on the relative importance of random connectivity and specific factors modulating it with Abeles' results and his synfire chain hypothesis (Sect. 1.3.1). It is then conceivable that such an interplay together with specific discharge patterns in the inputs could separate different classes of inputs patterns in the way illustrated in Fig. 4C.

This type of connectivity might have several interesting consequences. Firstly, the fact that group Ia afferents from a particular muscle may excite these interneurones in random combinations with Ia afferents from other muscles "precludes the possibility of primarily autogenetic feed-back, unless there were some highly specialized presynaptic control" (Harrison and Jankowska 1985b; p. 412). Secondly, a function of the "input fractionation" might be to enable two (or more) input systems, when activated in different combinations, to selectively activate different groups of interneurones. Depending upon the particular instantaneous input pattern, this selected set of interneurones could then exert a specific task, e.g. inhibit a certain combination of motoneurones. For this purpose, interneurones with different patterns of convergence should inhibit different motoneurones, but there is not yet enough evidence to support that idea (Harrison and Jankowska 1985b). One and the same interneurone should probably be envisaged to be integrated in different cell groups which carry out different tasks at different times according to the prevailing conditions. This *group selection hypothesis* is reminiscent of Abeles' multifunction of synfire chains (Sect. 1.3.1), of the "plurifunctional subpopulations" of reticular neurones participating, under different conditions, in different distributed systems (Schulz et al. 1985; Sect. 1.4.2) and of Loeb's "task group" hypothesis (Sect. 5.1.7). The fractionation of interneuronal pools may also occur in other systems, though probably to different degrees. Although the reciprocal Ia inhibitory interneurones have not been investigated with the same quantitative precision as the nonreciprocal ones, they appear to be less fractionated (E. Jankowska, pers. comm.).

The main point is that multimodal input patterns will be processed through a "fine grain" interneuronal interface which, after performing a "pattern analysis", generates specific output patterns. It is pertinent to emphasise that similar operations could happen in other interneuronal systems, such as the reciprocal Ia inhibitory system, which also receives convergent input from many sources (Baldissera et al. 1981), and the interneuronal system intercalated in pathways from spindle group II afferents (Schomburg and Steffen 1985). Fine features of discharge patterns on the input lines and within these interneuronal systems (such as correlations) are probably of importance in the actions and in the temporary stabilisation of these networks. For example, Fetz et al. (1979) proposed that early discharges evoked in Ia fibres by impulses on skeletofusimotor (β-) fibres (most likely of the fast contracting type; Sect. 2.5.2) might contribute to the inhibition of homonymous α-motoneurones by the Golgi tendon organ discharge that results when the related muscle unit contracts. If this were a significant physiological mechanism, it would indeed depend on a fine feature of afferent discharge. For, on repetitive activation of motor units at physiological firing rates, Golgi tendon organs tend to fire with one or two pulses in response to each motor unit contraction; particularly fast contracting motor units may "drive" tendon organ firing (Jami et al. 1985a). Since early discharges may also originate from spindles coupled in series to motor units, small subsets of Ia and Ib afferents can indeed be expected to fire in close correlation in response to motor unit contractions, signalling, in a way, the "event" of contraction, not its entire time-course and strength.

Another important point which probably plays a significant role in pattern analysis of the above networks is the fact that the last-order interneurones investigated by Jankowska and co-workers are mutually inhibitory (Brink et al. 1983; Harrison and Jankowska 1985a); this intrinsic connectivity thus parallels the mutual inhibition among Renshaw cells and among Ia inhibitory interneurones.

2.9.1.2 Oligo- and Polysynaptic Connexions to γ-Motoneurones

Oligo- and polysynaptic connexions from muscle afferents to γ-motoneurones are only briefly summarised. Muscle spindle Ia afferents have weak polysynaptic (not monosynaptic) excitatory effects, and Golgi tendon organ afferents stronger inhibitory effects on homonymous γ-motoneurones. Secondary spindle (or other muscle group II) afferents polysynaptically excite extensor γ-motoneurones, predominantly of the dynamic type (Ellaway and Trott 1978; Ellaway et al. 1979, 1981; Ellaway and Murphy 1980a; Noth and Thilmann 1980; Noth 1983; Appelberg et al. 1977; Appelberg et al. 1982). In general, however, the patterns of input distribution are complex and differ in various respects from the flexor reflex pattern to α-motoneurones (Appelberg et al. 1983a, b, c).

Appelberg and co-workers (1983a, b, c, d) have extensively investigated diverse segmental input systems to γ-motoneurones. The results have been briefly reviewed by Johansson (1985; see also Hulliger 1984). Johansson emphasises the complexity of the organisation of segmental reflexes to γ-motoneurones: "For a particular γ-cell (1) different effects were often obtained with stimulation of different inputs of the same type (e.g. stimulation of group II afferents of different nerves), and (2) group II muscle, group III muscle, skin and joint receptor afferents rarely give the same type of effects, and, for populations of γ-cells, (3) the patterns of effects elicited from muscle group II and group III, skin and joint receptor afferents regularly differ from each other" (1985; p. 299).

According to Ellaway et al. (1982), γ-motoneurones also receive closely coupled excitation from group III muscle afferents with low mechanical threshold. These afferents are probably distinct from the group of nociceptive afferents and react to muscle contraction and pressure. Ellaway et al. (1982) suggest that ". . .contraction sensitive Group III units are concerned with signalling the event rather than the degree of contraction. They have tended to give single impulses to twitch contractions and they fire at the onset of a tetanus with no appreciable sustained, or tension-related, discharge during the contraction" (p. 496). Homonymous (and heteronymous) γ-motoneurones react to muscle contractions in much the same way, i.e. as if "driven" by partially synchronised afferent volleys, with a short central latency suggesting one interposed interneurone, at the most. A problem Ellaway et al. (1982) face in interpreting their results arises if it is assumed that these group III discharges elicit the respective γ-motoneurone firing. The close coupling of the γ-motoneurone discharges to certain contraction phases suggests either that discharges in group III units are evoked as one or more coherent volleys or that synchronous excitation of only a few group III units is sufficient to fire a particular γ-motoneurone. Despite insufficient evidence, Ellaway et al. suggest that one or more synchronous volleys in group III units be generated.

"This would fit with the relatively common finding that two (sometimes three) peaks of γ-motoneurone discharge occur in response to contraction . . . compared with the predominantly single response to nerve stimulation" (Ellaway et al. 1982; p. 491). It remains to be established, however, whether the group III units signal activity of a restricted number or type of motor units repetitively activated at low physiological rates, since the single strong muscle twitch contractions used by the above authors are not exactly physiological. If they do, they would signal the event of motor unit contraction, not its time-course and strength, as may happen with Golgi tendon organs (see above).

2.9.2 Some Spinal Interneuronal Circuits Influencing Correlations

In the previous sections, by describing correlations between discharge patterns of motoneurones, muscle afferents and the possible effects of the latter on the former, we have closed the classical stretch reflex loop. However, some other spinal interneuronal systems, more or less closely related to these basic components of the stretch reflex, must also be considered in the present context, because they are likely to be important for signal transmission through and within the spinal cord, although, admittedly, their function is still poorly understood.

2.9.2.1 Broad-Peak Synchronisation Generated by Renshaw Cells?

First, it may be recalled that the broad-peak synchronisation between cat intercostal α-motoneurones was hypothesised by Kirkwood et al. (1982b; 1984) to be generated by synchronised interneuronal activity (see Sect. 2.1.2). But which interneurones are involved is not yet known. It is of particular interest that the above authors proposed Renshaw cells of the thoracic cord as "candidate neurones for the generation of the broad-peak synchronisation" (p. 131). Renshaw cells have indeed often been incorporated into schemes of networks producing synchrony, but there has been substantial controversy, as to whether their action was to correlate or de-correlate α-motoneurone activity (see Sects. 3.4.1 and 4.2.1.2). Possibly, these alternatives are not mutually exclusive because the action of Renshaw cells on α-motoneurone synchronisation might depend upon the dynamic state in which the whole neuromuscular central system is temporarily involved (see Sect. 3.4.1). In fact, the broad-peak synchronisation of respiratory α-motoneurones changes with changes in descending control (Kirkwood et al. 1982b), indicating that system states are variable.

Renshaw cells could act in the same way on γ-motoneurones as on α-motoneurones, since recurrent inhibition of γ-motoneurones via Renshaw cells now seems to be well established (Ellaway and Murphy 1980 b; Appelberg et al. 1983 d; Sect. 4.3.8). Thus, the broad bases of the CCH peaks reported for γ-motoneurones (see Sect. 2.3.1) could at least partially result from Renshaw cell action. It should be stressed, however, that the interneuronal system responsible for broad-peak synchronisation of α-motoneurones and similar synchronisation of γ-motoneurones has not yet been identified.

2.9.2.2 The Effect of Presynaptic Inhibition on Ia-α-Motoneurone Signal Transmission

Another interneuronal system influencing signal transmission through specific synapses is that producing presynaptic inhibition. In a series of papers (short review: Rudomín 1980), Rudomín and co-workers have studied the variability of signal transmission from Ia fibres to homonymous α-motoneurones. They found that fluctuations of successive monosynaptic reflexes recorded simultaneously in two ventral root (L_7) filaments in response to repetitive Ia afferent stimulation were correlated positively. The source of this correlation was supposed to be comprised of two parts: one independent component inherent in the separate neuronal (motoneurone) populations, and one common component due to common input to the two populations. The authors surmised that the common part of correlated fluctuations results presynaptically from a mechanism responsible for PAD (primary afferent depolarisation). The following account closely follows the description of Rudomín (1980, pp. 140 ff.).

The interneurones mediating presynaptic inhibition of Ia afferent terminals could correlate the latter's excitability variations and, hence, their synaptic efficacy. In order to show this, simultaneous intracellular recordings were made from two cat gastrocnemius α-motoneurones as illustrated in Fig. 19 A. Fine branches of the gastrocnemius nerve were then stimulated repetitively to produce monosynaptic Ia EPSPs in the two motoneurones. These EPSPs exhibited considerable fluctuations in peak amplitude superimposed on a varying baseline. These fluctuations could have two sources: presynaptic variations in the transmitter release mechanism, and postsynaptic noise due to other inputs. If postsynaptic noise were important, the smaller EPSPs in a series should decay faster than the larger ones. As shown in Fig. 19 B, however, this was not the case since the smaller EPSPs were scaled versions of the large EPSPs. The continuous lines in Fig. 19 C and D show averaged EPSPs elicited by Ia afferent stimulation in two other motoneurones. The hump on the decay of one of the EPSPs could have resulted from monosynaptic group II connexions or from polysynaptic Ia fibre connexions. In E

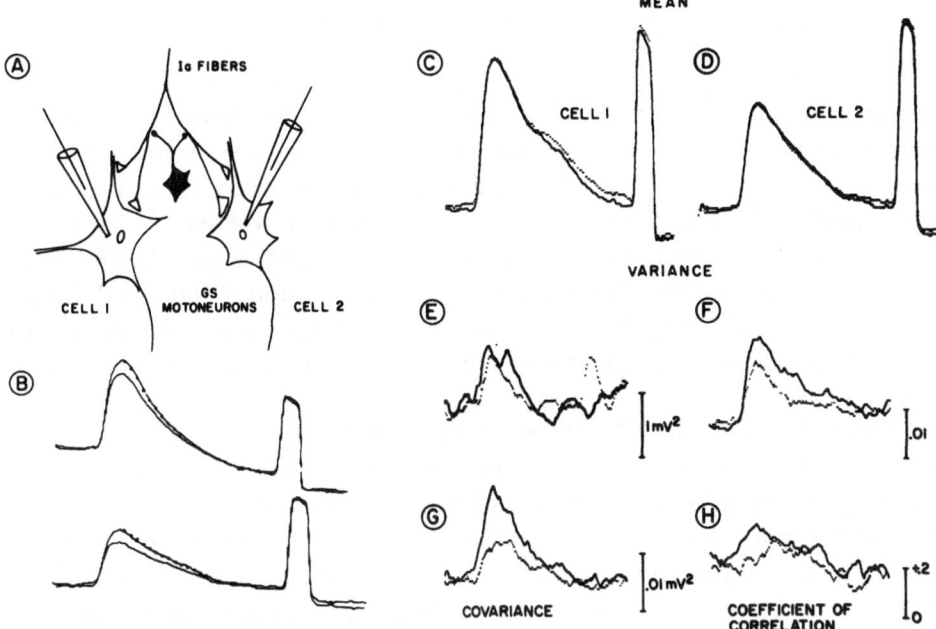

Fig. 19 A–H. Correlated fluctuations in Ia fibre EPSPs recorded simultaneously from two α-motoneurones. **A** Experimental arrangement: Small branches of the medial gastrocnemius nerve were stimulated with stimuli supramaximal for group I fibres, and the elicited monosynaptic EPSPs were recorded simultaneously from two α-motoneurones. The postulated source of the correlated fluctuations is symbolised as an interneurone (*black*) synapsing with the Ia fibres; **B** The averages of large and small Ia EPSPs collected during maximal group I fibre stimulation at 1/s. For each cell the series of EPSPs was separated into two parts, one set comprising all the EPSPs with peak amplitudes smaller than the mean, and the other set including all EPSPs larger than the mean. The EPSPs equal to the mean were not included in any set. The mean for each set was computed separately. The *stars* on the "larger" mean EPSP show the time course of the "smaller" mean EPSP after proper scaling for peak amplitude of the vertical axis. The time courses of the two sets of means were essentially identical. Scaling factors were 1.1 for one motoneurone (*upper set*) and 1.34 for another motoneurone (*lower set*). Calibration: 1 mV; 1 ms; **C** and **D** Mean Ia EPSPs recorded simultaneously from two medial gastrocnemius α-motoneurones (cells other than those illustrated in **B**), averaged from about 200 samples. Calibration: 2 mV; 1 ms; **E** and **F** Corresponding variances for each series of EPSPs also as a continuous function of time; **G** Covariance between the EPSPs; **H** Correlation coefficient between the EPSPs. *Continuous lines* denote data derived from control series; *dotted lines* are from EPSPs conditioned with stimulation of the biceps-semitendinosus nerve (two shocks at group I strength, 40 ms before test stimuli to medial gastrocnemius). (With permission from Rudomín 1980; his Fig. 7)

and F, the variance of the two EPSPs is plotted as a continuous function of time. In both cells the variance increased above a baseline in parallel with the EPSP time-course. The covariance and coefficient of correlation for the EPSPs in these two α-motoneurones are shown in Fig. 19 G and H. The correlation coefficients were calculated in essentially the same manner as for two afferent spike trains (see Sect. 2.5.2 and App. C), i.e. time-related to the input stimuli. The Ia EPSP fluctuations were positively correlated, suggesting that a common and synchronous influence was exerted on the Ia afferent terminals impinging on the two motoneurones recorded. When a conditioning stimulus was applied to the group I afferents of the biceps-semitendinosus muscle nerve 40 ms prior to the test stimulus, the average EPSPs of both cells were hardly affected, as indicated by the dotted lines, whereas the variance, covariance, and correlation coefficient were reduced at the EPSP peak. Hence, the conditioning stimulus changed the correlation of the EPSP fluctuations without altering their mean shapes and amplitudes. The change in correlation will affect the degree of correlated firing of the two α-motoneurones.

This is illustrated in Fig. 20 for the discharge patterns of two α-motoneurones recorded from ventral root filaments in response to repetitive stimulation of gastrocnemius afferents. Graph C plots the coefficient of correlation (for synchronous firing) of the two motoneurones (denoted by A and B) as a function of the sum of their firing indices (probabilities), $P(A_1)$ and $P(B_1)$ without conditioning cutaneous stimulation (open circles), and with such stimulation (closed circles). It can be seen that the correlation was decreased by the conditioning stimulus. An evaluation of these results in terms of some basic notions of information theory is presented in A, B, and D. Before any afferent test stimulus, the response (discharge or not) of a motoneurone is uncertain. The amount of uncertainty removed once the response of the cell is known is taken as a measure of the information gained, and is defined as follows:

$$H(A) = - [P(A_1).\text{ld } P(A_1) + P(A_0).\text{ld } P(A_0)], \qquad (7)$$

where $P(A_1)$ is the probability that cell A fires, $P(A_0)$ is the probability that cell A does not fire, and ld is the logarithmus dualis. With two completely independent channels (motoneurones) A and B (Fig. 20 A), the information transmitted through both channels is

$$H(A, B) = H(A) + H(B). \qquad (8)$$

If the two channels are partially correlated, the following inequality holds:

$$H(A, B) < H(A) + H(B). \qquad (9)$$

The relationship between $H(A) + H(B)$ and $H(A, B)$ was calculated for the two cells used in Fig. 20 C, and is plotted in Fig. 20 D. Such a plot would have a slope of 1 if the channels were independent [Eq. (8)]. The higher the correlation between the channels (motoneurones), the lower the slope. The open circles in Fig. 20 D were derived from the control series of afferent stimulation; the solid circles from the conditioned responses (stimulation of

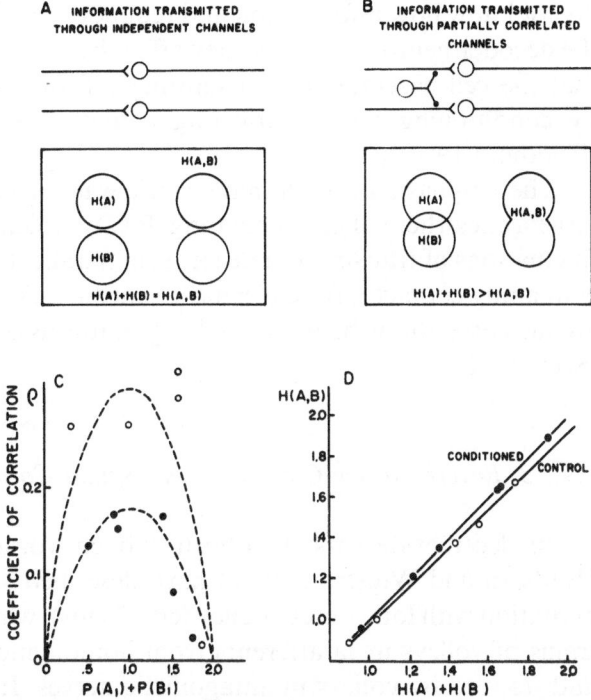

Fig. 20 A–D. Effects of conditioning cutaneous volleys reducing the variability of monosynaptic gastrocnemius reflexes on the information transmitted by a two-motoneurone ensemble. **A** The information transmitted through two independent channels is equal to the sum of the information transmitted through each independent channel; **B** Activity in the interneurones mediating PAD introduces partial correlation in the Ia fibre-motoneurone pathway. As indicated in the diagram, the information through partially correlated channels is less than the sum of the information transmitted through independent channels; **C** Correlation coefficient (ϱ) between monosynaptic responses of two α-motoneurones versus the sum of their firing indices $P(A_1) + P(B_1)$. *Open circles*: control series obtained with graded stimulation of group I gastrocnemius afferents; *Filled circles*: during conditioning stimulation of the sural nerve at twice threshold (30 ms before the test stimulus); **D** $H(A, B)$ versus $H(A) + H(B)$ curves derived from the same units used to obtain the data in **C**. Lines are best fits calculated with the least-squares method. The difference in the slopes is significant at the 0.01 level. Further explanation see text. (With permission from Rudomín 1980; his Fig. 8)

the sural nerve at 2 × threshold). Under control conditions the slope is 0.93; with conditioning sural stimulation 0.99, the difference being significant.

Rudomín's (1980) results indicate that the calculated information transmitted through *pairs of motoneurones* is increased during conditioning stimulations which reduce population variance. A strong correlation between pairs of units (Fig. 20 C) signifies redundancy of information about the source of the correlating input. Redundancy in parallel channels probably plays an important role in neural information transmission because it can compensate for the possible loss of information in single channels. "Since we do not know how information and redundancy of information are utilized by the decoder neurons, or how they affect behavioral readout, we can say only that the cell ensemble is transmitting information which can be modified by conditioning volleys affecting activity through the PAD pathways" (Rudomín 1980; p. 146).

The emphasis in Rudomín's work was on the effects of intraspinal interneurones (here those mediating PAD) which may introduce correlated fluctuations of transmitter release from parallel terminals of Ia afferents that commonly contact different α-motoneurones. Another source of such effects are the correlations between discharge patterns of the Ia afferents themselves (Sect. 2.5.2).

2.9.2.3 Reverberatory Circuits in the Spinal Cord?

In decerebrate cats, Hultborn and co-workers (Hultborn et al. 1975; Hultborn and Wigström 1980) have described states of long-lasting motor excitation with long onset latency (60–70 ms), which can be triggered by brief trains of volleys in Ia afferents from homonymous and synergistic muscles and stopped by volleys in antagonistic nerves. It is important to appreciate the long latency because it indicates a concealed buildup process. These excitatory states do not require intact reflex loops and were supposed by Hultborn and Wigström (1980) to be due to "reverberating" activity in intraspinal (segmental) interneuronal circuits (see Fig. 21 A). (Such closed intraspinal loops were already proposed in 1948 by Tönnies and Jung.) Recently Hounsgaard et al. (1984) presented evidence for an alternative explanation. They found that α-motoneurones in decerebrate cats (and in spinal cats treated with 5-HTP) showed "all-or-none" long-lasting depolarisation with ensuing maintained discharge, which could also be elicited by short depolarising current pulses applied through the intracellular electrode. They assigned this behaviour to an intrinsic motoneurone membrane property exhibiting bistability. Whether any long-lasting changes of activity within pre-motoneuronal cell networks occur in parallel remains to be investigated.

Mori et al. (1982) showed that the extensor muscle tone in decerebrate cats can be *set* to continuously variable levels by impulses descending from the brainstem (review: Mori 1987). There are two brainstem systems, from which the postural tone can be either increased (ventral pontine tegmental area corresponding to the rostral portion of the nucleus raphe magnus: Mori 1987) or decreased (dorsal pontine tegmental area corresponding to the caudal portion of the nucleus centralis superior: Mori 1987). Repetitive stimulation of the dorsal system reduced or stopped the pre-existing soleus EMG activity with a latency of several tens of milliseconds; impaled soleus α-motoneurones *progressively* hyperpolarised during the stimulation, and

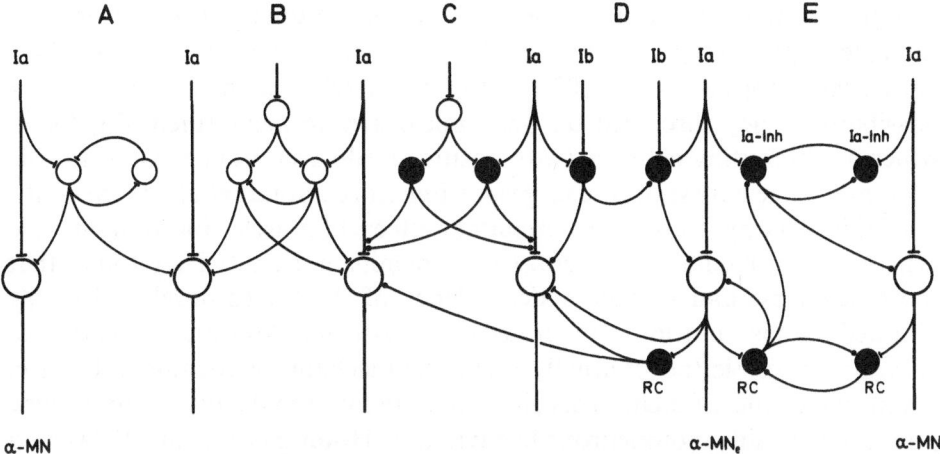

Fig. 21 A–E: The spinal interneuronal systems possibly involved in the correlation or de-correlation of α-motoneurones. *Open circles* symbolise excitatory, *solid circles* inhibitory interneurones. The different systems are separated in the different parts of the figure *(A–E)*. *α-MN$_e$* stands for extensor α-motoneurone, and *α-MN$_f$* for flexor α-motoneurone. *Ia-Inh* is the label for Ia inhibitory interneurones mediating reciprocal inhibition between antagonist motoneurone pools. For fuller description see text

soleus Ia afferents and Renshaw cells decreased their discharge rates; all these depressive effects outlasted the stimulation for long time periods (Mori 1987). Stimulation of the ventral system had opposite effects. Mori (1987) concludes that "withdrawal of group Ia facilitation is the most probable source of the long-lasting suppression" (p. 175) and vice versa. But this cannot be the only factor because the reduction of Ia firing must also come about somehow, e.g. by a decrease of fusimotor bias. So this would have to be explained. Anyway, the features of these "settings" are reminiscent of the "integrating property" of the interneuronal circuits originally proposed by Hultborn and Wigström (1980). Again, these features could also be accounted for by intrinsic motoneurone membrane properties. However, Mori (1987) showed an intracellular recording from a soleus α-motoneurone (his Fig. 11) which could be depolarised to different levels in a *graded* fashion by repeatedly stimulating the ventral tegmental area. There is no sign of a bistability of membrane behaviour. If, according to Mori (1987), this graded depolarisation came about by a corresponding graded increase in Ia discharge, the bistable membrane behaviour would have to reside in the γ-motoneurones, if it were to be saved. This appears unlikely. So interneuronal systems generating long-lasting reverberatory activity are not yet out.

Rymer and Hasan (1981) studied, in decerebrate cats, the prolonged excitation in extensor α-motoneurones following a tonic vibration reflex (see also Pompeiano and Wand 1980). They provided evidence "that longitudinal tendon vibration initiates processes with prolonged time course that are best explained by invoking prolonged discharge of Ia interneurons with excitatory autogenetic projections" (p. 102). As the most likely explanation among four possibilities, they discussed an anatomical arrangement (their Fig. 6), in which a self-exciting network (with positive feedback) could account for the prolonged time course of discharge (see Fig. 21 A). Stuart et al. (1986) continued this work by studying the excitatory effects of muscle vibration on close synergists (cat medial and lateral gastrocnemius muscles). They found that the time-course and frequency dependency of the forces developed in the vibrated agonist and in (nonvibrated) synergist muscles differed. They discussed six possible (not mutually exclusive) mechanisms for this difference, among which the differential distribution of Ia input to the two motoneurone pools, the bistable motoneurone behaviour of Hounsgaard et al. (1984), and again excitatory Ia interneurones (possibly incorporated into a self-excitatory loop) were considered the most important. The latter mechanism was held indispensable because the former two were unable to explain all the features of the results. Rainey et al. (1984) described lumbar interneurones responding with long-lasting discharges to Ia input. It is worth re-emphasising that this long-lasting excitation was initiated by high-frequency synchronised Ia input which might appear to be unphysiological and hence of no importance. While this possibility does exist, Rymer and Hasan (1981) conclude their discussion by stating: "Finally, while prolonged central effects have been documented here for vibratory stimuli only, it remains possible that prolonged interneuronal activity may also be important in more physiological forms of spinal activation" (p. 111). It appears less likely that the tonic vibration reflex observed in normal human subjects with a slow onset and rise and decay should result from motoneurone bistability. At least, many researchers have interpreted these results in terms of polysynaptic spinal or even supraspinal circuits intercalated between Ia afferents and α-motoneurones (Granit 1970; Matthews 1972; Hagbarth 1973). Such interneuronal circuits could well contain closed loops as suggested by Rymer and Hasan (1981), although direct evidence for their existence in the spinal cord is still lacking, whilst an exception has been reported for cerebellar circuits (see Tsukahara 1981).

Similar long excitations outlasting the triggering input can be obtained in certain cat preparations (decerebrate and spinal cats treated with L-Dopa) by stimulation of segmental afferents other than Ia fibres (see, e.g. Kniffki et al. 1981). Also, in spastic human patients, long-lasting muscle contractions can be elicited by multimodal, e.g. cutaneous, inputs. In these patients, the tonic state can change into more or less rhythmic "automatisms". An equivalent phenomenon can happen in spinal cats treated with L-Dopa where tonic activity can subside into locomotor-like activity patterns. This suggests the possibility that the same interneuronal networks may operate in different modes depending on circumstances not yet well understood.

The preceding discussion of intraspinal neural networks generating maintained activity patterns may gain some plausibility by reference to equivalent invertebrate systems. The advantage of such preparations is that neurones can be reliably identified and their action in small networks can be readily analysed. An example is provided by the work of Getting and Dekin (1985) on *Tritonia diomedea*. Swimming in this marine mollusc is episodic and consists of a series of alternating dorsal and ventral flexions initiated by a sensory stimulus. The locomotor pattern is generated by a pre-motoneuronal network of interneurones, which have been partly identified: two "cerebral cells" (C2), six "dorsal swim interneurones" (DSI), two "ventral swim interneurones" of type A (VSI-A) and two ventral swim interneurones of type B (VSI-B). Moreover, Getting and Dekin (1985) postulate the existence of two inhibitory interneurones (see their Fig. 15). The synaptic interactions between the various types of interneurone are complicated. But there are monosynaptic excitatory interconnexions among the DSI neurones, which establish positive feedback loops and which apparently are needed to prolong the basic depolarisation of the neurones within the network necessary for swimming. Whether this principle has survived in the vertebrate spinal cord, remains to be shown. Reciprocal inhibitory connexions between the DSI and VSI-B cells obviously are responsible for the alternation of activity in antagonistic motoneurones. Other reciprocal interconnexions usually are of opposite sign or mixed nature. Interestingly, the network appears to exist in at least two different activity states. That is, it is not "hardwired" to produce only a stereotyped swimming pattern. On the contrary, the multiplicity of interconnexions may enable the network "... to generate several different activity patterns for use in different behaviors" (Getting and Dekin 1985; p. 479). Although some skepticism is indicated as to whether these invertebrate networks can be regarded as layouts for vertebrate ones, which are much more complicated, yet a comparison of the two may promote our intuitive understanding of what is going on in complicated networks.

Figure 21 schematically summarises some of the spinal interneuronal systems alluded to above. For simplicity, they have been disentangled and displayed in parallel. The hypothetical reverberatory circuit fed by Ia fibres is depicted under A (Rymer and Hasan 1981). Part B shows a circuit of interneurones providing diverging excitation to α-motoneurones and, possibly, in turn receiving common interneuronal input or influencing each other so as to produce (broad-peak) synchronised activity (Kirkwood and co-workers); they may receive afferent input from Ia fibres as well as from other sources (unspecified input at top). In part C the system of neurones mediating presynaptic inhibition and correlating the Ia inputs to α-motoneurones (Rudomín and co-workers) is shown; many inputs may in turn influence these interneurones (Rudomín et al. 1983; Jankowska 1984) and thus correlate them. Part D (upper) shows the interneurones which mediate Ib-Ia "non-reciprocal inhibition"; they may also inhibit each other (Jankowska and co-workers) and send axon collaterals rostrally to inhibit DSCT cells; in contrast to these interneurones, those mediating presynaptic inhibition of Ia afferent terminals on motoneurones and DSCT cells appear to be organized in separate pools (Jankowska 1984). In E (above) are displayed the Ia inhibitory interneurones mediating reciprocal inhibition between antagonists; they may inhibit each other (for references see Pompeiano 1984) and receive recurrent inhibition from Renshaw cells as shown. Finally, included at the bottom in parts C through E is the recurrent inhibitory system which has a certain spatial distribution that extends farther than the probably excitatory direct interconnexions between α-motoneurones (illustrated in D); Renshaw cells belonging to antagonist motoneurone pools (see E) mutually inhibit each other (Ryall 1981). The Renshaw cell circuit will be discussed more extensively in Chap. 4. Several systems have been omitted, among them the interneuronal system which could provide for alternating activation of flexors and extensors due to its "half-centre organisation" (see Jankowska and Lundberg 1981). – All these interneuronal systems, singly and in conjunction, can contribute to correlating or de-correlating motoneurone discharges.

2.9.3 Summary

There are several interneuronal networks at segmental spinal levels which complicate and modify signal transmission from muscle (and other) afferents to motoneurones. For instance, signals in Ia afferents are not only transmitted to homonymous and heteronymous synergistic α-motoneurones via the excitatory monosynaptic pathway, but also via several excitatory and inhibitory oligo- and polysynaptic pathways. Signal propagation through the latter from a particular set of Ia afferents would be expected to be enhanced if the inputs are synchronised. Also, the postsynaptic effects exerted in α-motoneurones by Ia afferent discharge depend upon the time pattern of input; e.g. if a set of synchronously firing Ia afferents tend to discharge in advance (by a certain mean interval) of another set of synchronously firing Ia fibres, and the synchronous firings of the first set are delayed by the same interval in passing through an interneuronal chain, they might arrive at a common motoneurone synchronously with the firings of the second set (see Sect. 2.5.2). If both effects are excitatory, the postsynaptic firing would be disproportionately increased; the reverse effect would be obtained if the oligo- or polysynaptic transmission has an inhibitory action, as in the case of nonreciprocal Ia inhibition. It is worth emphasising again that these constellations yield various possibilities of synaptic modification and modulation, as suggested by von der Malsburg (Sect. 1.3.2; also Sect. 1.3.3). Direct synaptic links to a postsynaptic cell could cooperate with indirect (oligo- or polysynaptic) links if the signals carried in both are correlated. This cooperation could lead to synaptic potentiation if the two presynaptic signals arrive synchronously at the postsynaptic cell and both have an excitatory effect, and to suppression if they arrive synchronously and one has an excitatory and the other an inhibitory effect, or if the two signals are correlated inversely and both are excitatory. These manifold interactions are complicated by the convergence of other types of afferents on the interneurones in the oligo- and polysynaptic pathways, such as Ib fibres, joint and cutaneous afferents whose firings may also be partially correlated by those mechanical events which correlate Ia afferent discharges. The patterns of convergence are partially random, but in part obey "specific rules" which may, in turn, have developed ontogenetically, dependent on the firing patterns of the participating input fibres.

The interneurones mediating nonreciprocal Ia inhibition are probably a paradigm of the random pattern of input convergence. Temporally varying input combinations may thus activate different groups of cells – "task groups" as it were (cell assemblies performing particular tasks) – which might differentially connect to motoneurones on which they exert different effects according to the prevailing conditions. In a sense, these groups of lamina V–VI cells constitute distributed systems, since their inputs originate from widely distributed sources, and their outputs distribute to widely divergent targets (various motoneurone pools innervating different muscles, and DSCT cells). Spatial (topographical) order is not excluded *a priori* since the activation of muscles also occurs in a complicated spatio-temporal order. Mutually inhibitory interactions might play an important role in defining and delimi- nating these cell groups dynamically. Whether fast processes of synaptic modulation are involved remains to be seen. Likewise, their "sense", i.e. their physiological role, remains to be established.

Three further systems have been discussed with respect to effects of interneuronal systems on motoneuronal synchronisation: the Renshaw cell recurrent inhibition, the interneuronal system producing PAD and mediating presynaptic inhibition, and possible reverberatory loops intercalated be- tween Ia fibres and motoneurones. The correlating or de-correlating effects of Renshaw cells on motoneurones are probably not independent of the state of correlation between the motoneurones from which they receive their input and will therefore be discussed again in Sect. 3.4.1 in the context of tremor, and recurrent inhibition will be compared and integrated with proprioceptive feedback from the muscle in Chap. 4. The interneurones which presynaptical- ly inhibit Ia afferents change the monosynaptic signal transmission from the latter to homonymous α-motoneurones by correlating the excitability fluctu- ations in Ia afferent terminals projecting to different motoneurones. This effect is in turn modifiable by muscle afferent and cutaneous inputs. Rudomín (1980) has nicely demonstrated that the correlated excitation of different motoneurones can be altered by changing the *degree of covariation of Ia EPSPs* without changing the mean EPSP amplitudes. The existence of reverberatory interneuronal circuits intercalated between Ia fibres and motoneurones at the segmental spinal level is still uncertain, and, at best, the evidence in favour is indirect. However, if these circuits exist, their activation is most probably maintained by circulating synchronised firing (closed "synfire chains"), such as may also exist at supraspinal levels (Sect. 1.3.1). The stabilisation of the reverberatory activity in positive (re-excitatory) loops could then be supported by processes similar to synaptic modulation (Sect. 1.3.3). The effects of these interneuronal systems on correlated motoneurone firing are complicated by the fact that they receive input from some of those afferent systems whose correlating effects on motoneurones they modulate. Furthermore, they have so far been investigated under relatively static and artificial experimental conditions; the input patterns to the above networks might be more differentiated under natural conditions (see remarks in Sect. 2.4).

3 Tremor States

"In Lebensfluten, im Tatensturm
Wall' ich auf und ab,
Webe hin und her!
Geburt und Grab,
Ein ewiges Meer,
Ein wechselnd Weben.
Ein glühend Leben,
So schaff' ich am sausenden Webstuhl der Zeit
Und wirke der Gottheit lebendiges Kleid"

J. W. v. GOETHE[6]

In this chapter we shall consider the possible mechanisms of various types of physiological tremor. These may provide paradigms for different functional states in the central nervous system, which may be associated with different types of synchronisation (or correlation) between parallel neuronal elements. The various elements, which were considered separately in the previous chapter, are now put together and viewed from a new angle, that of their ability to produce various types of rhythmic activity patterns. We shall see that the control of these different states of activity is a tremendous task for the nervous system.

3.1 Frequency-Displacement Amplitude Relations for Normal Hand Tremor

Several investigators observed an inverse relationship between frequency and displacement amplitude of physiological tremor (briefly reviewed by Stiles 1976). Stiles (1976, 1980) systematically studied this relationship. He asked (normal) subjects to keep their right or left hand extended horizontally for 15 to 60 min. The forearm was supported on a table, and the metacarpal and finger joints were immobilised by a plastic mould weighing about 100 g. The tremor movement developing at 16 cm from the wrist joint was measured by an accelerometer. Successive records lasting 16 s each were analysed for mean frequency content and root-mean-square (rms) deviation (as a measure of tremor amplitude). It turned out that within the first few minutes the predominant tremor frequency remained approximately constant at ca. 8–9 Hz. At the same time, tremor amplitude slowly increased but was on average below roughly 100 μm. Subsequently, the amplitudes of the tremor increased while its frequency decreased. Stiles emphasised "the general find-

[6] Goethe JW: Faust; First Part. Lines 501–509 (Ghost)

ing that between certain displacement amplitude levels the frequency and the displacement of the tremor changed very rapidly. These rapid changes in frequency and displacement amplitude generally occurred over a time period of less than 16 s" (1976; p. 47).

Several points are of interest. (1) For rms amplitudes below about 100 μm, relatively little change in the tremor frequency occurred when displacement amplitudes increased. (2) For rms amplitudes above about 100 μm, the tremor frequency systematically decreased when the displacement amplitude increased. (3) The relationships between tremor frequency and amplitude appeared to indicate that physiological hand tremor oscillations with rms displacement amplitudes between 1 and 4 cm will occur at frequencies between 3.75 and 6 Hz. (4) There appear to be "discontinuities" in the frequency–amplitude relations separating various relatively "stable" regions. The first break-point delimits the low-amplitude region (up to around 100 μm) and high-frequency (8–12 Hz) region. Another discontinuity is intercalated between medium- and large-amplitude tremor. (Note, however, that this is the present author's view, which Stiles himself disagrees with; he instead stresses the continuous variation of amplitudes and frequencies; Stiles 1976).

Stiles (1976, 1980) suggested that low- and larger-amplitude tremors (with their related frequencies) result from different mechanisms. In the low-amplitude form (< 100 μm), oscillations of the whole stretch reflex do not occur. On the contrary, neural feedback might attenuate tremor excursions. The basic mechanism of this form is suggested to be the mechanical oscillation of a viscoelastic-mass system (muscle-limb mechanics) which is forced by a broad-band input due to essentially asynchronous motor unit contractions. "Another problem is that the amplitude of the tremor will be affected by the spectral characteristics of the internal perturbations (asynchronous motor-unit contractions, the pressure pulse)" (Stiles 1980; p. 55). This connects to the topic of the next section.

The larger-amplitude tremor probably involves reflex oscillations, i.e. modulation of motor unit activity (with partial "broad-peak" rhythmical synchronisation) by muscle afferents whose discharges are in turn modulated (and partly synchronised) by the tremor movements. As mentioned above, the afferents predominantly but not exclusively responsible for this neural feedback may be Ia afferents. But assuming that the neural feedback does have a sizeable influence on the generation and maintenance of larger-amplitude tremor, the problem remains as to how the wide range of frequencies (from 9 down to 4 Hz) could come about since the tremor frequency should, in the reflex oscillation hypothesis, be primarily determined by the delays around the loop. This problem is not resolved by Stiles. All he states is that the interaction of mechanical and reflex factors allows a limb of a normal subject to generate tremors of many different frequencies and displacement amplitudes. "Certainly a system producing these oscillations cannot be viewed as existing in a relatively few states of stability (or instability)" ... "Overall, the results suggest that many different equilibrium conditions can exist in which internal and external perturbations, muscle-load mechanics, and neural feedback effects determine the major tremor oscillation of the hand" (Stiles 1980; p. 58).

Hagbarth and Young (1979) and Young and Hagbarth (1980) reported findings qualitatively similar to those of Stiles (1976, 1980) though in a less quantified way. They additionally recorded presumed muscle afferent discharges from the trembling muscles (wrist flexors) by microneurography. They also concluded that low-amplitude physiological tremor is not exclusively dependent on motor unit synchronisation. Instead, the basic mechanism lies in the unfused contractions of asynchronously discharging motor units. The stiffness of the contracting muscles and the inertia of the moving parts then act as a mechanical filter, influencing the resonant frequency of the system (Hagbarth and Young 1979). "Enhanced" physiological tremor, however, occurring after some minutes of maintained hand position (as in Stiles' experiments), was proposed to be sustained by oscillation in the segmental stretch reflex, the afferents primarily responsible being the Ia fibres. In recordings of muscle afferents and EMG from the trembling muscle they found rhythmic modulations in which peak activity in the afferents preceded peak activity in the EMG by about 20 ms, a time interval compatible with spinal reflex and conduction latencies. This "enhanced" tremor might be a limit-cycle oscillation because the amplitude-dependent sensitivity of the primary spindle endings ("gain compression nonlinearity"; see Hulliger 1984) and the delays in the spinal monosynaptic pathway would damp larger amplitudes. Since Hagbarth and Young (1979) did not wait long enough to observe the very large amplitudes (of the order of cm) and low frequencies (4–6 Hz) as investigated by Stiles, they were not compelled to find an explanation for the latter tremor. It would have had to transcend their model of "enhanced tremor". In order to distinguish between these two forms, we shall henceforth call the "enhanced" tremor of Hagbarth and Young "medium-amplitude tremor" and the low-frequency tremor "large-amplitude".

It is important to emphasise in the present context that low-amplitude and enhanced physiological tremors represent two functional states of the spinal segmental neuromuscular system which are associated with two forms of correlation between α-motoneurone (and probably Ia fibre) discharge patterns. In low-amplitude tremor there is no apparent sign of gross "grouping" of α-motoneurone discharges ("broad-peak" synchronisation) related temporally to the tremor movements, whereas there might be some tendency towards short-term synchrony. In contrast, Hagbarth and Young (1979) reported "an obvious grouping of afferent spindle discharges from the contracting wrist and finger flexor muscles, with bursts of spindle impulses occurring mainly during the stretch phases of the tremor cycles and relative silence during the shortening phase" (p. 512). However, these inferences were drawn from multi-unit recordings with one electrode so that the single units were not precisely characterised as to receptor type. But assume that all units were Ia afferents. Then the afferents were probably contained in a small nerve fascicle and may have originated from a restricted muscle region. This is also suggested by the observation of "another characteristic feature of the afferent stretch discharges: their 'intensity' is not directly proportional to the amplitude of any given tremor cycle. Even the smallest tremulous movements, barely discernible in the goniometer trace, were often accompanied by distinct stretch discharge" (Hagbarth and Young 1979; p. 512). This finding might imply that the afferents recorded the contractions of a small set of motor units whose contributions to the tremor excursions varied from cycle to cycle. Then, afferent synchronisation would show a spatial pattern. But note that microneurographic mass recordings (and to some degree also goniometer recordings, as they may be determined by a complex interaction between synergists and antagonists) are notoriously difficult to interpret (Prochazka and Hulliger 1983).

The results of Stiles (1976, 1980), and Hagbarth and Young (1979) are partially at variance with recents results of Burne et al. (1984) who maintain that motor unit synchronisation does already occur with "normal" hand tremor in the frequency range of 8–12 Hz.

With a different experimental paradigm, namely driving the human ankle joint with sinusoidal torques, Agarwal and Gottlieb (1977) found the resonance of the active gastrocnemius-soleus muscles to lie near 6 Hz. Figure 22 A shows some of their findings. In this case, the ankle was driven at 10 Hz, the upper trace (labelled τ) depicting the internal motor torque, the second trace (labelled θ) the joint angle, the third trace the EMG of the anterior tibial (AT) muscle, and the lower trace the gastrocnemius-soleus (GS) EMG. It is apparent that most of the input at 10 Hz is not transferred into ankle movements of equal amplitude and sinusoidal shape at the same frequency (as expected of a linear system), except during brief periods indicated by horizontal arrows beneath the angle recording. More often the input is transferred to distorted periodic movements at about half the input frequency. At this frequency (here ca. 5 Hz) the gastrocnemius-soleus EMG bursts are synchronised and large. The sudden changes between these two dynamic, obviously discrete states (faithful sinusoidal transfer and "frequency halving") suggest the existence of (at least) two oscillators whose interaction could account for similar results obtained by Brown et al. (see below). The "clonic oscillation" (Agarwal and Gottlieb) near 6 Hz tended to go on self-sustained after the machine driving stopped (Fig. 22 A), particularly after prolonged driving, as if "warming up" occurred, perhaps due to post-tetanic potentiation. Figure 22 B shows the dependence of ankle joint compliance on

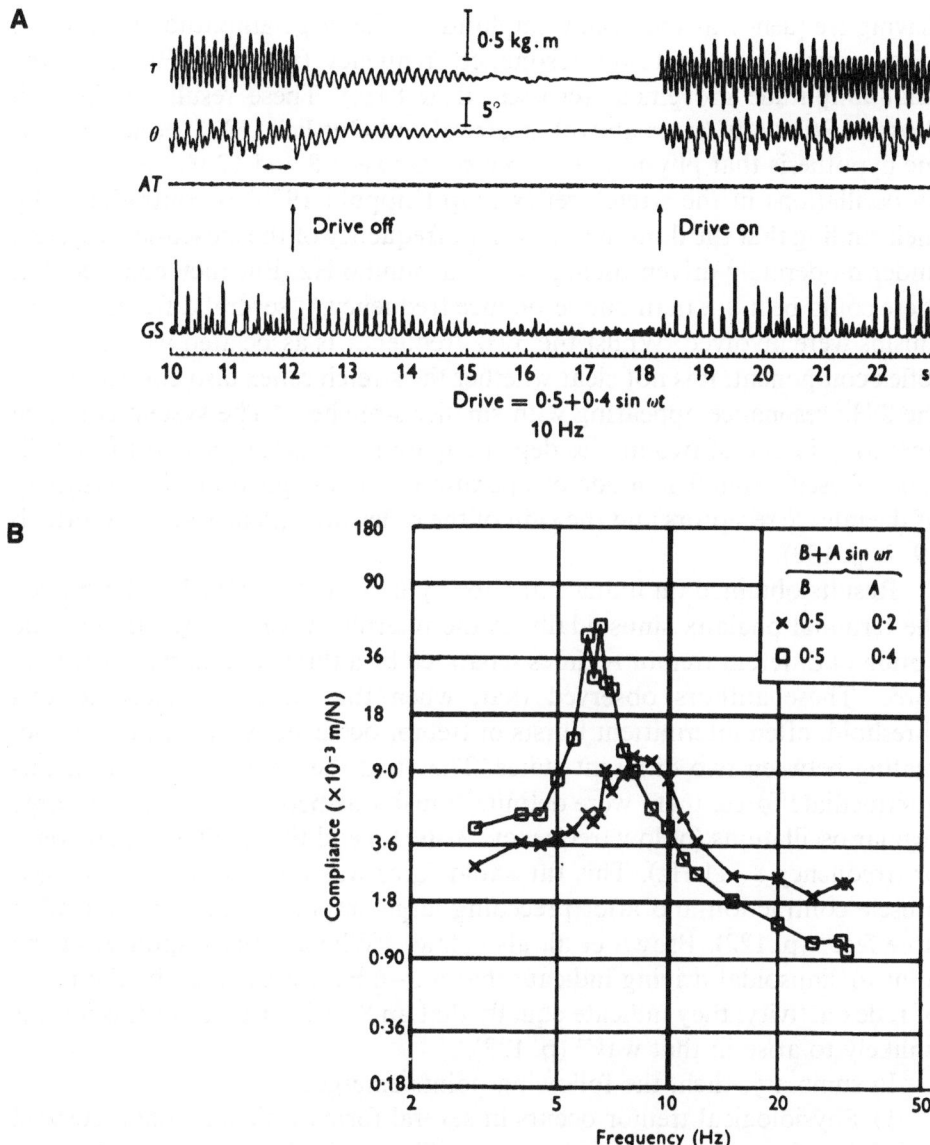

Fig. 22 A. The forced oscillation at a drive frequency of 10 Hz. The *arrow* indicates the time when the modulation signal of the motor was turned off. The self-sustained oscillation of the foot continued for several seconds near 6.35 Hz as measured from the Fourier transform analysis. As the modulation signal was turned on again, the nonsinusoidal wave-form developed rapidly. Recurrences of 10 Hz oscillation in between the nonsinusoidal response are indicated by *horizontal double arrows* beneath the angular rotation curve. (With permission from Agarwal and Gottlieb 1977; their Fig. 4); **B** Effective compliance in metres/Newtons as a function of the drive frequency. The motor bias voltage was kept constant at 0.5 V. The amplitude of the modulation signal for the two cases was 0.2 (X) and 0.4 (square) volts. The gastrocnemius-soleus muscle was tonically active against the motor bias producing an average torque of about 0.2 kg.m to maintain the zero angular foot position. (With permission from Agarwal and Gottlieb 1977; their Fig. 8)

driving frequency at two input amplitudes. The large-amplitude movement (open squares) has a lower resonance frequency (near 6.25 Hz) than the small-amplitude movement (crosses: near 8 Hz). These results fit in with those of Stiles (1976; see above). Agarwal and Gottlieb (1977) conclude that the hypothesis that physiological tremor (between 8 and 12 Hz) is produced by oscillations in the stretch-reflex loop (Lippold 1970) is contradicted by their finding that the dominant resonant frequency of the closed-loop system under moderately driven oscillations is around 6 Hz. But they concede that there could be more than one resonance frequency, if several different mechanisms were involved. Whilst the 6 Hz frequency is associated with a strong reflex component, it is not clear whether the stretch reflex also contributes to the 8 Hz resonance appearing with smaller stretches. "The system could be operating in one of two modes depending on the bias torque and the amplitude of oscillation. The mode of operation would depend on the sensitivity of the muscle receptors and the gain of the reflex arc" (Agarwal and Gottlieb 1977; p. 170).

Results obtained on human thumbs by Brown et al. (1982), who moved the terminal phalanx sinusoidally at the interphalangeal joint, support the notion of different tremor regimes separated by a threshold, namely a critical force. These authors observed that, when the force was close to this threshold, often intermittent bursts of tremor occurred with the thumb fluctuating between two different states. They had the impression that, at this intermediate force, there were definite transitions between the state of large regular oscillations (of low frequency: 3–6 Hz) and the small irregular tremor (frequency 8–11 Hz). This threshold force was lowered after prolonged muscle contraction and after preceding large-amplitude tremor: "warming up effect" (p. 122). Brown et al. also state: "Whereas the responses of the joint to sinusoidal driving indicate that a 3–6 Hz tremor may be the result of reflex activity, they indicate equally that an 8-11 Hz tremor of this joint is unlikely to arise in that way" (p. 122).

In summary, then, the following points emerge:

1) Physiological tremor occurs in several forms which are characterised by distinct frequency-amplitude regions. Tremor in the *low-amplitude, high-frequency region* is sustained by unfused contractions of motor units firing nearly asynchronously and does not involve oscillations of the entire spinal stretch reflex, whereas the *higher-amplitude, lower-frequency tremors* (above referred to as medium- and large-amplitude tremor) are at least partially based on such reflex oscillations.

2) Tremor progresses from the former into the latter region by crossing an *amplitude threshold*. Any higher amplitudes therefore bear the risk of reflex oscillations (larger-amplitude tremor).

The transition between the two states of low-amplitude and "enhanced" tremor poses a special problem which will be considered later. But first, the stochastic nature of (constant) muscle contraction will be dealt with, which is essential for understanding low-amplitude tremor and the intricate mechanisms by which it is sustained at least over some time.

3.2 Muscle Contraction as a Stochastic Process

Constant muscle contraction is essentially a stochastic (random) process which has bearings on the development of tremor on the one hand, and the control of muscle contraction by the nervous system on the other. It is far too little recognised that the control of this stochastic process is an enormous task for the nervous system, requiring both intricate mechanisms and possibly adequate plant models. It is not unlikely that the way the nervous system handles this task is a sort of paradigm which may reveal insights into mechanisms of signal processing in the central nervous system in general. The stochastic process of muscle contraction will therefore be dealt with in some detail because it is easier to observe it than maybe similar processes in the central nervous system. For comparison, however, the latter will later be discussed in the light of the spinal recurrent inhibitory system (see Chap. 4).

We deal first with a linear stochastic model of muscle contraction which elucidates the basic mechanism of physiological tremor, and then consider some nonlinearities in this stochastic process.

3.2.1 The Basic Mechanism of Small-Amplitude Physiological Tremor

The dominant mechanism of small-amplitude physiological tremor was lucidly described by Christakos (1982) using mathematical analysis and computer simulation. Similar suggestions were previously made by Marshall and Walsh (1956), Allum et al. (1978), Cussons et al. (1979) and Clark et al. (1981). Christakos showed that tremor is an unavoidable consequence of the normal activation of skeletal muscles. His population stochastic model of skeletal muscle (based on data from the human first dorsal interosseus muscle) is able to explain the following features of the tremulous force oscillation during steady, constant-strength muscle contraction: (a) its apparent regularity; (b) its amplitude variations; (c) its relatively low frequency compared to the firing rates of the majority of the active motor units; (d) the increase in tremor amplitude with increasing mean muscle force; (e) the relative stability of the dominant tremor frequency in the face of varying levels of contraction, and the increase of this frequency at very high force levels (Christakos 1982).

These features result from the way in which motor units work during skeletal muscle contraction.

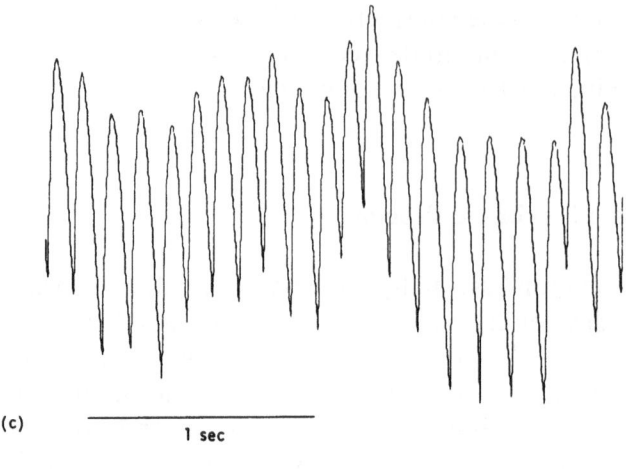

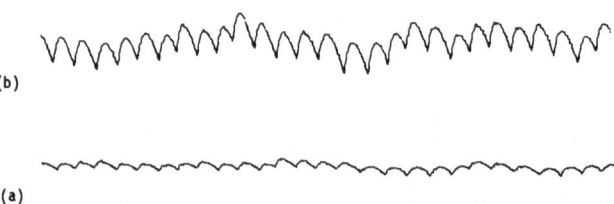

Fig. 23 a–c. Computer-simulated force oscillations. **a** shows the force contribution of a small, **b** that of a medium-sized, and **c** that of a relatively large motor unit to a sustained muscle contraction at 17% maximal contraction. Amplitude and time scales are the same in **a** to **c**. The differences in amplitude are of particular importance; these differences become large especially at high force levels since larger motor units are then recruited. (With permission from Christakos 1982; his Fig. 6)

1) With increasing (steady) force, successively larger (stronger) and faster motor units are recruited and begin to fire at approximately the same low rate (ca. 6–10 pps). Figure 23 demonstrates the force fluctuations contributed by motor units of different size firing at slightly different rates at a low level of muscle contraction. In the model previously recruited smaller motor units increase their firing rate roughly linearly with increasing contraction force, and this has indeed been neatly demonstrated (Milner-Brown et al. 1973).

2) The discharge pattern of each steadily firing motor unit is semi-regular with a roughly Gaussian interval distribution and with low variability (coefficient of variation of the order of 0.1 to 0.2). In the low-amplitude region of physiological tremor, motor units fire relatively independently of each other, i.e. in an uncorrelated fashion (with the possible exception of a tendency towards short-term synchrony, especially in weight-lifters (Milner-Brown et al. 1975; see Sect. 2.1.1).

3) The force-producing system of each motor unit can be modelled as a damped second order system, with a low cut-off frequency for slow motor units and a damping ratio that increases with increasing sustained discharge rate (inherent motor unit nonlinearity). These low-pass characteristics imply that the power spectrum of the force developed by a newly recruited motor unit has a strong low-frequency component, in the band of 0–2 Hz, and a local peak at the mean discharge rate λ_i, above which the slope declines steeply with power up to 15 or 20 Hz (depending on the type of motor unit).

Thus, at each level of muscle contraction, there exists a relatively small group of newly recruited motor units (decreasing in number with increasing mean force), which fire near their recruiting rate and hence, by their unfused contractions, contribute force fluctuations to the overall force output. At most force levels, except for the very lowest, most of the active motor units fire well above their recruiting rates; the slower the units, the more their contractions are fused, whereby the earlier recruited units contribute less to force fluctuations (i.e. to tremor), the earlier they have been recruited. This mechanism together with experimental evidence for it were already clearly described by Allum et al. (1978). The superposition of the force fluctuations contributed by the motor units recruited last yields a force fluctuation showing regularities which wax and wane as illustrated in Figs. 24 and 25 (upper traces). Thus, the local peak in the tremor spectrum at 8–10 Hz (see Fig. 25, lower trace) results from the unfused contractions of the newly recruited motor units. This explains the five features of physiological tremor listed above (items a through e).

The apparent regularity (item a) and the random variations in amplitude (item b) of the tremor oscillation as exemplified by the simulated muscle force waveforms in Figs. 24 and 25, result from the following two phenomena. The semi-regularity of the motor unit discharge pattern, i.e. the strong spectral component at the firing rate in the random force output of each unit; and the fact that for uncorrelated units the individual auto-spectral components add linearly (upon superposition of motor unit activity) to yield a similarly strong spectral component in the overall muscle force waveform within the same narrow frequency band. This latter component indicates a prominent, fairly regular oscillation of randomly varying amplitude in the muscle force.

It merits attention that an extension of this theory to the case of partially correlated processes has recently been presented (Christakos 1986).

Cross-correlated rhythmic motor unit activity at similar rates implies significant cross-spectral components at about the same frequency, in addition to the auto-spectral components mentioned in the preceding paragraph. All these components combine in such a way as to produce a large peak in the spectrum of the overall process, again indicating a fairly regular oscillation in the muscle force signal. This explains the regularity of tremor when a fraction of motor units are pairwise correlated, possibly to different de-

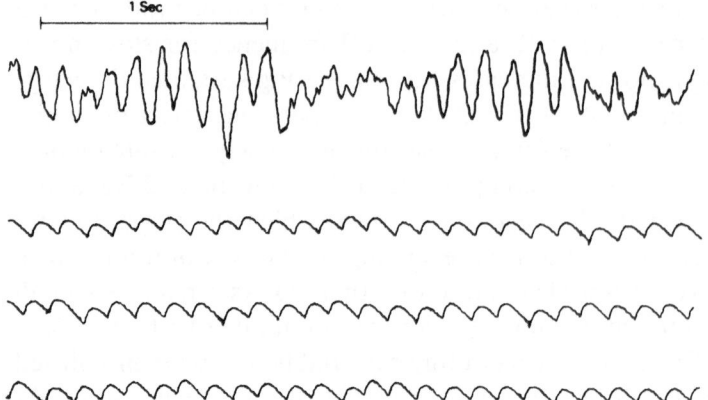

Fig. 24. The effect of superimposing the force contributions of motor units firing semi-regularly at similar rates (about 9 Hz). The *lower traces* show the force fluctuations of three motor units of similar size and discharge rate. If 30 such uncorrelated processes are added, a force fluctuation, as shown in the *upper trace*, results. (Courtesy of C.N. Christakos)

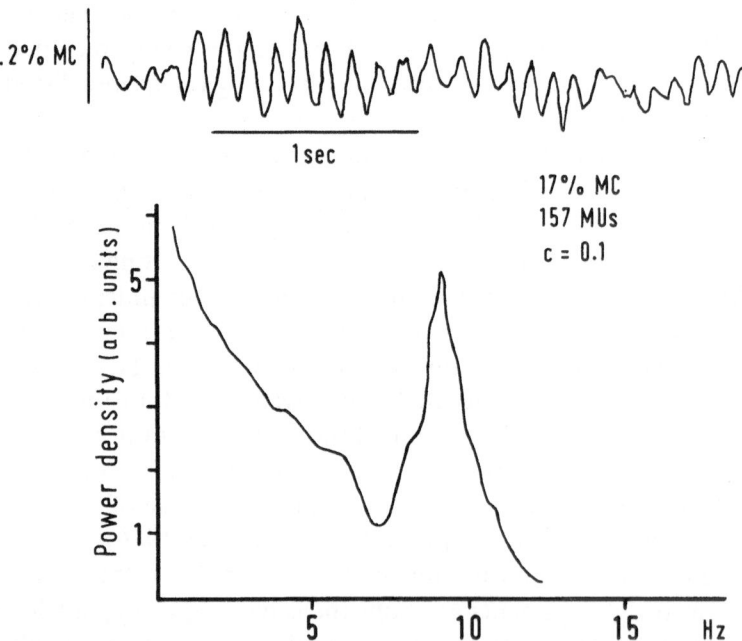

Fig. 25. Computer-simulated force tremor oscillations (*upper trace*) and their power spectrum (*lower plot*) for a steady contraction of the human first dorsal interosseus muscle at a level of 17% of maximal contraction (*MC*), at which 157 motor units are assumed active; motor units discharge semi-regularly with a coefficient of variation of 0.1. (With permission from Christakos 1982; part of his Fig. 3)

grees, ranging from very weak phase relations to complete synchrony, and for different time constellations. This mechanism applies also to the case of large-amplitude tremors to be considered in the following sections. The underlying theory can also be applied to rhythms at other levels of the central nervous system, such as the α-rhythm in the EEG.

Intuitively, the emergence of regular oscillations by superimposing uncorrelated semi-regular processes can be grasped as follows. The local peak in the power spectra at 8–10 Hz represents the most conspicuous and prominent oscillation in the force spectra (except for the low-frequency components). As noted above, it is due to the asynchronous activity of a small group of newly recruited motor unit firing semi-regularly at or near their recruiting rates. At any instant, there will be some motor units among this group which happen to fire nearly synchronously (by chance). Since each motor unit of this subgroup produces interspike intervals of relatively low variability, the chance is great that at least some of them again fire almost synchronously after a mean interspike interval (see App. E). At this new instant, other motor units may have entered the subgroup of synchronously discharging motor units. Thus, because of the semi-regularity of motor units discharging at about equal rates, there occurs a chance "grouping" of events (see A. Taylor 1962). In addition, it should be noted that the requirements on "synchrony" (for rhythmic force generation) are not so very high due to the long contractions of motor units (as compared, for example, to postsynaptic potentials). The groupings of motor unit action potentials (and subsequent contractions) wax and wane, and this phenomenon readily explains the waxing and waning of tremor amplitude.

It must be pointed out that the above model accounts for the fluctuations in muscle *force*, from which displacement fluctuations result through a mechanical filtering. The filter is provided by the visco-elastic properties of limb tissues (including muscles) and the limb masses. Since these filters can vary widely, they may influence the displacement characteristics for different muscles and/or joints rather differently.

3.2.2 Nonlinearities of Motor Unit Behaviour

As argued in Sect. 3.1, the fluctuations of force and the internal muscle length fluctuations during small-amplitude tremor may have to stay below a certain critical amplitude threshold in order not to cast the entire stretch reflex into oscillation (Stiles 1976; Hagbarth and Young 1979). That is, any larger-amplitude excursions which may occur by chance would increase the risk of instability.

Such larger-amplitude chance excursions may result from several mechanisms. (1) In steady muscle contraction, successive activation of individual motor units usually occurs at relatively regular intervals. Not infrequently, however, small interspike intervals are interspersed (Andreassen and Rosen-falck 1980; Bawa and Calancie 1983; Sect. 3.2.2.1). The ensuing temporal overlap of the related contractions would cause a larger excursion by them-selves, but in addition, the second contraction may be nonlinearly enhanced (potentiated; for review see Burke 1981; also Parmiggiani et al. 1982; Sect. 3.2.2.1). (2) The independent semi-regular activation patterns of the population of freshly recruited motor units cause temporary chance synchro-nisations of subsets of them (Christakos 1982; App. E), thereby producing periods of relatively regular force oscillations of enhanced amplitude inter-rupted by irregular force fluctuations. In Christakos' model (1982; Sect. 3.2.1), these fortuitously synchronised twitches add linearly. However, this does not hold in general (Sect. 3.2.2.2).

In order to assess the risk of stretch reflex instability, it is essential to understand the way in which the above force fluctuations are picked up by muscle spindle afferents, particularly by Ia fibres from primary spindle end-ings. It has been amply shown that these afferents sense the unfused contrac-tions of even single motor units (e.g. Binder and Stuart 1980a; Schwestka et al. 1981; Christakos and Windhorst 1986b), and that this information is transmitted to homonymous α-motoneurones (W. Koehler et al. 1984a, b). Therefore, following a discussion of the two features of motor unit behaviour outlined in the preceding paragraph, the response patterns of muscle spindle afferents to those features are illustrated subsequently (Sect. 3.2.3). In gener-al, these responses are nonlinear, and this nonlinearity possibly contributes significantly to the damping of tremor.

3.2.2.1 Nonlinearities in Single Motor Units

Whether nonlinearities appear in the behaviour of single motor units depends very much on the specific stimulus pattern used (see Burke 1981). They can however be expected to play a significant role under physiological conditions. We have stimulated single motor units in cat medial gastrocne-mius and soleus muscle with random (Poisson) patterns, and applied an analysis technique described in App. D (see also Windhorst et al. 1983).

Figure 26 (upper row) shows a typical result obtained in about 67% of the medial gastrocnemius motor units. The right plot (D) displays the mean twitch tension response of the unit averaged with respect to all the stimuli delivered to its axon in a ventral root filament. Part A is a three-dimensional plot of the tension responses to selected *pairs* of stimuli, the mean interval δ between these stimuli being plotted on the z-axis. That is, the first stimulus is the reference event (at $\tau = 0$ ms), and the second follows around $\delta = 60$ ms later for the curve in the foreground, and at successively briefer intervals for

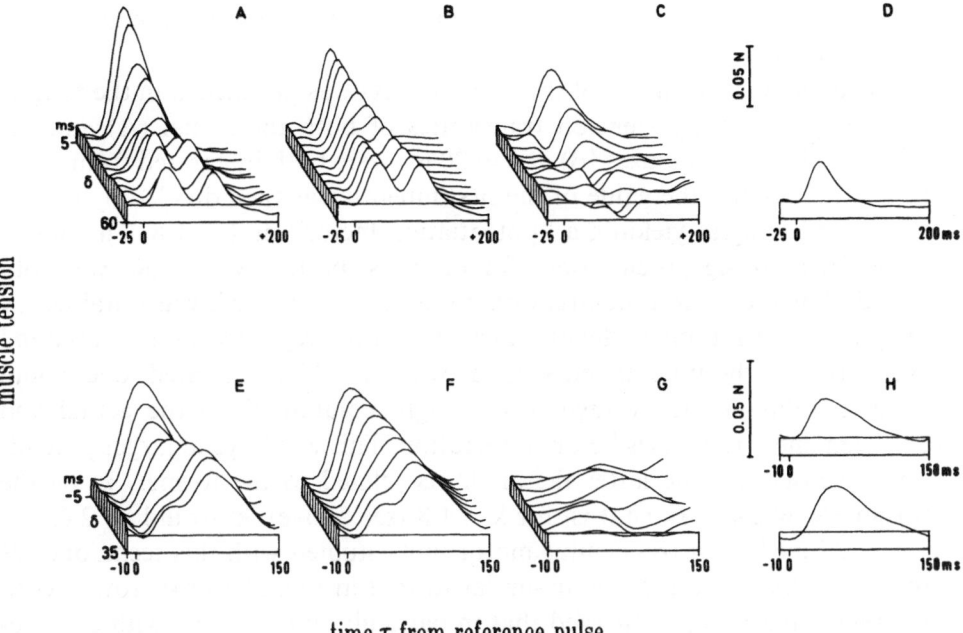

time τ from reference pulse

Fig. 26 A–H. Comparison of the nonlinear tension summation evoked by stimulation of one medial gastrocnemius motor unit (*upper row*: **A** through **D**) and by stimulation of two medial gastrocnemius motor units in parallel (*lower row*: **E** through **H**) with random stimulus patterns. For the *upper row*, the overall twitch response of the motor unit averaged over all the stimuli delivered is shown in **D**. For the plots in **A**, pairs of stimuli separated by selected mean intervals δ (oblique axis) were sorted out from the stimulus train to the motor unit, as illustrated schematically in Fig. 60 (App. D). The medial gastrocnemius tension fluctuations were averaged over these pairs of stimuli and plotted for various δ's in a pseudo-three-dimensional manner. The plots in **B** represent the expected linear responses obtained by superimposing twice the overall twitch (**D**) at appropriate intervals δ. The plots in **C** are the differences between those in **A** and those in **B**, and show the potentiation of the second twitch at short intervals after the first stimulus. The mean stimulus rate was 6.1 pps. The number of reference events varied from 23 to 31. In the experiment underlying the *lower row*, three medial gastrocnemius motor units were stimulated independently at rates of 7.7, 9.3 and 6.4 pps. The interaction of the first two is depicted. Pairs of stimuli from the two stimulus trains were selected, as illustrated in Fig. 60 (App. D). **E** actual responses; **F** expected linear responses; **G** differences between **F** and **E**, i.e. the depression appearing in **E** as compared to **F** (for δ's between 0 and 15 ms) is plotted as upward deviations from the zero plane; **H** twitch trajectories averaged with respect to all stimuli in either of the two stimulus trains. Number of reference events between 45 and 63. (With permission from Niemann et al. 1986; their Fig. 2)

the curves following towards the background. Part B shows the responses which would have been expected had the two subsequent twitches superimposed linearly; they were obtained by summing the curves in part D at appropriate intervals. The differences between the curves in part A and in part B are displayed in part C, and it appears that the closer a second activation of a motor unit follows upon a first, the more the second twitch is potentiated.

A rough quantification of the above effects was performed in the following way. The average tension trajectories of both the actual (A) and the expected (linear) responses (B) at $\delta = 20$ ms were both integrated from $\tau = 5$ to $\tau = 120$ ms and the value of the first integral was then divided by that of the second integral yielding a "potentiation factor" of $k = 1.8$. This was a value for a strong potentiation; for the most part, lower values were obtained, down to 1 (no potentiation). Note that the factor k was obtained by integrating over both twitches which were thus regarded as a compound event. If only the potentiated second twitch is to be evaluated, one would have to subtract the average twitch (right column) from the actual and expected linear responses before integration. This would yield larger potentiation factors, k'. The latter factor, k', is related to the former, k, by the formula $k' = 2k - 1$. That is, for $k = 1.8$ (see above), k' would be 2.6.

Of 27 medial gastrocnemius motor units studied with this technique, 18 (ca. 67%) showed a behaviour similar to that in Fig. 26 (upper row). As to the remaining 33%, it emerged that in particular motor units with contraction times (10% to 90% tension rise) greater than 40 ms did not potentiate.

In order to look specifically at S-type motor units, the same technique as described above was applied to motor units of the soleus muscle. Nonlinearities showed up much less frequently: 12 out of 16 single soleus units behaved linearly; in the remaining 4 cases there was a slight depression at short intervals of activation (from $\delta = 10$ to $\delta = 25$ ms), i.e. the actual response was slightly smaller than the expected linear response.

Our results are essentially compatible with those of Parmiggiani et al. (1982) who used the same stochastic stimulus patterns as we did, but investigated whole slow or fast muscles instead of single motor units. Thus, in this respect, our results are an extension to the motor unit level. However, it is not a priori self-evident that whole muscles would display the same behaviour as motor units of the same contraction speed since the synchronous activation of many motor units in the former case might cause substantial nonlinear interactive effects (see Demiéville and Partridge 1980; and below). The main point to be emphasised here is that very often a second motor unit twitch following a preceding one at a short interval can be more or less strongly potentiated and that this tends to generate larger force (and length) transients.

This result is probably of importance during natural muscle contractions. Andreassen and Rosenfalck (1980) recorded, in humans, the discharge patterns of tibialis anterior motor units during sustained muscle contractions. They quite often observed "double discharges"; the criterion for a double discharge was that the interspike interval was less than half the mean interval. Double discharges often occurred when a unit was recruited. But in about

30% of the 20-s recordings, double discharges also occurred during maintained contraction. Double discharges often preceded small increases in torque or followed immediately after a small decrease in torque. The latter small decrease in torque must be expected to occur as a random fluctuation due to the stochastic activation patterns of motor units (see Sect. 3.2). It is highly probable that this decrease, which was superimposed on a high background torque (see their Figs. 9 and 10), was sensed by a small subset of muscle afferents whose discharge patterns were thereby partially correlated. This fluctuation in afferent discharge (probably a local transient increase in Ia firing) then produced the motoneurone double discharge. Andreassen and Rosenfalck discuss several mechanisms that may enhance the dynamic sensitivity of α-motoneurones to such input fluctuations. Among these are afterhyperpolarisation, recurrent inhibition via Renshaw cells, and Ia and Ib feedback. The relative importance of these mechanims is unknown, but they all may affect the pattern of motoneurone discharge in roughly the same way. "The functional value of these effects lies in the increased dynamic sensitivity of the motor neuron pool. The motor neuron pool thereby enhances the higher frequencies in the synaptic input which partially compensates for the pronounced low-pass characteristics of muscle" (p. 904; see Sect. 4.2). However, the above results on single motor unit twitch potentiation suggest that this sensitisation is more pronounced for fast than for slow motoneurones. (It is noteworthy that α-motoneurones possess a dynamic sensitivity even when recurrent inhibition and proprioceptive feedback are inoperative or very weak; see Baldissera et al. 1982). Another recent study by Bawa and Calancie (1983) on the human flexor carpi radialis muscle showed that such doublets may also occur repetitively, particularly at the onset of slow motor unit recruitment. This phenomenon was seen in a large number of units of all types.

3.2.2.2 Nonlinear Interaction Between Several Motor Units

Whereas the predominant motor unit nonlinearity resulting from interaction of twitches in a single channel is potentiation (previous section), that resulting from interaction of twitches of two (or more) motor units – if present – is depression. These opposite responses are contrasted in Fig. 26.

The "in parallel" nonlinearity is illustrated by a typical example of a pair of medial gastrocnemius motor units in the lower row of Fig. 26 (E through H). For comparison, the lower plots have the same basic structure as the plots in the upper row (A through D), which show the potentiation effects in a single medial gastrocnemius motor unit. For the lower row, three medial gastrocnemius motor units were stimulated in parallel, but the interaction of only two of them is shown. The overall twitches averaged over all stimuli in the respective stimulus trains are illustrated in H. Part E depicts the tension trajectories averaged over various time constellations of activations of two motor units, the time interval δ ranging from -5 to 35 ms. The expected linear trajectories are shown in part F. Comparison of E and F indicates that the tension actually developed is less than expected from linear behaviour for time constellations from about $\delta = 0$ to $\delta = 15$ ms. The difference between E and F is plotted in G, but – for graphical reasons – values above the zero plane this time imply that the actual response is less than the expected linear one, in contrast to the corresponding plots in Fig. 26 C.

Quantifying this depression in the same way as the potentiation described in the previous section, i.e. dividing the tension integrals at $\delta = 10$ ms (maximal depression) from $\tau = 5$ ms to $\tau = 130$ ms, yielded a depression factor of 0.67 for the unit in the lower row of Fig. 26.

Of 36 pairs of medial gastrocnemius motor units investigated in this manner, 10 (28%) showed a nonlinear summation of their twitches as illustrated in Fig. 26 (lower row), albeit on average slightly less pronounced, with a "depression factor" between 0.75 and 0.85. The rest (26 pairs) behaved rather linearly, although this statement is subject to detection errors arising from statistical variation. Thus, only depression factors of 0.85 or less were here considered significant. As argued below, much weaker depressions in many motor units may sum up to considerable effects.

As before, soleus motor units behaved much more linearly than medial gastrocnemius motor units, also in the case of parallel interaction. Of 15 pairs investigated, 13 summed their twitches linearly for all time constellations of stimuli, whereas two showed a small depression when activated nearly simultaneously.

In many muscle models in which skeletal muscle is composed of many motor units in parallel, it is implicitly assumed that the (independent) twitches of the various motor units sum linearly at the tendon (e.g. Coggshall and Bekey 1970; Christakos 1982; see Sect. 3.2.1). However, as shown here for the cat medial gastrocnemius muscle, this does not hold true in general. After completing this series of experiments, we became aware of similar results obtained by Demiéville and Partridge (1980). Although their report was entitled *Probability of peripheral interaction between motor units and implications for motor control*, these authors actually investigated the interactions of two *entire* frog gastrocnemius muscles coupled mechanically either in parallel or in series. They extensively discussed the degree and possible consequences of mechanical interactions between motor units. Some of their conclusions are very important in the present context and are therefore reproduced here somewhat extensively.

In muscles composed of many units, as contrasted to two-unit systems, the force production of each unit, contributing to the overall muscle output, can be modified by more than one other motor unit. Imagine a muscle at a recruitment level, at which n motor units are active. The total force output produced by the activity of the i-th motor unit includes the output force, which that unit would produce without interacting with the remaining $n - 1$ units, $f_{i,0}$, plus all of the indirect (cross-) effects on the force production of the other $n - 1$ active units, resulting from action of the i-th unit. Such an indirect effect of unit i on unit j may be denoted by $\Delta f_{j,i}$. Then the total output force, F_i, resulting from activity in the i-th unit

$$F_i = f_{i,0} + \sum_{j=1}^{n} \Delta f_{j,i}; \quad j \neq i \tag{10}$$

As the interaction appears to be nonlinear, $\Delta f_{j,i}$ is not constant, but it will vary with the total number, $n - 1$, of other motor units active and, hence, with the level of recruitment. At a low level of recruitment, the i-th individual unit would contribute a large part to the total muscle force, and large interactions will occur. With increase in muscle force due to the recruitment of new motor units into activity, the nonlinear cross-effect of the previously regarded i-th motor unit might be assumed to decrease because its force contribution becomes relatively less and, hence, the nonlinear interactions diminish. However, this decrease does not vary exactly inversely with the number $n - 1$ of the other active units because the usual recruitment order adds successively larger motor units. The amount of nonlinearity arising in natural muscle contractions is therefore very complicated. According to Demiéville and Partridge (1980), the nonlinear interaction of motor units could even lead to the situation that a motor unit contributes negative force to the overall muscle force if the second term in Eq. (10) becomes negative and absolutely greater than the first term.

Demiéville and Partridge (1980) also add some interesting quantitative considerations. In a muscle with many active motor units, any particular nonlinear interaction is likely to be unmeasurably small, even when the summed effect of all interactions is large. For instance, consider the situation where the force reduction due to interaction in a whole muscle is 50%. If that muscle is composed of 100 motor units interacting nonlinearly, the average force reduction resulting from one of the 9900 one-way interactions would amount to only about 0.01% of the whole-muscle force. "Consequently, in muscles with a large number of active units, the importance of interaction lies in the cumulative effect, rather than in the unitary effect emphasized in the model" (Demiéville and Partridge 1980; p. R133). In this light, our 30% of nonlinear interactions found may well be an appreciable underestimate. Smaller interactions that we did not consider significant due to noisy records might still be physiologically significant.

As to the mechanisms responsible for nonlinear interaction, Demiéville and Partridge (1980) suggest the well-known length–force and force–velocity relations. In their experiments, both the length and velocity of the test muscle could be altered by activity of the interacting muscle because they used a nonisometric arrangement. The length–tension and force–velocity effects produce alterations in the response of a muscle to a neural input, particularly if it is allowed to move. In any nonisometric system, therefore, a change in force will result in changes in acceleration, velocity, position, power, and work. But, due to their internal anatomical structure, even isometrically contracting muscles almost always move (mostly shorten) internally. "Thus a part, perhaps all, of the observed interaction can be explained by these two well-known muscle properties" (p. R132). It may be added that interaction might include a spatial component insofar as motor units whose territories overlap considerably are more likely to interact mechanically. In our experiments, we tried to localise the muscle units in the muscle by stimulating the motor units at high rates, causing fusion, and drawing a sketch of the muscle surface area which showed strong dimpling. However, this technique is rather inaccurate, since it makes inferences on the overlap of motor unit territories from the surface projection and cannot really determine the degree of overlap in the depth of the muscle. Therefore, we cannot tell with certainty that strong nonlinear mechanical interaction of motor units is related to their spatial proximity in the muscle. It should also be emphasised that these interactions probably depend on the particular load against which the muscle is working and on the time constellations of motor unit activation.

As specifically pointed out by Demiéville and Partridge (1980), the above mechanical interaction between contractions of parallel motor units poses a particular control problem to the nervous system. The mechanical response elicited by activation of a motor unit can be "known" to the motoneurone (or, more generally, to the CNS) only in a statistical sense (as an average or so), or post factum, i.e. as reported by sensory feedback. "It follows that the central nervous system cannot organize motor responses by calling up the appropriate sum of responses, choosing from a catalogue of well-defined and fixed unit responses" (Demiéville and Partridge 1980; p. R133). It is not easy to imagine how the central nervous system should have an a priori knowledge of what may happen to its motor commands in the periphery, since the actual motor output depends strongly upon prevailing mechanical conditions. It should therefore have an accurate picture of this situation delivered by sensory feedback. But this would probably not suffice since, in order to appreciate the sensory messages, the nervous system should also have a central model with which to compare them. This topic is dealt with more explicitly in the next two chapters, the sensory messages that might play a role being addressed first.

3.2.2.3 Summary

When motor units of the isometrically constrained cat medial gastrocnemius or soleus muscles are activated randomly, the temporal pattern of the input is an important determinant of the force produced at the common tendon.

1) Many single motor units activated with Poisson-like stimulus pulse trains react sensitively to the precise temporal sequence of subsequent stimuli. About 67% of the medial gastrocnemius motor units show a conspicuous potentiation of the second of two twitches in response to pairs of stimuli separated by 5 to 50 ms. The remainder, as well as soleus motor units, exhibit no conspicuous nonlinearities. This type of twitch potentiation is considered an important mechanism for rapid force production at the onset of fast muscle contraction. However, small interspike intervals also occur during sustained muscle contraction in which they may then produce force (and related internal length) fluctuations of larger amplitudes that may in turn augment the risk of reflex oscillation. The latter effect, however, depends on the way spindle afferents respond to potentiated motor unit twitches.

2) The mechanical interactions in force production of parallel motor units have not yet been investigated very extensively, but are important for the force development of the whole muscle. In our study, about 28% of *pairs* of medial gastrocnemius motor units produce conspicuously less twitch tensions than expected for linear summation if activated nearly simultaneously. This figure can give only a very rough idea of the degree to which motor units interact in natural muscle contractions in which many motor units are active in parallel. As argued by Demiéville and Partridge (1980), the cumulative effect of the immense number of possible mutual interactions can amount to considerable loss of tension available at the load. This poses a severe control problem to the nervous system which will be discussed later. In the present context of tremor, this nonlinear interaction of parallel motor units would appear to mitigate to some extent the effects of chance synchronisation of motor unit twitches.

3.2.3. Effects of Motor Unit Contractions on Spindle Afferent Discharge

The previous two sections analysed the system between motor axon and force output (muscle unit). The following sections add another subsystem, namely the muscle spindle with its afferent axons. For the following, it is essential to note some peculiarities of signal transmission from skeletomotor efferents to spindle afferents, particularly an aspect termed "sensory partitioning" by Cameron et al. (1981).

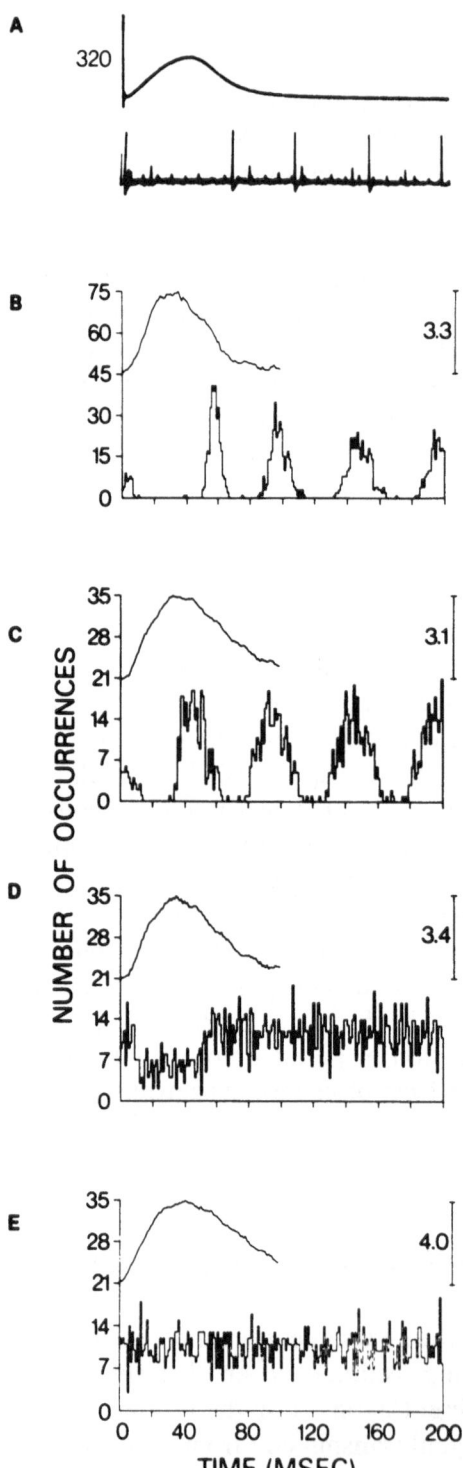

Fig. 27 A–E. Responses of a single Ia afferent to the twitch contractions of the whole muscle and four different motor units. **A** response to single whole-muscle twitch contraction (320 g) at muscle length L_0-2 mm. Initial deflection of the afferent trace is a stimulus artefact, and the first afferent spike was produced by the stimulus shock delivered to the muscle nerve; **B – E** cross-correlation histograms compiled for the Ia afferent and four different, randomly stimulated motor units. Average motor-unit contraction profile is superimposed on each correlogram with the average tension indicated by the *vertical bar on the right*; **B** displays strong coupling between the afferent and motor unit, coupling index = 1.44. Correlogram based on 327 stimuli; **C–E** display progressive decreases in afferent-motor unit-coupling (coupling indices are 0.94, 0.24, 0.09, respectively). Correlograms **C – E** based on 325, 96, and 139 stimuli, respectively. Note the similarity in the time-courses and average twitch tensions of the four different motor units. (With permission from Binder and Stuart 1980a; their Fig. 1.)

A good impression of some statistical aspects of signal transmission from α-motoneurone axons to spindle afferents is given by a paper by Binder and Stuart (1980a). In the underlying experiments, skeletomotor axons of motor units belonging to the cat tibialis posterior muscle (held isometric) were repetitively stimulated at a low rate (about 2 pps), and the discharge patterns of spindle afferents from that muscle were recorded from dorsal root filaments. An example for the results obtained is displayed in Fig. 27. The reaction of a Ia fibre to a single twitch of the whole muscle is shown in Fig. 27A; it exhibits a typical "spindle pause". Parts B to E show averaged Ia fibre responses (poststimulus-time-histograms: PSTHs) to repetitive activations of different motor units of about equal strength and time course of contraction as judged from their averaged twitches. Contractions of the first three motor units (B–D) provoke a typical "unloading response" in the spindle, although to different degrees in the three cases. The spindle does not respond to contractions of the fourth motor unit (E). The diverse strengths of the responses of the afferent to contractions of the four motor units are indicative of variations in the degree of mechanical coupling between the two elements. Independent evidence has been accumulated that the different degrees of mechanical coupling as expressed in the depth of PSTH modulations result in part from the spatial relations between spindles and motor units (Cameron et al. 1981; Schwestka et al. 1981).

Whereas in Fig. 27 motor units of about equal contraction strength were chosen to demonstrate their differential effect on a spindle afferent, strong motor units on average influence afferent discharge more strongly than do weak motor units. This is an important aspect in the present context and is again illustrated by the results obtained by Binder and Stuart (1980a). The authors categorised the spindle responses to motor unit contractions in two ways: qualitatively and quantitatively. The qualitative characterisation was performed by visual inspection of the PSTHs using criteria such as the duration of the post-stimulus spindle pause and variability of successive peaks and troughs. Thus, PSTHs were categorised as showing "null", "weak", "intermediate" and "strong" coupling. (For example, the PSTH in Fig. 27E would be judged as expressing a coupling index of 0, whereas the PSTHs from D upwards would receive values around 1; see Binder and Stuart 1980a). For quantifying the degree of spindle responsiveness to motor unit contraction, the PSTH was computed with 40 5-ms bins (extending over a 200-ms analysis period). The mean and standard deviation of the 40 bin contents were calculated, and their quotient – the coefficient of variation – served as a numerical "coupling index" (see legend to Fig. 28).

In order to relate the responses of Ia and spindle group II afferents quantitatively to the twitch amplitudes of the motor units stimulated, the latter were divided into three classes: those lower than 1% total muscle twitch amplitude, those between 1% and 3%, and those greater than 3% whole

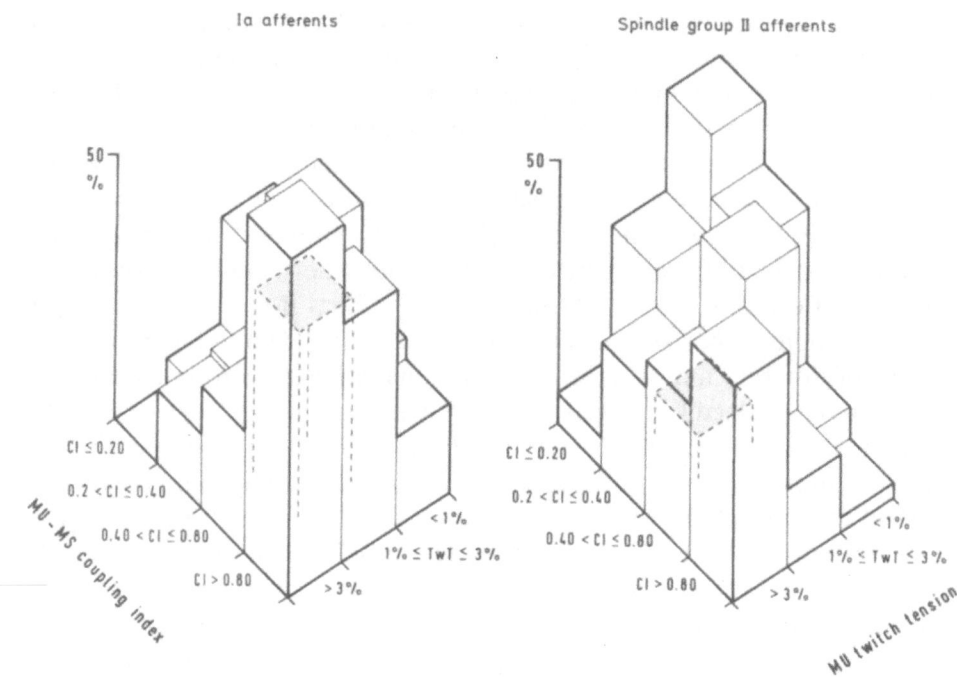

Fig. 28. Pseudo-three-dimensional representation of the relationship between motor unit twitch force and spindle afferent responses, *on the left* for Ia afferents, *on the right* for spindle group II afferents. The twitch forces of cat tibial posterior motor units, measured under isometric conditions and expressed in percents of whole muscle twitch force, are divided into three classes and plotted on the *right abscissae* (*MU twitch tension:* TwT). The responsiveness of afferents to contractions of single motor units was assessed qualitatively from PSTHs such as those in Fig. 27, but expressed quantitatively by a "coupling index" (*CI*) ranging between 0 and somewhat above 1 (see Fig. 27), such that 0 implied no coupling (no spindle response to motor unit twitch) and 1 a very strong coupling indicated by deep modulation of the corresponding PSTH. These coupling indices are divided into four classes and plotted on the *left abscissae* (*MU-MS coupling index*). The number of motor unit-spindle afferent combinations falling into each (two-dimensional) class is plotted on the *ordinate* ,such that all afferent responses within each class of motor unit twitch force sum up to 100%. The columns with *dashed sides* and *shaded upper surfaces* are hidden from visibility by the front columns. (Based on data from Binder and Stuart 1980a)

muscle twitch force. Figure 28 shows the relation between the above two measures: "coupling index" and normalised motor unit twitch force (plotted from Table 1 of Binder and Stuart 1980a). The entries (in both plots) are the proportions of motor unit-afferent combinations in percentages of all combinations studied. The figure implies that:

1) Group II afferents from secondary spindle endings (right distribution) are generally less sensitive to motor unit contractions than are Ia fibres (left distribution), as expressed by the higher percentages of null responses and the

lower percentages of strong responses (see also Christakos and Windhorst 1986 b).

2) For both Ia fibres (left distribution) and group II spindle afferents (right distribution), the probability of null responses ($0.2 \leq$ CI) decreases and the probability of strong responses (CI > 0.8) increases with increasing motor unit contraction strength. This implies that *strong motor units affect a larger number of spindle afferents than do weak motor units*. Also, motor unit force is positively correlated with spindle response, as expressed by the coupling index (corr. coeff. 0.44; $P < 0.001$; see Binder and Stuart 1980 a).

3) It follows that during natural constant muscle contractions of what ever strength those motor units which have been recruited last, i.e. which are the strongest and contribute most to physiological tremor (Sect. 3.2.1), exert the strongest influence on spindle discharge.

The modulation of spindle afferent discharge by motor unit activity has mostly been studied using time domain analysis, i.e. by cross-correlation methods. In regard to tremor it is of interest to see the effects in the frequency domain, too. We have recently applied this type of analysis to spindle afferent discharges (Christakos et al. 1984; Christakos and Windhorst 1986 b; for the method see App. F). Figure 29 shows typical examples of auto-spectra computed for a soleus Ia afferent whose discharge was modulated by the activation of one (in A) or two (in B) soleus motor units with semi-periodic stimulus patterns (see legend). The stimulus train had a fundamental frequency of $\lambda_m = 5.5$ Hz and a mean rate of pulses of 8 Hz. Two distinct components can be seen in the afferent spectrum, one at λ_m (indicated by single arrow labelled MU 1) and a smaller one at the first harmonic of λ_m (11 Hz). These components reflect the effects of the twitch contractions of the motor unit on the afferent discharge and indicate that the Ia afferent spike train carries information on the rhythm of activation of the unit. The prominent spectral peak at 23 Hz in A represents, at least to some extent, the receptor's own rhythm corresponding presumably to the rhythm of the dominant pacemaker of the sensory ending. Spectral components above about 15 Hz may partially be due to nonlinear interactions between frequency components originating from the input rhythm and those due to the pacemaker rhythm.

The example of Fig. 29 A is representative of cases of moderate coupling between a (slow) motor unit and a primary spindle ending. Similar features were observed in cases of weak coupling, but the modulating components were small as compared to the rest of the spectrum.

An example for the case of parallel stimulation of two motor units is given in Fig. 29 B. The auto-spectrum of the discharge of the same soleus primary is shown for (independent) semi-periodic stimulation of two motor units, one at 5.5 Hz as in A, the other at 7 Hz. The two distinct spectral components at $\lambda_{m1} = 5.5$ Hz and $\lambda_{m2} = 7$ Hz (note the extended frequency scale for better distinction of the two spectral components) and the one at the first harmonic

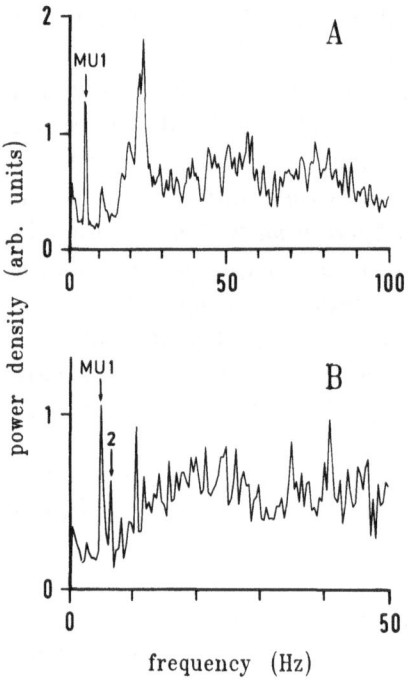

Fig. 29 A, B. Typical examples of auto-spectra computed for a soleus Ia afferent when one (**A**) or two (**B**) motor units were stimulated with semi-periodic stimulus patterns. These stimulus patterns consisted of a strictly periodic component at a fixed rate with interspersed random stimulus pulses elevating the mean stimulus rate slightly. The auto-spectrum of such a stimulus train shows a fairly regular succession of narrow spectral peaks at the fundamental frequency and its harmonics up to high frequencies, superimposed upon a constant level of power corresponding to the randomly injected pulses. The soleus muscle was held isometric. **A** Auto-spectrum of the discharge of the primary ending during semi-periodic stimulation of a single soleus motor unit (*MU 1*) moderately coupled to the receptor. Note the distinct spectral components at the fundamental rate of stimulation of the unit, 5.5 Hz, and its first harmonic, indicative of the modulation of the afferent train by the activity of the motor unit. Also note the prominent spectral peak around 23 Hz; **B** Auto-spectrum of the discharge of the same primary ending during parallel, semi-periodic stimulation of two soleus motor units (*MU 1,2*) moderately coupled to the receptor. Two distinct spectral components can be seen at the fundamental rates of stimulation of the units, 5.5 Hz and 7 Hz, and one at the first harmonic of 5.5 Hz. Note the different sizes of these components, indicative of the strength of coupling between the individual motor units and the spindle. Power density is in arbitrary units. The frequency scale is different in the bottom plot. (Christakos and Windhorst, unpublished)

of λ_{m1} (11 Hz) show that the primary discharge pattern contains information on the rhythms of activation of both motor units. The different sizes of these components, on the other hand, reflect the differences in mechanical coupling between the motor units and the spindle, which in fact were also detectable in the time domain (PSTHs; not shown here). At higher frequencies the spectrum is practically flat. These results (for further details see references above) indicate that spindle afferents, Ia fibres in particular, are well suited to sense and convey information on the firing rates and unfused contractions of motor units.

3.2.3.1 Effects of Successive Motor Unit Contractions on Spindle Discharge

If spindles are strongly coupled to a motor unit, their afferent responses to closely spaced activations of that motor unit are usually nonlinear in the manner illustrated in Fig. 30. Part D (lower right) shows the overall response of a Ia fibre (upper curve; mean firing rate subtracted), averaged over all stimuli applied randomly to a medial gastrocnemius motor unit whose average twitch is shown below. The responses of the afferent and the motor unit

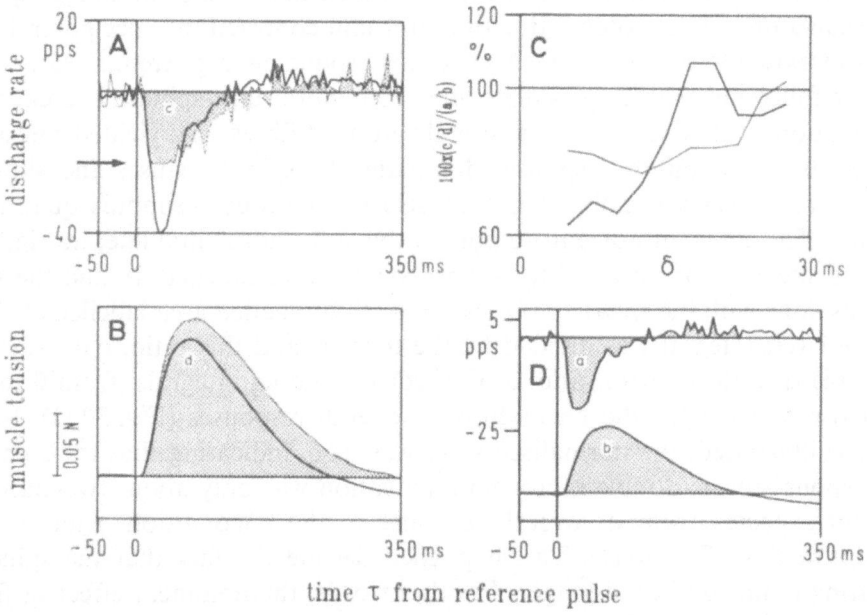

time τ from reference pulse

Fig. 30 A-D. Quantified effects of nonlinear successive contractions of a medial gastrocnemius motor unit on the discharge of a Ia fibre. **D** shows the motor unit twitch (*lower trace*) and Ia fibre rate change (*upper trace*) averaged over all stimuli delivered (number of reference events: 2296). The *thin curve* in **B** shows the actual motor unit response to pairs of stimuli separated by an average of $\delta = 10$ ms; the *thick curve* is the expected linear response resulting from superposition of the lower curve in **D** at the appropriate interval δ. **A** contains the corresponding curves for the spindle response (number of reference events for **A** and **B**, actual reponses: 227); the *horizontal arrow* on the ordinate indicates zero firing rate. The various responses were integrated as shown by the *shaded areas* labelled by *small characters*. Then, ratios c/d were determined for various δ's, normalised by dividing by the control ratio a/b (given in percents), and plotted as a function of δ in **C** (*thick line*). Another example evaluated in the same way is represented by the *thin line* in **C**. (Number of reference events for curves in **C** between 237 and 289.) (Niemann, Windhorst and Meyer-Lohmann, unpublished)

to *pairs* of stimuli separated by a mean interval of $\delta = 10$ ms (see App. D) are shown by the *thin curves* in Fig. 30 A and B, respectively. The *thick curves* in A and B represent the responses that would have been expected if both systems behaved linearly (see Sect. 3.2.2.1). The motor unit response is slightly potentiated (B). In contrast, the spindle response is less than the expected linear response, mainly because the firing rate cannot be reduced below zero rate; but the pause does appear to last a bit longer than that in the linear response.

Nonlinear spindle responses, such as that in Fig. 30 A, *do not require* a nonlinear motor unit response (potentiation) like the one shown in Fig. 30 B. That is, even if the motor unit behaves linearly regarding the summation of subsequent twitches, a spindle strongly coupled to it would usually still show the nonlinearity just described (see also below). Nevertheless, it is interesting to look for the quantitative relation between this type of nonlinear spindle behaviour and the potentiation of motor unit contraction, since potentiation is supposed to be one of the mechanisms producing large tremor excursions. The following analysis was performed on 6 motor unit-spindle afferent combinations (with five Ia and one spindle group II fibres) and yielded consistent results. Only spindle responses dominated by spindle pauses and showing almost no relaxation discharge were selected in order to simplify quantification. The actual motor unit response (above the horizontal line) in Fig. 30 B was integrated from $\tau = 0$ to 340 ms (shaded area labelled d), and the same was done with the spindle response (Fig. 30 A: shaded area labelled c). Then the latter integral was divided by the former, and this ratio, c/d, was normalised to the control ratio, a/b, of corresponding integrals (from 0 to 260 ms) calculated for the unconditioned overall responses (Fig. 30 D). In the case illustrated, the normalised ratio was 0.66, indicating that the spindle's response to the double motor unit activation was only about two-thirds of that expected from its overall response to the unconditioned motor unit contraction (Fig. 30 D). This happened despite the fact that the spindle's firing reduction lasted longer. In other words, the nonlinear effect of firing limitation to zero frequency outweighed the prolongation of the spindle pause. If such normalised ratios are determined for various values of δ, a relation as illustrated in Fig. 30 C (thick line) results displaying the time-course of this nonlinearity. Another relation for a further motor unit-spindle combination (from another experiment) is represented by the thin line. The minimal factor was 0.77. This value was within the range (0.70−0.80) found for another four combinations of motor unit and spindle afferent in which the motor units potentiated and the spindle afferents showed predominantly a spindle pause without relaxation discharge.

These results indicate that the amplitudes of the force (and internal muscle length) fluctuations due to paired motor unit activation and ensuing potentiation are not fully represented in firing rate changes of spindle afferents strongly coupled to the respective motor units. This effect would damp any tendency towards reflex oscillation caused by motor unit potentiation.

3.2.3.2 Effects of Chance Motor Unit Synchronisations on Ia Afferent Discharge

As discussed above, even small-amplitude physiological tremor shows segments of relatively regular force oscillations interrupted by irregular periods (Figs. 24 and 25). Both phenomena are manifestations of the same random process which involves chance synchronisation of small groups of semi-regularly firing (and contracting) motor units during the fairly regular periods. Chance synchronisation can impair the maintenance of stability in the segmental reflex loop, since the occurrence of reflex oscillation appears to depend on displacement amplitude. The question then is how these events are signalled back to homonymous motoneurones via Ia fibres.

Two further features of the spindle response to motor unit contraction (in addition to that shown in Fig. 30), which are important for the following discussion, are illustrated by the peristimulus time histograms (PSTHs) in Fig. 31. In the underlying experiment three motor units of the cat medial gastrocnemius muscle were stimulated in parallel via their ventral root axons with independent patterns. Part A displays the average tension response (isometric recording) of one motor unit. The average response of a medial gastrocnemius Ia fibre is shown in C. It consists of the familiar reduction of firing during motor unit tension rise, followed by a vigorous *relaxation discharge*. The other response feature is displayed by the same afferent in response to another medial gastrocnemius motor unit (Fig. 31 B and D). Here the afferent fires an *"early spike"* (peak, in D, near the onset of motor unit contraction).

Let us first take a closer look at the "early discharge". In principle, as pointed out in Sect. 2.5.2, this feature can result either from β-innervation of certain spindles (Fig. 12 C) or from the spindle being localised functionally in series with the activated motor unit. This type of response also represents a nonlinearity. For example, Fig. 31 H shows the spindle response effected by a time constellation of motor unit activations in which that motor unit (B) causing the early discharges was activated slightly later than the other motor unit (A) producing a "spindle pause". The expected linear response resulting from superposition of PSTHs C and D is displayed in I. Comparing the actual response H with the expected linear one I shows that the early discharge has been suppressed (difference between H and I in J). The corresponding motor unit twitches show no nonlinearity, as evident from comparison of E and F (difference in G). This sensitivity of early discharges to contractions of parallel motor units is quite a common finding (see also

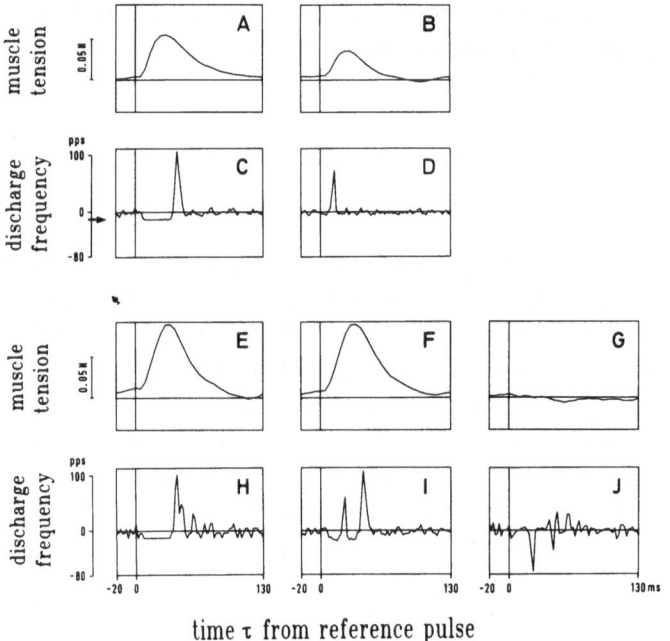

time τ from reference pulse

Fig. 31 A-J. Linear and nonlinear behaviour of muscle tension and discharge of a Ia fibre from the medial gastrocnemius muscle. Three motor units were stimulated in parallel with independent random patterns at mean rates between 8.1 and 10.6 pps. The interaction of two motor units in influencing the Ia afferent discharge is displayed. **A** and **B** show the average twitch responses of each motor unit under study, **C** and **D** the corresponding spindle afferent responses (note that, as in Fig. 30, the mean rate was subtracted from the spindle responses, so that zero firing rate in **C** occurs during the spindle pause as indicated by a *small horizontal arrow* on the *ordinate*); **E** and **H** show tension and spindle responses, respectively, to *pairs* of stimuli, consisting of a stimulus to the first motor unit (occurring at $\tau = 0$) followed within 5 to 15 ms by a stimulus to the second motor unit. **F** and **I** display the responses expected with linear superposition of the single responses in **A** and **C**, and **B** and **D**, respectively, at an appropriate time delay; **G** and **J** present the differences between **E** and **F**, and **H** and **I**, respectively. Number of reference events: for **A** and **C**, 563; for **B** and **D**, 643; for **E** and **H**, 56. (Niemann, Windhorst and Meyer-Lohmann, unpublished)

Windhorst and Schwestka 1982). Their possible significance in the context of tremor mechanisms will be discussed in Sect. 3.3.

The effects of near-synchronous activations of two motor units on the response of a Ia afferent displaying spindle pause and relaxation discharge are illustrated in Fig. 32. In this case, the Ia afferent reacted similarly and strongly to contractions of two medial gastrocnemius motor units as shown by the two PSTHs labelled A and B, which were computed with respect to all the stimuli delivered to one motor unit (PSTH A) or the other (PSTH B). These responses indicate that this Ia afferent had a strong mechanical coupling to both motor units and that it was probably located well within the

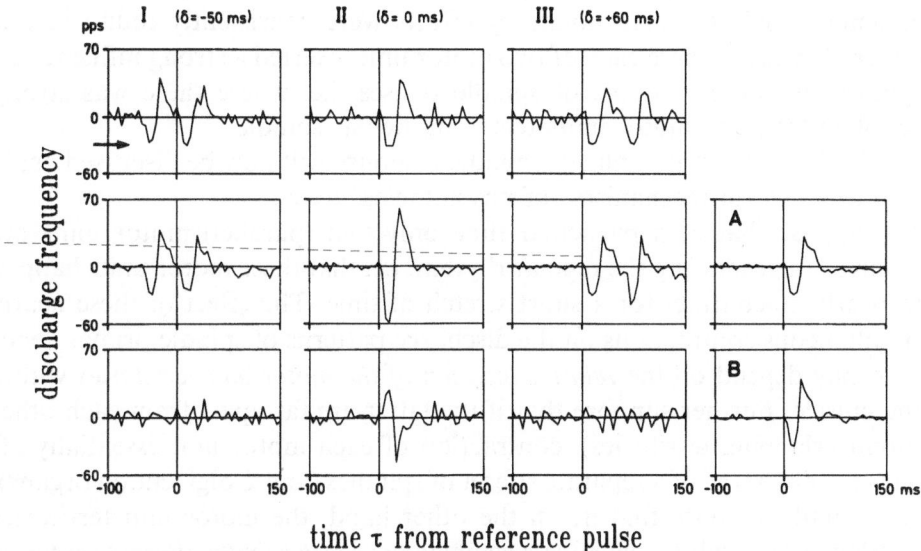

time τ from reference pulse

Fig. 32 A-B, I-III. Effects of selected time constellations of activations of two medial gastrocnemius motor units on the discharge of a Ia fibre. Three motor units (1 through 3) were activated with independent stochastic stimulus patterns at mean rates of 6.8, 5.7, and 7.7 pps, respectively. Shown are the interactive effects of two motor units (1 and 3). *Parts A and B (right column)* display PSTHs calculated for the afferent discharge with respect to all stimuli delivered to motor units 1 and 3, respectively. *Columns I through III* show the actual afferent responses (*upper row*) to defined time constellations of stimuli; e.g. in *I* stimuli to motor unit 1 (occurring at τ = 0 ms) were preceded by stimuli to motor unit 3 by an average of δ = 50 ms (within a window of 10 ms width centred around this mean), and so forth, as indicated on *top of each column*. Zero firing rate of the afferent is indicated by a *horizontal arrow* on the *ordinate*. The *second row* contains the expected linear responses, the *third row* the differences between actual and linear responses. Number of reference events: *A*: 2041; *B*: 1714; *I*: 124; *II*: 110; *III*:109. (With permission from Niemann et al. 1986; their Fig. 5)

overlapping territories of both motor units. The compound spindle reaction to contractions of both motor units depended very much on the time constellation of the motor unit activations, as shown in columns I through III. In column I (upper row) the motor unit that caused response B was activated on average 50 ms prior to activation of the motor unit causing response A. The compound response looked very much like that below it (second row) which is the linear superposition of the two responses in A and B. This is confirmed by the difference of the actual and linear responses displayed in the third row (column I). Essentially the same happens when the time relation of activation of the two motor units was reversed, as demonstrated in column III. However, the actual response was much less than the expected linear response would have to be when the two motor units are activated nearly simultaneously (column II). In this case the difference (lower row) showed large consistent deviations from random fluctuations otherwise found in columns I and III. Such nonlinear effects were consistently found in our material in cases where each of two motor units exerted a strong influence on spindle discharge in terms of spindle pauses, i.e. where there was strong coupling from the motor units to the particular spindle.

The latter type of nonlinear spindle response will now be discussed again in the context of mechanisms of physiological tremor.

Suppose that at a particular time only two (parallel) motor units are involved in generating the regular rhythm, i.e. that these motor units happen to nearly co-contract for a short stretch of time. The effect of these nearly simultaneous contractions on the discharge patterns of spindle primary endings may depend on the *relative location of the motor unit territories* within the muscle. Further suppose that if the latter are far apart from each other (nonoverlapping territories), contraction of each motor unit essentially affects the discharge of a separate subset of spindles (and Golgi tendon organs). And finally suppose that if, on the other hand, the motor unit territories overlap considerably, *two effects combine to reduce spindle afferent response.* Firstly, in this case, the subsets of spindles influenced by the adjacent motor units are not distinct, but they may have common elements. This results in an *occlusion* that restricts the central efficacy of signal transmission from this particular muscle region. And secondly, the effect resulting from the nonlinear spindle responses described in the preceding paragraph adds to this restriction. For if each of the two motor units on average strongly modulates the discharge of a particular Ia fibre, almost synchronous contraction of the two motor units exerts an effect on the spindle discharge that is a "sublinear summation" of their separate effects, as exemplified in Figs. 31 and 32 B and C. Even if two motor units, both influencing a spindle, sum their near-synchronous twitches linearly, the spindle response might still be sublinear. If the twitches sum nonlinearly, such that the force output is less than expected in the linear case, the spindle nonlinearity might be reduced (this being almost impossible to test experimentally since the nonlinear spindle response to the linear motor unit summation is unknown). Be that as it may, the occurrence of near-synchronous motor unit contractions leads to damping, either mechanically by nonlinear twitch summation, or neurally by nonlinear spindle response.

In conlusion, then, the effect of the co-contraction of two adjacent motor units on motoneurone excitability should be weaker than that of two remote motor units and should therefore also bear less risk of instability. This "spatial factor" related to the topography of motor units and spindles within a complex muscle may be of particular importance in the context of similar factors in afferent reflex connexions (see Sect. 3.3). Note that, for simplicity of the argument, we have assumed motor units of about equal strength. This is justified by the fact that the most recently recruited motor units, which contribute most to physiological tremor (see Sect. 3.2.1), belong to the same category. These motor units are also the strongest (amongst those already recruited) and should therefore exert the strongest actions on spindle afferents, particularly since their contractions are not yet fused, whereas the contractions of the motor units recruited earlier not only are more completely fused, but they also contribute less force anyway.

3.2.3.3 Effects of Motor Unit Contractions on α-Motoneurone Membrane Potential

Before pursuing the above line of argument, it would be nice to know whether the information on motor unit contractions carried by spindle and Golgi tendon organ afferents (and possibly further muscle afferents) is actually transmitted to homonymous α-motoneurones. This has indeed recently been demonstrated by W. Koehler et al. (1984a, b).

The experimental paradigm was as follows. In anaesthetised cats, medial gastrocnemius α-motoneurones were impaled with micropipettes to record membrane potential fluctuations. Three medial gastrocnemius motor units were activated in parallel by electrically stimulating their axons in ventral root filaments with different random stimulus trains at mean rates between 10 and 18 pps. Muscle tension was recorded with the muscle isometrically constrained. Figure 33A presents a 1 s section of the relevant data. The upper trace (labelled PSP) represents the a.c. component of the membrane potential fluctuations, the middle trace (labelled T) the muscle tension fluctuations, and the lower three traces (labelled S1, S2 and S3) are dot displays of the stimulus patterns to the motor units. Averaging (over a 2-min period) the PSP and the tension with respect to stimuli in train S1 resulted in PSP transients (upper trace labelled PSP → S1) and T transients (lower trace labelled T → S1), respectively, as shown in Fig. 33B. Corresponding averages with respect to stimulus trains S2 and S3 are shown in Fig. 33C and D. It can be seen in B that, although motor unit 1 produced a well-developed (relatively fast) average twitch, this was scarcely reflected in PSP transients exceeding the noise level. The situation is different in C where a hyperpolarising PSP transient accompanied the rise of tension in motor unit 2 (relatively slow) and where a slight but long depolarising transient occurred during the relaxation phase. Finally, motor unit 3 twitches (in D) evoked an initial depolarising transient (at the onset of contraction) followed by a hyperpolarising transient during tension rise, followed by two further oscillations during relaxation.

The sequence of hyperpolarisation and depolarisation in C could most easily be interpreted as resulting from decreased muscle spindle discharge and increased Golgi tendon organ discharge during tension rise and increased muscle spindle discharge during relaxation. In D the early depolarising phase

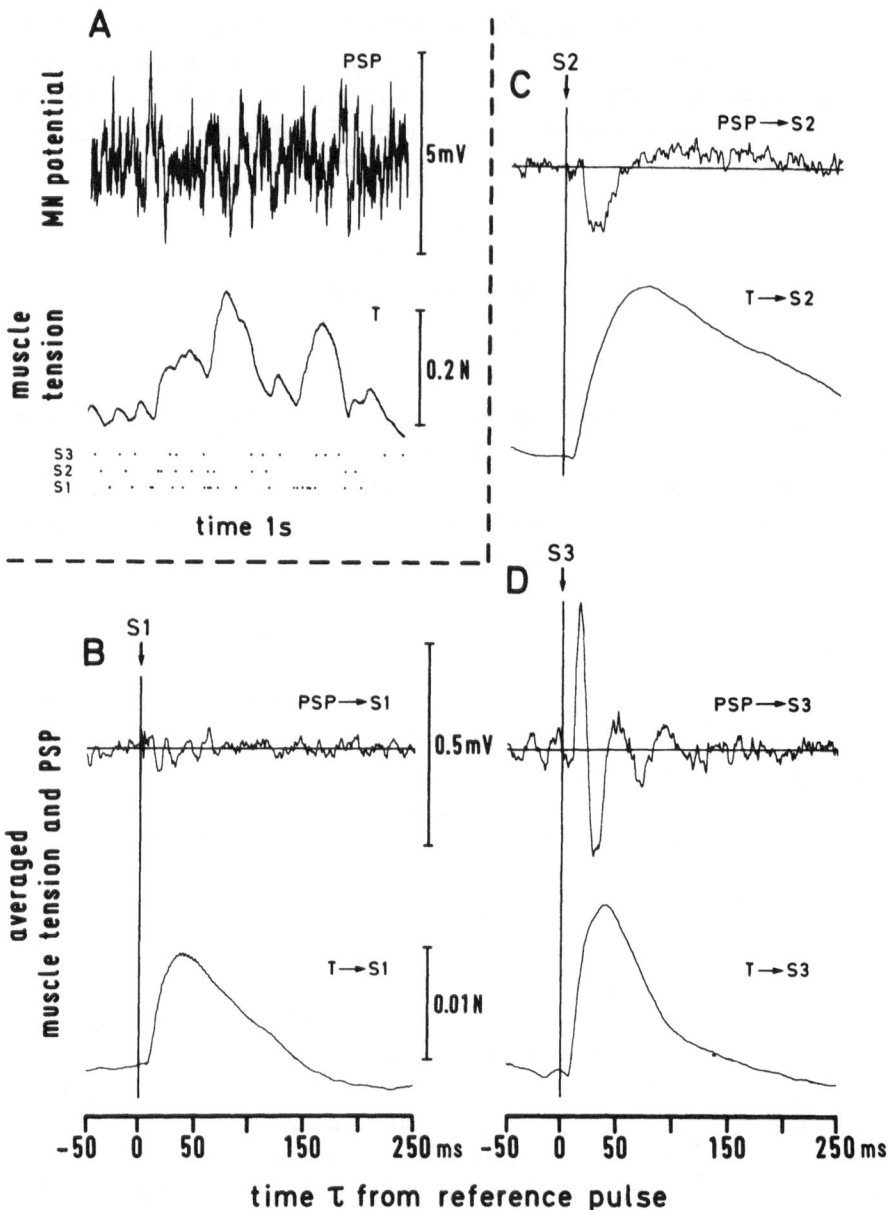

Fig. 33 A-D. Medial gastrocnemius muscle tension (*T*) and α-motoneurone (*MN*) postsynaptic potential (*PSP*) fluctuations in response to in-parallel activation of three homonymous motor units with random stimulus patterns. **A** Samples of a.c.-recorded motoneurone potentials (*upper trace*: *PSP*; depolarisation upwards) and of muscle tension fluctuations (*middle trace*: *T*; tension increase upwards) in response to the three stimulus patterns (*lower traces*: *S1* through *S3*) depicted as *dot displays*. The fourth second (3–4) of sampled data is shown; **B – D** *PSP* (*upper trace*) and *T* (*lower trace*) trajectories averaged over all stimuli of the respective trains (**B** 2026 reference stimuli, **C** 1538 reference stimuli; **D** 1147 reference stimuli). Recording duration: 113.5 s. *Horizontal straight lines* represent the mean *PSP* of the sampled data. Calibrations at the lower right of **B** apply to **B–D**. (With permission from W. Koehler et al. 1984a; their Fig. 1)

could have been caused by "early" muscle spindle discharges in response to motor unit 3 contractions followed by "spindle pause" and rather synchronised relaxation discharges. (These oscillations depend upon dynamic mechanical conditions in the muscle, because they may be prominent when the motor unit contraction is superimposed on a low background force, and may be small or absent when the twitch is superimposed on a high background force.) All these effects probably involve more or less synchronous discharges of several affected spindles (and possibly Golgi tendon organs) coupled to the particular motor unit. Positive correlations between muscle afferent discharges would be expected to be responsible for the early and late (oscillatory) depolarisations in Fig. 33 D.

It is also of importance to note that, in the same motoneurone, the effects of contractions of different motor units are reflected in very different ways, both quantitatively and qualitatively. This suggests the existence of *differential feedback via muscle afferents to motoneurones* from certain motor unit territories (see Sects. 2.8.2.3 and 3.3).

It is also of interest in the present context to know the degree to which the spindle nonlinearities in response to contractions of single (Sect. 3.2.3.1) and several motor units (Sect. 3.2.3.2) show up at postsynaptic levels, particularly in homonymous α-motoneurones. Yet scanty data are currently available on this point. Only recently some relevant data were provided by W. Koehler et al. (1984b). An example of the motoneurone membrane response to different time constellations of activations of a motor unit is shown in the upper part of Fig. 34 and compared to corresponding responses of a Ia fibre (lower part; from another experiment). The upper part is adapted from W. Koehler et al. (1984b, their Fig. 2), and is based on the same data as described above. Part A shows the membrane potential trajectory averaged with respect to all the stimuli delivered to one motor unit. Columns I, II and III show the effects of selected time constellations of stimuli (as indicated on top), the thin lines representing the actual motoneurone reponses, the thick lines the expected linear responses calculated by superimposing the responses in A at appropriate intervals δ. The lower row contains the differences between actual and linear responses. In the lower part of the figure, the corresponding responses of a primary spindle afferent to stochastic activation of a single gastrocnemius motor unit are displayed in the same way (except for slight differences in the time windows used). It is apparent that the nonlinearities in spindle and motoneurone membrane potential responses show very similar time-courses, so that one might conclude that the motoneurone nonlinearity could result from nonlinearities in Ia fibre behaviour. However, there were other instances of linear behaviour of the motoneurone membrane potential which led W. Koehler et al. (1984b) to suggest that "... the afferent pathways mediating the mechanical events in the muscle to a motoneurone behaves rather linearly (at least in this small-amplitude range) or might even have a

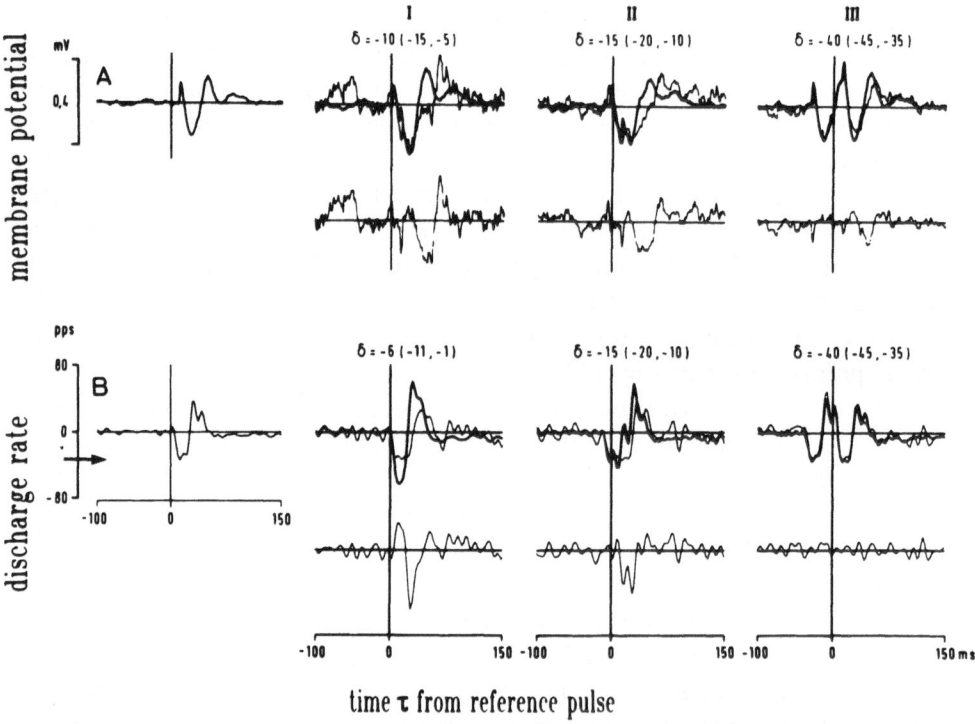

time τ from reference pulse

Fig. 34 A, B, I–III. Comparison of the nonlinear behaviour of a medial gastrocnemius Ia afferent and a medial gastrocnemius α-motoneurone to selected time constellations of medial gastrocnemius motor unit activations. The *two lower rows* show the behaviour of a Ia fibre that responded to motor unit contractions with a pause followed by a relaxation discharge (see **B**). Zero firing rate is indicated by a *horizontal arrow* on the *ordinate* in plot **B**. The *second row from below* exhibits superimposed plots of the actual responses (*thin curves*) and the expected linear responses (*thick curves*); the *bottom row* shows their differences. **B** shows the overall spindle response to all the stimuli to the motor unit. Number or reference events: **B** 2041; *I* 114; *II* 122; *III* 143. The *two upper rows* show an example of a nonlinear response of a medial gastrocnemius α-motoneurone (intracellular recording) to time constellations of medial gastrocnemius motor unit activations. The membrane potential fluctuations were averaged with respect to stimulus events in the same way as the Ia afferent discharge in the *two lower rows* (see W. Koehler et al. 1984a, b). The *horizontal lines* indicate mean membrane potential. Number of reference events: **A** 1010; *I* 22; *II* 77; *III* 99. (With permission from Niemann et al. 1986; their Fig. 4)

linearizing effect" (p. 387). At present, the verdict then is that simply more data is needed before a definite conclusion can be reached.

Even less material is available regarding the effects on motoneurone membrane potential of various time constellations of activations of two motor units (W. Koehler et al. 1984b). The few instances examined showed more or less satisfactory linear superposition of the average effects of each single motor unit. Hence, unfortunately no reasonable comparison between motoneurone responses and muscle spindle responses is possible at present.

3.2.3.4 Summary

1) The discharge patterns of muscle spindle afferents are modulated by the unfused contractions even of single motor units. Ia fibres are more sensitive in this respect than spindle group II afferents. Strong motor units have a stronger effect on spindle discharge than do weak motor units, both in terms of the number of spindle afferents influenced ("projection frequency") and of the size of the effect on each single afferent ("gain"). Thus, in natural steady muscle contractions, in which the strongest motor units recruited last dominate the process of physiological tremor generation (Sect. 3.2.1), these motor units also exert the strongest modulatory influence on spindle afferents.

2) Muscle spindle afferents usually do not respond as vigorously to *pairs* of twitches separated by short intervals as they do to twitches, which are separated by longer intervals. Muscle force (and internal length) fluctuations resulting from motor unit "doublets" are therefore not fully represented in the afferent signals of those spindles, which are strongly coupled to the motor units. This would diminish the risk of reflex oscillation.

3) If two motor units both produce a spindle pause and possibly a relaxation discharge in a muscle spindle afferent, i.e. if they are both strongly coupled to it, the near-synchronous activation of these units produces discharge variations that are smaller than expected for linear summation. If one motor unit produces an early discharge, simultaneous contraction of another motor unit, arranged in parallel, often prevents this. Thus, any larger-amplitude muscle force (and internal length) fluctuations resulting from chance synchronisation of motor units with overlapping territories are not fully represented in the discharge variation of spindle afferents, which are strongly coupled to both. This effect is enhanced by the fact that overlapping motor units influence common subsets of spindles (occlusion), whereas remote motor units influence distinct subsets. These factors may also contribute to damp tendencies towards reflex oscillation. This is particularly valuable since motor units sometimes show a tendency towards synchronous discharge, the extent of which is correlated with the mean amplitude of physiological tremor (Dietz et al. 1976).

4) Unfused contractions of motor units of the medial gastrocnemius muscle are reflected in membrane potential changes of homonymous α-motoneurones. The time-course of the averaged potential in the same motoneurone to contractions of different motor units varies; hyperpolarisation during motor unit tension rise may be followed by depolarisation during motor unit twitch relaxation, or an initial depolarisation may precede the above sequence. An α-motoneurone may show qualitatively and quantitatively different responses to different motor units, suggesting a differential coupling of the afferents from different muscle regions to the motoneurone. Nonlinear

potential summation may occur in response to successive contractions of one motor unit resembling the corresponding response of some spindle afferents.

The interpretation of the above results as to their significance for tremor is subject to some reservations, which have to be eliminated by further studies. Firstly, in the small-amplitude range considered here, the sensitivity of Ia afferents is reduced strongly by static and mildly by dynamic γ-fusimotor action (Goodwin et al. 1975; Hulliger and Sonnenberg 1985; see also Sect. 4.3.11), the latter effect being particularly prominent with low-frequency (4 Hz) and small-amplitude muscle stretches superimposed on a constant muscle length (see Hulliger and Sonnenberg 1985). Thus, depending on the amount of fusimotor spindle input, Ia afferents might not respond as sensitively to motor unit contractions as shown above, although in most of our experiments a residual amount of fusimotor action probably persisted because we endeavoured to leave intact as much of the ventral roots as possible. Also, present knowledge does not permit to predict the fusimotor effect on the response of afferents to successive motor unit twitches (item 1) (M. Hulliger, pers. comm.). In general, however, spindle afferents are known to respond to small-amplitude irregular movements during muscle contraction (Hagbarth and Young 1979; see also Stiles 1980), but this might well be the response to near-simultaneous contractions of several adjacent motor units. In summary, static fusimotor action might mitigate stability problems, which arise from the response of Ia afferents to unfused motor unit contraction. Secondly, it was shown above that early discharges appear to be an elusive phenomenon in that they are sensitive to contraction of surrounding motor units. They may also be so to fusimotor spindle input. Further factors determining their frequency of occurrence are discussed below.

3.3 Localised Stretch Reflexes Acting as a Tremor-Suppressive Mechanism?

We shall now try to integrate, in a single model, the data on the spatial organisation of monosynaptic Ia fibre connexions to α-motoneurones (Sect. 2.8.2.3) and the data set presented in the preceding discussion, thus closing the reflex loop.

3.3.1 Muscles with "In-parallel" Structure and Their Monosynaptic Reflex Organisation

We shall discuss this issue first with respect to an idealised model muscle that is built according to data on the cat medial gastrocnemius muscle. This model muscle is assumed to have the simple spindle-like structure depicted in Fig. 35. In the cat medial gastrocnemius muscle, motor unit territories are

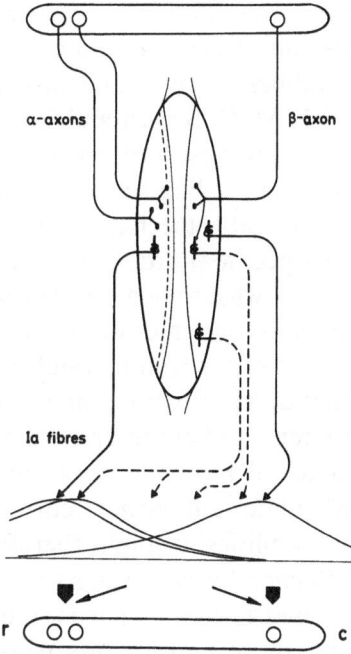

Fig. 35. Schematic model of an idealised simple muscle structure and its topographical Ia monosynaptic reflex connectivity. For clarity, the spinal motoneurone columns are represented twice (*at top and bottom*); *r* rostral: *c* caudal. The territories occupied in the muscle by three motor units are delineated by *thin lines* (*solid* or *dashed*). The β-axon at the right gives off synapses to extrafusal *and* intrafusal muscle fibres. The Ia fibres represented by *solid lines* are assumed to produce spindle pauses and relaxation discharges *without* early discharges. These fibres are proposed to project in an ordered manner to the spinal cord such that their muscle origin is topographically related to the level at which they enter the spinal cord (see text). From their entry point they project slightly rostrally to the motoneurone pools onto which they distribute their synaptic effects in a weighted manner as indicated by the *continuous lines below the arrowed* Ia fibre ends (see Windhorst 1979a; Lüscher et al. 1980). This weighting is represented once again in a discontinuous way by the *arrows of different widths* above the lower motoneurone pool. The Ia fibres producing early discharges (of whatever origin) are represented by *dashed lines* and assumed to project in an unordered manner to the spinal cord. (Modified from Windhorst 1984; his Fig. 1)

restricted to parts of the whole muscle area (e.g. Burke and Tsairis 1973; Burke et al. 1977; Cameron et al. 1981; own observations; the same applies to the lateral gastrocnemius, see English and Weeks 1984), and their intramuscular position is grossly related to the rostro-caudal position of their respective motoneurone somata in the motoneurone columns of the spinal cord, such that rostrally located motoneurones have their muscle fibres situated near the dorsal margin of the medial gastrocnemius (Swett et al. 1970; Burke and Tsairis 1973; Burke et al. 1977) and so forth, as illustrated in Fig. 35. (For the cat lateral gastrocnemius see Weeks and English 1982; for other muscles in other species, and for the possible mechanisms involved in the ontogenetic development of these topographic patterns see review by Landmesser 1980, and also Bennett and Lavidis 1982; Booth and Brown 1983; M. C. Brown and Booth 1983; with further references).

Let us first look for further mechanisms (in addition to those discussed in the preceding sections) which could reduce the risk of reflex oscillation. One possible mechanism could make use of spatial (topographical and "species-specific") order, if present, within the monosynaptic stretch reflex system. It was argued above that the nearly synchronous contractions of two spatially separate motor units would be reflected more strongly in the discharge of the spindle (and probably Golgi tendon organ) afferent ensemble than those of overlapping motor units. Due to the diverging projection of Ia afferents to homonymous α-motoneurones, these signals would be distributed to all motoneurones, and they would, in a qualitatively similar way, modulate and, hence, correlate their membrane potentials and discharge probabilities. It might therefore be advantageous to restrict or localise their influence by weighting the monosynaptic Ia-motoneurone coupling in an ordered fashion. That is, the Ia afferents should be most strongly coupled with those motor units whose contractions most strongly modulate their discharge patterns (Windhorst 1978 b; Binder and Stuart 1980 b). Or, conversely, the subset of Ia afferents which are influenced by a particular (their "own") motor unit should produce weaker EPSPs in motoneurones of spatially separated motor units than it does in their "own" motoneurones. This idea is illustrated for the model muscle in Fig. 35, where, for graphical clarity, a topographically weighted Ia-motoneurone connectivity is assumed. Such a connectivity pattern would prevent synchronous contractions of remote motor units from being reflected equally powerfully in all motoneurones. This pattern of connexions and couplings would be the basis for "localised stretch reflexes", and would help, at least initially, to restrict and localise any fortuitously originating reverberatory tendencies in the monosynaptic reflex system. Remember that topographical connectivity patterns are not the only possibility of establishing localised reflex connexions, "species-specific" patterns based on "cell recognition" would also do it. This appears to be partially the case in the cat medial gastrocnemius muscle, the lateral gastrocnemius, the semimembranosus and the biceps femoris muscle (see the end of this section).

Another means of counteracting the tendency towards cyclic excitability modulation in the reflex loop would be to generate anti-cyclic, out-of-phase discharges. By this we mean the "early discharges" which occur at the onset of motor unit contractions (see Fig. 31). These discharges are anti-cyclic in the sense that they occur at the moment of (or soon after) the decrease in firing rate of other Ia fibres in response to motor unit contractions. It would seem reasonable to *distribute the effects of these early discharges monosynap-*

tically but nonpreferentially to all the homonymous motoneurones in order to make the more excitable motoneurones discharge in advance (out of phase) of those which caused the early discharges (Windhorst and Schwestka 1982). Such an unordered projection of Ia fibres carrying early discharges (dashed lines in Fig. 35) would blur the topographical order possibly established by those Ia fibres not exhibiting them (solid lines in Fig. 35). This functional interpretation of spindle "early discharges" is new and in particular gives a novel meaning to β-innervation, but it is subject to the reservations raised in the preceding summary (Sect. 3.2.3.4) and to further qualifications outlined below.

Carried to the extreme, the above suggestion is that the spindle Ia afferents from a muscle can be divided into two distinct populations characterised by their different projection patterns. Now this is certainly an exaggeration because Ia afferents responding with early discharges to motor unit contractions will also be unloaded to a greater or lesser degree. Thus, the two populations may overlap to some extent. This extent and the frequency of early discharges depend on a number of anatomical and functional factors, such as: (a) the structure of the muscle of concern (because the length of the muscle fibres relative to the overall fascicle length co-determines the frequency of in-series arrangements of muscle fibres and spindles and, hence, of early discharges; see Richmond et al. 1985; Loeb et al. 1987); (b) the composition of the muscle of the various types of motor unit (because, for instance, static β-innervation involves fast motor units; see Jami et al. 1982); (c) the recruitment level (for the same reason as under (b), and because the activity of parallel motor units will influence the occurrence of early discharges; see Sect. 3.2.3.2); (d) the patterns of discharge of the active motor units (synchronised or not); (e) the number and pattern of distribution of spindles in the muscle (co-determining the frequency of in-series arrangements of muscle fibres and spindles); (f) the overall frequency of static β-innervation; (g) the amount of fusimotor bias to the spindles; (h) muscle length during work (because muscle length influences the mechanical coupling of muscle fibres to spindles). This complexity renders impossible any attempt to generally define the quantitative significance of early discharges for tremor suppression.

Thus, the suggestion of a *division of the Ia afferent population into functional subclasses* which would have different monosynaptic connectivity patterns to homonymous α-motoneurones is a heuristic simplication. Yet, it could provide one hypothetical explanation for the lack or paucity of spatially ordered Ia-motoneurone connexions in some muscles (see below). For example, Swett and Eldred (1959) found that the Ia afferent projection from the cat medial gastrocnemius muscle appears to be much less ordered topographically than the efferent projection (Swett et al. 1970; Burke and Tsairis 1973; Burke at al. 1977). But the "species-specific" connectivity pattern might also turn out to be stronger if it could be tested experimentally not by electrically stimulating Ia afferents in a muscle nerve in an undifferentiated fashion, but by selectively activating functionally different subpopulations. This is extremely difficult, particularly as long as such subpopulations have not been defined or even conceived of.

The cat *medial gastrocnemius muscle* is not the ideal prototype for the discussion of our model. Its internal structure is more complicated than presumed in the sketch of Fig. 35 (see Burke and Tsairis 1973). It is organised in "compartments" which are defined by branches of the medial gastrocnemius nerve and are not arranged strictly in parallel to each other (see Cameron et al. 1981). Indeed, Lucas and co-workers (Lucas and Binder 1984; Lucas et al. 1984) demonstrated the localised monosynaptic reflex connexions of Ia afferents onto homonymous motoneurones correctly with respect to this "internal coordinate" system. Interestingly they presented evidence that this sort of localised Ia connectivity appears to be more prominent on large than on small motoneurones. The same was suggested by Munson et al.: "... analysis of the interactions of FF motor units with pairs of primary afferents revealed a tendency for the afferent generating the larger EPSP to be more unloaded by activation of the motor unit" (Munson et al. 1984; p. 1268). Since muscle force tremor is known to increase with increasing contraction level based on the recruitment of stronger motor units (e.g. Allum et al. 1978; Christakos 1982), and since larger motor units exert stronger average effects on spindle afferent discharge than do small motor units (see Fig. 28), this dependence on motor unit type appears to be physiologically reasonable and consistent with the hypothesis presented here. Another feature fits into this scheme, namely that dynamic β-fibres belong to S-type motor units, and static β-fibres to F-type motor units (Jami et al. 1982; Murthy 1983). It is the static variety of β-motoneurones that must be expected to easily produce early discharges.

Partitioned monosynaptic Ia-motoneurone connexions have also been demonstrated for the following parallel-fibred muscles.

Lateral gastrocnemius muscle: A differential feedback of Ia afferents from different nerve branches (compartments) of the cat lateral gastrocnemius muscle onto α-motoneurones projecting to these compartments was demonstrated by Vanden Noven et al. (1986). The general pattern is again that of Ia afferents having their most powerful monosynaptic connexions with motoneurones belonging to "their" compartment.

Biceps femoris muscle: Another example was documented by Eccles and Lundberg (1958) and by Botterman et al. (1983a) who investigated the cat biceps femoris muscle. It is innervated by three nerve branches, one to the anterior biceps (hip extensor) portion, one to the middle portion and one to the posterior biceps (knee flexor) portion. Composite monosynaptic Ia input to anterior biceps α-motoneurones from the anterior biceps nerve branch is stronger than from posterior biceps Ia fibres. Conversely, posterior biceps α-motoneurones receive strong posterior biceps Ia input but little from anterior biceps. Thus, "homonymous" Ia input is strongest from their "own" Ia fibres. The data of Eccles and Lundberg (1958) already suggested that the α-motoneurones to the middle portion could perhaps be classified into two functional groups: an extensor and a flexor group. Botterman et al. (1983 a) reported that middle biceps motoneurones in turn receive stronger Ia input from their own compartment than from anterior biceps or posterior biceps branches. But also within the groups of extensor or flexor cells, partitioning was evident in that anterior biceps (extensor) Ia input was stronger to "homonymous" motoneurones than to middle biceps-extensor cells, and posterior biceps (flexor) Ia input was stronger to "own" motoneurones than to middle biceps-flexor cells. Thus, both topography and species specificity play a role in this EPSP partitioning.

Semimembranosus muscle: The cat semimembranosus muscle is divided into a short and a long head, the former inserting on the distal femur and the latter on the tibia (Sacks and Roy 1982). The muscle is innervated by two nerve branches, one to the anterior and one to the posterior portion. Again the EPSPs elicited by Ia afferents from any portion are larger in the α-motoneurones supplying the same muscle region, i.e. in their "own" α-motoneurones (Hamm et al. 1985a). As above, the two factors, topography and species specificity, both contribute to this type of organisation.

McKeon et al. (1984) were unable to demonstrate somatotopic projections of muscle afferents to homonymous α-motoneurones in the human *tibialis anterior muscle*. But the evidence to support this view is rather indirect. In contrast, in the cat anterior tibial muscle, there is a somatotopic relation between spinal motoneurone level and location of their muscle fibres within the muscle, suggesting an efferent compartmentalisation (Iliya and Dum 1984). In this context, it is of interest that, in the same muscle, the *fusimotor system* appears to show a spatial organisation. Huhle (1985) demonstrated that γ-efferents to spindles tend to leave the spinal cord at about the same level as the related Ia afferents enter it. Moreover, of 30 muscle spindles located in the distal part of the muscle, 29 had Ia- and γ-fibres in the more caudal segment L7; 72% of these muscle spindles were found in the more caudally lying middle third of this segment. For 70% of the 35 muscle spindles located in the proximal muscle region, Ia- and γ-fibres were found in the caudal part of the more cranial segment L6 and the cranial part of L7 (Huhle 1985). Interestingly, this topographical arrangement is more of the serial type and thus similar to the oblique arrangement of muscle compartments (Iliya and Dum 1984). In general, Huhle's results suggest that fusimotor innervation may not be as diffuse as one might suspect. But clearly, more studies of this sort are highly desirable.

3.3.2 Muscles with "In-series" Structure and Their Monosynaptic Reflex Organisation

Muscle structures differing considerably from the idealised model dealt with in the preceding section are found in certain (cat) muscles which have recently attracted increased attention.

The *semitendinosus muscle* consists of two separate compartments in series (see Fig. 36 A). Ia afferents from one compartment exert on average equally strong monosynaptic actions on motoneurones whose muscle fibres are situated in either compartment (Nelson and Mendell 1978; Botterman et al. 1983 b). But neither the motoneurones projecting to the two compartments nor the Ia fibres originating from them are separated clearly with respect to their rostro-caudal location within the ventral horn or their entry point (Schwestka 1981), as is the case for cat neck muscles (see below). This somewhat "chaotic" situation is represented schematically in Fig. 36 A.

If one semitendinosus compartment is activated by electrical stimulation of its nerve branch, spindles in this compartment produce spindle pauses (except perhaps for early discharges), whereas spindles in the adjacent (in series) compartment behave like Golgi tendon organs, i.e. with an increased discharge during muscle tension rise (Botterman et al. 1983 b; Schwestka and Windhorst unpublished observations). With single motor unit activation, spindle behaviour is similar, though spindles in the adjacent compartment do not usually produce bursts of action potentials but only single spikes at contraction onset, i.e. a kind of early discharge (Schwestka et al. 1981; Windhorst and Schwestka 1982). In models of several coupled control loops composed of linear elements, such "in-series" arrangements seem to be less stable than "parallel" arrangements (W. Koehler and Windhorst 1980, 1981), although the situation may be somewhat different in a population stochastic model of reflex muscle control which incorporates the discharge statistics of the neuronal elements. If early discharges are beneficial for stabilising the system (as suggested in the preceding section), then they are particularly so in the semitendinosus muscle because of their frequency of occurrence.

Stretch reflexes of the (cat neck) *splenius and biventer cervicis muscles* are strongly localised. These muscles are partially compartmentalised. For instance, the splenius is incompletely divided by two tendinous inscriptions into essentially three compartments which are innervated by segmental nerves C_2 to C_4 (except for a small C_1 branch: see Brink et al. 1981; Fig. 36 B). Ia afferents in any of these branches exert their most powerful monosynaptic effects on motoneurones of the same segment and have weaker connexions to motoneurones, the further away these are situated. This holds for projection frequency as well as for amplitude of (aggregate) EPSPs (Brink et al. 1981). The weighted distribution is strongly asymmetrical, with the Ia fibre-motoneurone connexions reaching further rostrally than caudally (see the

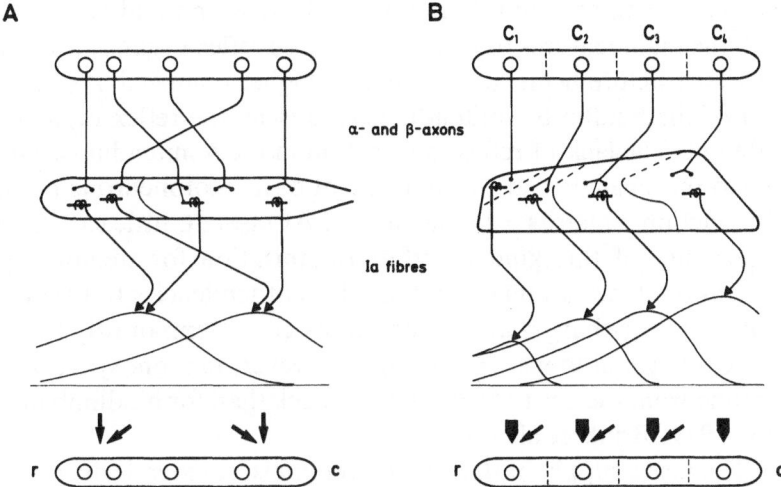

Fig. 36 A-B. Muscles with in-series arrangements of compartments and various monosynaptic reflex organisations. **A** Scheme of the cat semitendinosus muscle which is composed of two compartments completely separated by a tendinous inscription (as indicated by a *thin line*); the efferent (motor) as well as afferent (sensory) innervation of these compartments is essentially unordered in that fibres leaving or entering the spinal cord at any height can innervate both compartments (nothing is known about a possible transverse order as in the case of Fig. 35, though such an order is unlikely); the Ia fibres are not divided into types here; their monosynaptic effects are assumed to be weighted in much the same way as in Fig. 35, i.e. to depend upon distance of the motoneurones from the Ia entry level as indicated by the *continuous curves below the arrowed* Ia fibre ends; **B** Scheme of the cat splenius muscle (redrawn from Brink et al. 1981), which is incompletely divided into three compartments by tendinous inscriptions represented by *thin continuous lines.* However, its segmental motor innervation divides the muscle into the four regions indicated by the *dashed* (plus the *solid*) *lines.* The Ia afferents in the spinal nerves C_1 through C_4 distribute their monosynaptic effects on homonymous motoneurones in a strongly (topographically) weighted fashion as indicated by the *continuous curves below the arrowed* Ia fibre ends. These curves were constructed from the discontinuous data presented by Brink et al. (1981), taking account of (composite) EPSP amplitudes and projection frequencies, and are fairly accurate except for spatial scaling of spinal segment lengths, which were assumed to be equal. (Modified from Windhorst 1984; his Fig. 2)

medial gastrocnemius muscle; Lüscher et al. 1980). Similar data were obtained by Brink et al. (1981) for the biventer muscle. Almost certainly, this topographical organisation underlies the "localization of proprioceptive reflexes in the splenius muscle of the cat" which was recently demonstrated by Bilotto et al. (1982) and Ezure et al. (1983).

The morphology of spindles and Golgi tendon organs in these muscles is more refined than in hindlimb muscles in that they tend to be associated in complexes such as tandem spindles, "dyads" (associations of spindles and Golgi tendon organs), etc. (Richmond and Abrahams 1975b). The splenius muscle is schematically represented in Fig. 36 B (based on Brink et al. 1981;

see also Richmond and Abrahams 1975a; Richmond et al. 1985) together with the main features of its monosynaptic reflex organisation. Although the muscle structure bears, apart from its higher complexity, some similarity to that of the hindlimb semitendinosus muscle, the reflex organisation is very different. The lack of reflex localisation in the semitendinosus may be partly due to the smaller rostro-caudal extent of its motoneurone pool, but probably functional factors also play a role. At present, little can be said as to the importance of this kind of reflex organisation for tremor suppression or generation. One can imagine, though, that tendencies towards synchronised motor unit discharges and contractions could temporarily be kept localised in the segmental loops. On average, however, Ia monosynaptic excitation of motoneurones seems to be weaker for neck than for hindlimb muscles (Rapoport 1979; Brink et al. 1981).

The nerve supply of the above neck muscles extends over 4 to 5 (cervical) segments. Even more extended is the motor control system of the *respiratory muscles* of the cat rib cage, which spans 13 segments. A Ia fibre entering one segment makes strong connexions to motoneurones of its own segment, weaker ones with motoneurones of the adjacent segments, and none at all with motoneurones of more remote segments (Kirkwood and Sears 1982a). In a way, therefore, the (external intercostal) respiratory motor system is organised in a similar (if exaggerated) manner as the above neck muscles. The intercostal muscles can also be envisaged as being arranged in series, the compartments here being separated by the ribs instead of tendinous inscriptions. In this view, the intercostal muscle and monosynaptic reflex system is simply spatially more extended (distributed) than that of the cat neck muscles. Thus, in both these cases, a clear topographical organisation exists.

It should be recalled that the synchronisation of cat intercostal α-motoneurones shows a spatial distribution indicative of localised inputs (Kirkwood et al. 1982a). The three types of synchronisation (see Sect. 2.1.1) were investigated for motoneurones of the same and separate segments. It turned out that, for all three categories, the degree of synchronisation declined with rostro-caudal separation of the α-motoneurones, although the rate of decline was more variable for broad-peak synchronisation than for the other two types. Short-term synchronisation was undetectable for distances exceeding three or four segments, but for high-frequency oscillation (h.f.o.) and broad-peak synchronisation clear peaks were still visible in the cross-correlograms across distances of five segments. The authors conclude that "the short-term synchronisation in these preparations is generated by the bulbospinal respiratory neurones and that the majority of their axons do not branch to make strong synaptic connexions to motoneurones over more than three to four segments" (p. 137). Thus, any presynaptic synchronisation would also affect the motoneurones in a spatially weighted distribution.

3.3.3 Summary and Limitations

Apart from "peripheral" factors (nonlinear motor unit-spindle interactions in the muscle) that, in conjunction with fusimotor spindle input (see Sect. 3.2.3.4), may contribute to reducing the risk of reflex oscillation, there may be "central" factors based upon the spatially ordered monosynaptic projections of Ia afferents to homonymous α-motoneurones. It is suggested that Ia fibres responding to contraction of a motor unit with a typical spindle pause and possibly a relaxation discharge connect most strongly to the related motoneurone (and functional neighbours), and least strongly to motoneurones innervating muscle fibres far apart from the spindle. This organisational scheme would provide for reflex localisation, whether based on topography or species specificity. Ia afferents producing early discharges in response to contraction of a motor unit should connect more diffusely to homonymous α-motoneurones. This would blur the pattern established by the former Ia afferents in tests using indiscriminate synchronous stimulation of all Ia fibres. However, the quantitative significance of early discharge is uncertain because this type of spindle response is subject to a number of anatomical and physiological factors. There is now some evidence of possible topographical order in the projection of γ-efferents to muscle spindles in the cat tibialis anterior muscle. However, to what extent the latter projection can be integrated into a cogent hypothesis of localised stretch reflexes remains to be elucidated.

The foregoing discussion concentrated on patterns of monosynaptic connexions from Ia fibres to α-motoneurones which might intuitively be expected to increase the stability of the stretch reflex. There are several limitations to this approach; three may be mentioned here.

Firstly, the pattern of central monosynaptic connectivity may be governed not only by stability respects but, in addition, by other functional principles, for example the use made of the muscle (see Sect. 5.1). Thus, the semitendinosus is a biarticulate muscle, its proximal compartment being regarded as a hip extensor and its distal compartment as a knee flexor. The medial gastrocnemius muscle also spans two joints (knee and ankle). If this functional diversity imposes certain (as yet unknown) requirements on central connectivity, the latter demands would concur and possibly compete with those originating from stability as outlined above.

Secondly, McKeon et al. (1984) maintain to have demonstrated an absence of somatotopy in the projections of muscle afferents onto homonymous α-motoneurones in the human tibialis anterior muscle. They suggested that reflex partitioning may occur only in muscles with complex physiological actions or with some degree of anatomical partitioning, which is associated with a partitioning of the motoneurone pool. They regard the tibialis anterior as a simple pennate muscle that crosses joints only at its distal end and has only one physiological action – dorsiflexion and eversion of the ankle. Spinal reflex partitioning should not apply to such simple muscles having single physiological functions. However, since they had previously demonstrated a sensitivity of spindle afferents of that muscle to local motor unit activity (McKeon and Burke 1983), the above considerations led them to suggest an interesting alternative, which transcends monosynaptic reflex connexions: "It remains possible that the importance of sensory partitioning lies not in spinal reflex connections but in supraspinal projections" (p. 193). This opens new perspectives (see below). Incidentally, tibial anterior function is not as simple as suggested by McKeon et al. (1984). These authors themselves state that the muscle produces dorsiflexion and eversion which need not be tightly linked to each other (see also Burgess et al. 1982). It is somewhat ironic that just in the tibial anterior muscle, though of the cat, there is evidence for compartmentalisation (Iliya and Dum 1984) and topographical organisation of γ-fusimotor input to spindles and of Ia afferent feedback to the spinal cord (Huhle 1985; see above).

Thirdly, the monosynaptic connectivity has been studied most extensively because it happens to be most easily accessible experimentally; it is not known which precise role specific oligo- and polysynaptic connectivity patterns play in tremor suppression. This role is very likely to be great. Be that as it may, one should endeavour not to overlook possible functions of reflex partitioning as long as a coherent theory of stretch reflex function is still lacking (see Chap. 5).

Some of the partitioned monosynaptic reflex connexions surveyed above may not appear to be of a significant quantitive strength. It should be taken into account, however, that the methods used to establish them (indiscriminate electrical stimulation of larger *sets* of Ia fibres or gross mechanical stimuli to excite sets of muscle spindles) are likely to fail to detect important further factors which co-determine signal transmission. Among these are the fine architecture of Ia fibre-motoneurone connexions, i.e. the relative location of synapses of many Ia fibres in the motoneuronal dendritic tree and the temporal characteristics of Ia discharge. The importance of these factors has repeatedly been emphasised throughout this book (see also Windhorst 1978 b) and ought to be taken into account in any model of stretch reflex function in which fine-grain information is likely to play a role.

In the preceding sections, emphasis was put on the possibility that the monosynaptic reflex connexions, if organised appropriately, may suppress rather than enhance physiological tremor. This idea is supported in part by experimental results obtained by Goodwin et al. (1978) on trained awake monkeys. The authors showed that in the monkey masticatory muscles tremor at low frequencies increased upon surgically opening the stretch reflex, and they concluded that the frequent emphasis on the instability of the intact stretch reflex was not supported. It is in this system – albeit in cats – that

Appenteng et al. (1978) provided evidence for a differential monosynaptic connectivity of muscle spindle afferents onto α-motoneurones. However, the masticatory muscles and their reflex organisation show some organisational peculiarities, and the results of Goodwin et al. (1978) were obtained under special experimental conditions, so that it is uncertain to which extent their conclusion can be generalised.

Within the framework outlined above, it is quite conceivable that for optimal functioning the stretch reflex can tolerate or even require a certain degree of partial synchronisations between its parallel channels, but that, beyond a certain threshold, it would fall into self-sustained oscillation. After crossing this threshold, the patterns of synchronisation, particularly between motor units, would shift to the rhythmical broad-peak type. Note that the preceding discussion has again stressed spatial aspects. Indeed the above transition process also has a spatial aspect insofar as a possible localised synchronisation may work as a "crystallisation focus", from which synchronisation, by diverging connexions of afferents, finally spreads throughout the parallel channels of the reflex system. This would probably require time-variable processes, however, most likely based on synaptic modification. These processes might then bear some resemblance to the "kindling" phenomena which eventually lead to epileptiform seizures and which appear to be associated with changes in synaptic efficacy (Sutula and Steward 1986).

3.4 Further Feedback Systems

3.4.1 Involvement of the Renshaw Cell System in Tremor

In the preceding sections population statistical events in the neuromuscular system were described in some detail to give an intuitive picture of the processes occurring in low-amplitude physiological tremor, and of the mechanisms which might prevent its quick transition into larger-amplitude tremor. One might argue that too much effort has been spent on processes which appear to be very specific to a system governed by rather long time constants (due to the sluggishness of muscle contraction): What, then, is "brain-like" about them? Apart from the fact that the CNS has to cope with the problems posed by the neuromuscular system, one can argue that the above processes are somewhat representative of processes occurring exclusively within the CNS. The brain will be shown to build internal dynamic models of the peripheral situation, and long time constants are nothing exceptional in these models. In the following chapter this will be argued to be the case for a specific network: the spinal Renshaw cell system. It is noteworthy that recurrent inhibitory networks exist at all levels of the CNS as kind of basic network structure. In this section the possible role of Renshaw cells in tremor mechanisms is briefly discussed.

Different dynamic states of neuronal activity patterns underlying different forms of physiological tremor were exemplified above in terms of a particular feedback system, viz. the classical monosynaptic reflex loop. Similarly, another feedback system, namely that set up by Renshaw cells in the spinal cord, may show different activity states in correspondence with the reflex system. A dispute of long standing concerns the role of recurrent inhibition via Renshaw cells in tremor processes. Some authors favour the idea that Renshaw cells de-correlate α-motoneurone discharge (Gelfand et al. 1963; Adam et al. 1978; Buahin and Rymer 1984) or otherwise reduce the tendency towards instability in the motor system (Stein and Oğuztöreli 1984). Others suspect Renshaw cells of enhancing or generating tremor. For example, Elble and Randall (1976) suggest that Renshaw inhibition-rebound causes the synchronous 8- to 12-Hz modulation in motoneurone firing that they observed in their experiments on human forearm muscles. They liken this mechanism of physiological tremor generation to the "inhibitory phasing mechanism" proposed by Andersen and Andersson (1968) for thalamic spindles. Andersen and Andersson (1968) had shown that an afferent volley to thalamic nuclei produces a series of rhythmic discharges in the thalamocortical fibres as a result of inhibition-rebound. The same phenomenon could be taking place in the motoneurone pool subjected to recurrent inhibition (Elble and Randall 1976). In contrast, Adam et al. (1978) stressed the de-correlating and, hence, tremor-suppressing function of the Renshaw cell network. The solution might be that both are true, but under different circumstances. Elble and Randall (1976) were probably not concerned with the low-amplitude form of physiological tremor, for which the suggestion of Adam et al. (1978) might hold.

The textbook version of Renshaw cells envisages them as cells typically producing long (up to 50–60 ms) and high-frequency (up to 1500 pps) bursts of discharge when excited via recurrent α-motoneurone axon collaterals. However, this response usually occurs only when many collaterals impinging on a given Renshaw cell are excited synchronously at a low input rate. Although Renshaw cells occasionally produce a burst in response to a spike of a *single* α-motoneurone (Ross et al. 1975; van Keulen 1981), they mostly respond with only a few discharges, particularly with asynchronous, higher-frequency inputs. Thus, in low-amplitude physiological tremor, where α-motoneurones fire more or less asynchronously, the asynchronous input must be expected to make Renshaw cells discharge relatively irregularly in a sequence of single or few grouped discharges. Each single discharge of a Renshaw cell, unless synchronised with discharges of other Renshaw cells, usually produces merely a weak hyperpolarisation in the contacted α-motoneurones (van Keulen 1981). Thus, in low-amplitude tremor where the Renshaw cells projecting to a given motoneurone pool can be expected to fire in a random and uncorrelated manner (except possibly for tendencies towards short-term synchronisation), the suggestion of Adam et al. (1978) may well be true.

Suppose, however, that substantial numbers of α-motoneurones fire more or less synchronously (broad-term, rhythmical), as in enhanced (medium-amplitude) tremor. The Renshaw cells they excite would then produce longer bursts in accordance with the rhythm of their input. And the discharge of α-motoneurones contacted by them would in turn be substantially inhibited during the Renshaw cell bursts. This would delay their next firing, causing it to approach the subsequent firings of those motoneurones that set up the Renshaw cell bursts. In this situation, Renshaw cells could not only sustain but promote the expansion of rhythmical synchronisation throughout a motoneurone pool.

Interestingly, there exists a parallelism between Renshaw cells and muscle spindles (and Golgi tendon organs) regarding spatial aspects of their input from α-motoneurones. As outlined in Sects. 3.2.3.2 and 3.3.1, the effect of any temporarily synchronised motor unit contractions on the discharge patterns of muscle spindles depends upon the relative location of the motor unit territories within the muscle. A similar statement holds for the effects upon Renshaw cells, with the qualification that the relative rostro-caudal location of the motoneurone cell bodies is the relevant spatial parameter. In the cat lumbosacral cord, an α-motoneurone (say of the medial gastrocnemius pool) contacts Renshaw cells distributed over a longitudinal range of (at most) 2 mm (for details see Sect. 4.3.1). (Remember that the longitudinal extent of motoneurone pools in the cat is of the order of 6–12 mm.) Conversely, any Renshaw cell can be excited by two α-motoneurones which may be separated by at most 2 mm (ca. 1 mm to either side). Thus, only motoneurones within this limited reach can, by their synchronised firings, influence the Renshaw cells in the way described in the preceding paragraph. Whether this influence remains initially localised depends, of course, also on the projection of Renshaw cells back to motoneurones. Indeed, there is a spatially weighted distribution of the inhibitory effects of any one Renshaw cell on α-motoneurones (Sect. 4.3.1), though these effects have a longer range than the excitatory effects of motoneurones on Renshaw cells. Thus, negative feedback via Renshaw cells is also localised in a certain sense. This spatial pattern also prevails in the respiratory motor system (Kirkwood et al. 1981), although the spatial dimension is larger than in the lumbar spinal cord.

The transition between the *two activity (or tremor) states* described above might again occur gradually or through a threshold crossing as in the case of the proprioceptive feedback loop (Sect. 3.1). This problem has not yet been investigated experimentally, and, because of its complexity, can possibly only be studied by computer simulation. A prerequisite would be the acquisition of precise data on the sensitivity of Renshaw cells to synchronous inputs, i.e.

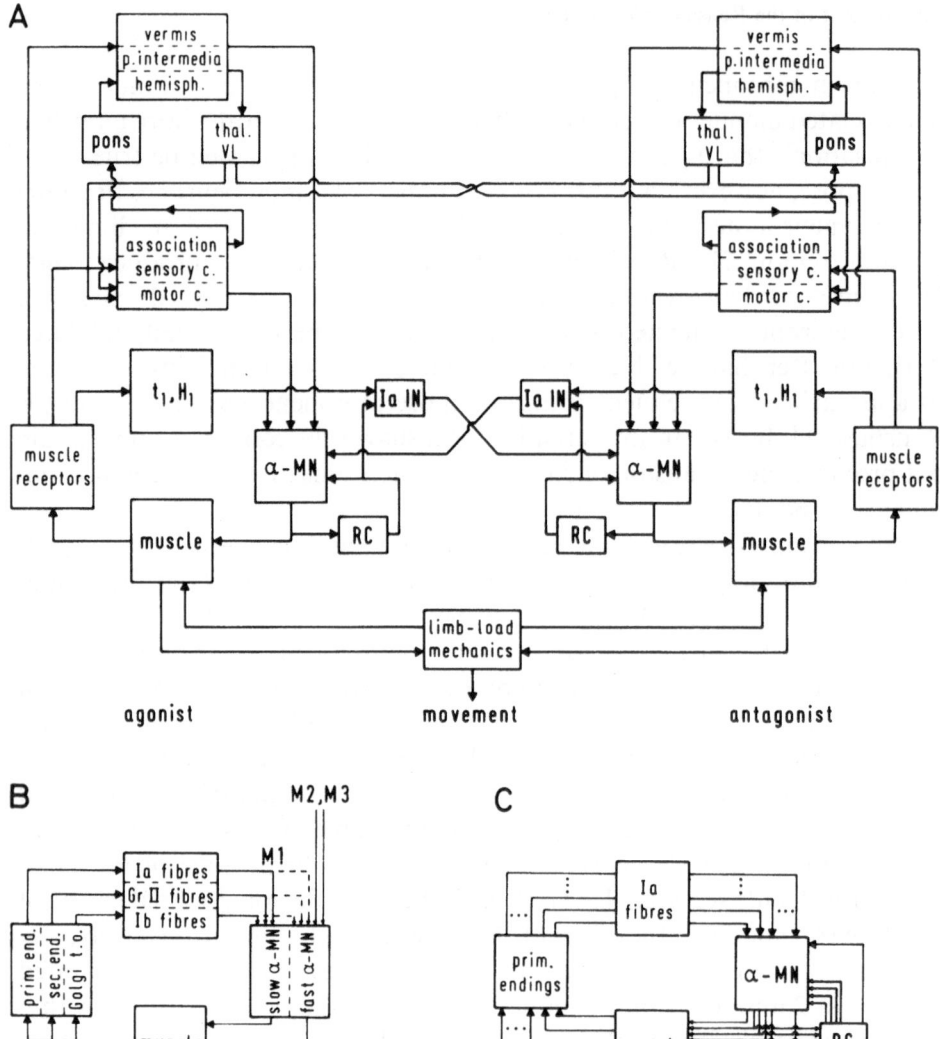

Fig. 37 A–C. Block diagram of the neuro-muscular control system emphasising multichannel structures and their interconnexions. **A** Scheme of the structures and connexions possibly involved in the genesis of cerebellar tremor. The two reflex systems of an agonist muscle (left) and its antagonist (right) are linked by the following connexions: (1) supraspinal (*upper horizontal*) connexions between the cerebral cortex [*block* divided into "association", "sensory" and "motor" cortex (*c*.)] and cerebellar cortex (*upper block* divided into "vermis", "pars intermedia" and "hemispheres") via "pons" and ventrolateral thalamus ("*thal. VL*"); (2) spinal (*middle horizontal*) connexions between antagonistic MN pools via Ia inhibitory interneurones ("*Ia IN*") which are inhibited by homonymous Renshaw cells (*RC*); (3) peripheral connexions (*below*) via muscle-load mechanics; **B** Division of the segmental (spinal) reflex loop into several "subloops" according to different functional classes of afferents (Ia fibres from primary muscle spindle endings, group II fibres from secondary endings, Ib fibres from Golgi tendon organs) and of motor efferents (slow and fast α-MNs). Bawa and Tatton (1979) proposed that slow motoneurones are recruited predominantly during the M1 (segmental) reflex component, and fast motor units during the M2 and M3 components, as indicated. This suggestion has however been modified by Calancie and Bawa (1985). **C** Further subdivision of one of the above "subloops" (for Ia fibres) and for the motoneurone-Renshaw cell system

on their ability to act as "coincidence detectors". From what is known about Renshaw cells, they seem to be well suited for this role.

A somewhat more complicated suggestion regarding the role of recurrent inhibition in suppressing large-amplitude tremor (or clonus in the pathological case) was recently made by Katz and Pierrot-Deseilligny (1984). However, this suggestion was related to the inhibition, by Renshaw cells, of the Ia inhibitory interneurones mediating reciprocal inhibition, and to the supraspinal control of Renshaw cell excitability (see Sect. 4.3.11). If, in patients with diseases of the "upper motoneurone", the central control of Renshaw cells is absent, the balance of reciprocal inhibition might be disturbed so as to facilitate clonus. This suggestion relates to the coupling of different reflex loops (here of antagonist muscles), which is dealt with in the following section (see Fig. 37).

3.4.2 Multiple Meshed Feedback Loops

The preceding discussion opens up a new problem: If both feedback pathways impinging onto α-motoneurones, one via muscle spindles and the other via Renshaw cells, play a role in setting up the two tremor states and in governing the transition from one to the other, how do they interact? Does one of them dominate over the other or are their properties "tuned" to each other in such a way as to yield a cooperative system of higher complexity? Whereas the latter possibility is, perhaps, suggested by certain similarities between the two feedback loops (see next chapter), the former possibility was preferred by Elble and Randall (1976) in assigning a dominant role to Renshaw cells for tremor genesis, particularly for the determination of tremor frequency. However, each of the two feedback systems on its own is complex enough to preclude simple prediction of its behaviour under various conditions. The coupling of the two systems adds another dimension of complexity. [In continuation of previous model studies, Stein and Oğuztöreli (1984) investigated the stability of a system of antagonist motoneurone pools and their muscles, which were peripherally connected to a common load and centrally coupled by reciprocal inhibition via Ia inhibitory interneurones. Increasing the muscle stiffness or autogenetic monosynaptic reflex gain as well as introducing reciprocal inhibition furthered the tendency towards oscillations, whereas recurrent inhibition via Renshaw cells decreased it.]

When considering these new dimensions, further feedback systems must be taken into account. In fact, secondary muscle spindle endings and Golgi tendon organs with their respective afferents are often, tacitly or not, disregarded in the discussion of tremor mechanisms. A simple reason is that considering these afferents implies discussing interneuronal systems, which, although they may be expected to be important in propagating correlated discharges between their inputs, have an input-output organisation of further complexity (see Sects. 2.9.1 and 2.9.2), so that their involvement in various states of tremor is completely obscure. Last but not least, there are long-latency reflex pathways (see reviews in Desmedt 1978) whose precise anatomical course and functional significance are a matter of considerable debate (for a recent discussion see, e.g. Matthews 1984).

3.4.2.1 A Reflex Model with Several Different-Latency Feedback Loops

The coupling of different feedback loops with different signal conduction delays can yield surprising behaviour, as shown by the computer simulations of Oğuztöreli and Stein (1976). In principle, their model was quite simple. They assumed that the various reflex pathways from muscle receptors only differ in the latency of the response, t_k, and the gain of the response, H_k. They considered three possible pathways (k = 1, 2, 3) with different latencies. "These three pathways correspond to the three pathways studied in the forearm of normal human subjects and will be referred to (Milner-Brown et al. 1975) as the *spinal* (t_1 = 30 msec), *cortical* (t_2 = 55 msec), and *transcerebellar* (t_3 = 85 msec) pathways" (p. 88). (In another frequently used terminology, the EMG responses to brief muscle stretches supposed to be mediated by these pathways are briefly denoted by M1, M2, and M3; see Desmedt 1978.) With a "standard parameter configuration" for the peripheral muscle-load system and the muscle spindles, Oğuztöreli and Stein performed computer simulations on the effects of different feedback constellations upon oscillatory system behaviour. The most interesting results in the present context can be summarised as follows.

1) If only the spinal pathway (t_1 = 30 ms) is operative and endowed with a sufficiently high gain H_1 ($H_1 > 0$; $H_2 = H_3 = 0$), a growing oscillation of frequency 10.1 Hz will ensue. With only the cortical loop active at a sufficiently high gain ($H_2 > 0$; $H_1 = H_3 = 0$), a growing oscillation emerges at a frequency of 7 Hz. With an active transcerebellar loop at a particular gain ($H_3 > 0$; $H_1 = H_2 = 0$), two independent oscillations, one decaying and one growing, with frequencies of 5.1 and 13.5 Hz, respectively, are generated (see Oğuztöreli and Stein 1976).

2) If a constant amount of gain sufficient to render each single loop unstable (as in 1) is divided among the three loops in a favourable way, growing oscillations can be converted to decaying ones, i.e. the system can be stabilised.

3) When the distribution of gain between the three feedback pathways is changed continuously ($H_1 + H_2 + H_3$ = const.), the frequency of oscillation often changes smoothly between the limits set by the "natural" frequencies

of each loop in isolation (see 1). Occasionally, however, there may be rather abrupt changes between high and low frequencies of oscillation.

4) "In conclusion, complex interactions are possible between the various reflex pathways, but under suitable conditions the presence of multiple pathways can be very effective in reducing the tendency for oscillation which will be inherent when any one pathway is present with a high gain" (Oğuztöreli and Stein 1976; p. 99). Hence, it is easily conceivable that if the fine balance of these "suitable conditions" is altered, e.g. by pathological increases in gain or destruction of one or the other pathway, "inherent" oscillatory tendencies may become manifest.

There are two particularly interesting aspects in the above simulation studies, namely the occasionally abrupt changes between higher and lower frequencies of oscillation and the occurrence of low-frequency oscillations due to long-latency reflex loops. Both aspects are reminiscent of Stiles' (1976) results (Sect. 3.1) and suggest that the very low frequency (3–6 Hz), large amplitude tremor is due to oscillation in a (now dominant) long-latency reflex loop, whatever its precise nature. This possibility would define a third tremor state, above the two dealt with above (and investigated by Hagbarth and Young 1979; Young and Hagbarth 1980; Sect. 3.1). Recently, Gottlieb and Lippold (1983) showed that the 3–6 Hz tremor may coexist with the 8–12 Hz tremor during sustained extension of the middle finger in normal humans. They also proposed that the slow tremor originated from oscillation in a long-latency reflex loop. Thus, complex phenomena can originate from the interaction of multiple and possibly nonlinear oscillators (see also Winfree 1980).

3.4.2.2 Cerebellar Tremor

Oğuztöreli and Stein (1976) tentatively assumed in their model that the longest-latency loop ($t_3 = 85$ ms) traversed the cerebellum (see above). Apart from anatomical considerations, this seemed also justified in view of the low-frequency cerebellar tremor arising in cerebellar disorders. Lesions of deep cerebellar nuclei, particularly of the dentate nucleus, produce "pendular" tendon reflexes and "intention" tremor that occurs when a goal-directed voluntary movement is approaching its end. These symptoms can be regarded as being signs of the general ataxia and most often concur with a hypotonia of the musculature. Quite successful primate animal models have been developed for these cerebellar symptoms (e.g. Vilis and Hore 1977, 1980) and give some insight into the pathophysiological mechanisms possibly involved in these phenomena.

An interesting proposal concerning the pathophysiology of cerebellar tremor has recently been derived by Vilis and Hore (1980) from their results obtained in cebus monkeys trained to grasp a handle, keep it in a constant position, and resist any disturbances injected by a torque motor. By means of implanted cryoprobes the nuclei interpositus et dentatus could be cooled to reversibly block their function. Handle (arm) position, EMG activity of biceps and triceps muscles (of the arm) and discharge patterns of precentral motor cortex cells were recorded in response to the injected torque pulses (of 40 ms width). The pulses displaced the handle in the direction of arm (elbow) extension. The return movements usually overshot and were occasionally followed by a damped oscillation at a frequency of 6–8 Hz. The lower the temperature, to which the cerebellar nuclei were cooled, the more the damping was reduced. The oscillations in response to torque pulses increased in amplitude and decreased in frequency until at 10 °C an undamped oscillation at about 4 Hz resulted. This sequence of events was reversed upon rewarming.

Simultaneous recordings of EMG and motor cortex cell activity led Vilis and Hore (1980) to the following conclusions:

1) If a limb is rapidly displaced from a position to be maintained, the return is effected through a series of reflexes and voluntary corrections initiated by afferent impulses from the muscle initially extended. The stretch reflexes of this muscle are not altered by cooling the cerebellar nuclei.

2) With *impaired cerebellar function* the sequence of events is as follows. The return movement initiated by contraction of the primarily stretched muscle (the agonist) in turn stretches the antagonist, initiating a series of contractile responses in this latter muscle. This again leads to a stretch of the agonist, and the whole sequence repeats itself. Cerebellar tremor can thus be viewed as a "series of alternating stretch reflexes" primarily involving the transcortical loop (see above), which explains its low frequency of 3–5 Hz. Thus, its basic mechanism is similar to that of (spinal) clonus, the only difference being the length of the loop involved.

3) *In normal subjects*, the series of alternating stretch reflexes in the transcortical loops is interrupted by the *predictive capabilities of the intact cerebellum*. On the basis of "peripheral" information concerning the perturbation of arm position and of "cortical" information regarding the subjects's "intention to move" (e.g. to resist the perturbation), the cerebellum calculates a prediction of the stretch and probable return movement of the agonist muscle. This prediction is sent to the motor cortex in order to initiate a quick and appropriate command to the antagonist to properly brake the return movement in time. The "damping effect" of an intact cerebellum is thus based on the fact that the "predicted" antagonist response is faster than, and hence phase-advanced in regard to, that initiated by the antagonist's stretch via a transcortical reflex. Cerebellar tremor in this interpretation comes about through an unmasking of oscillatory tendencies in the transcortical reflex loops when the predictive capacities of the cerebellum are eliminated

(see also Hore and Vilis 1984; Hore and Flament 1986). (Another possibility for involving the cerebellum in phase-advancing the motor cortex response with respect to simple stretch reflexes was also discussed by Vilis and Hore 1980.)

It becomes apparent from the above that long loops may operate in a more intricate manner than assumed in the simple model of Oğuztöreli and Stein (1976); yet the significance of meshed parallel loops for stability is nicely illustrated by this model. The complicated interaction of parallel reflex loops may occur not only between long-latency loops, but also – mutatis mutandis – between loops at lower levels.

3.4.3 Remarks on the Possible Significance of Physiological Tremor

In the preceding sections, different tremor states were discussed at some length as paradigms for physiological states of activity in the nervous system, in which different forms of correlation between parallel neuronal channels play a role. One might argue that tremor was thereby given undue honour, hampering as it does the precise execution of motor acts. This view, while partially correct, may yet be too restricted. There is no question that various tremors, particularly pathological ones, interfere with precise movements. And physiological tremor, too, might well be the undesired side-effect of properties of the neuromuscular system that were designed for other purposes. Still, a second look from a different vantage point may be enlightening.

For example, Stein and Oğuztöreli (1976) argue that it may be desirable to have a small amount of oscillation. They base their argument on results obtained by Joyce et al. (1969), who found that with a partially fused tetanus the muscle behaved quite differently when it was stretched or allowed to contract at very slow velocities. This type of nonlinearity, which results from the small-range elastic properties of cross-bridges (Hill 1968), could be abolished by the superposition of a small vibration on the stretch or contraction. Stein and Oğuztöreli (1976) suggest that, in the monosynaptic reflex loop, the velocity sensitivity of Ia muscle spindle afferents may compensate for the sluggishness of muscles (see also Stein 1974). At high gain, this may entail a tendency towards instability (physiological tremor), but on the other hand, the resulting small oscillations may serve to linearise muscle properties.

Hagbarth and Young (1979) put forward a similar hypothesis: "In practical applications of System Control Theory, a rapid low-amplitude alternating movement (dither) is often added to a position servo to permit graded or proportional control of non-linear or discontinuous systems. Whether physiological tremor can serve a similar purpose for the motor control of body parts remains to be demonstrated" (p. 522).

For the large-amplitude tremors, it is important to note that their frequency falls within the range of the fastest alternating voluntary movements. Stiles (1976) reminds us of the fact that similar extension-flexion oscillations of the hand can be achieved voluntarily. He observed that certain of his subjects "were at times unsure whether the hand oscillations that occurred at these large-displacement amplitudes were being performed voluntarily or involuntarily" (p. 53). Although Stiles subsequently gives reasons to believe that the large-amplitude tremor movements were indeed involuntary, a close relation between the latter and voluntary alternating movements is suggested by his volunteers' statements.

Such a relation is also proposed by Freund (1983), who points out that tremor represents the fastest involuntary movement showing conspicuous similarities and analogies with the fastest voluntary or reflex movements. With regard to these fastest alternating movements, Freund (1983) states that a subject asked to perform alternating finger movements as fast as possible would do best by using tremor-like finger or hand oscillations. There is also an inverse relationship between amplitude and frequency in that it is so much the easier to achieve the maximum rate, the smaller the movement amplitude (or force alterations in case of isometric contractions) is. On the other hand, alternating finger and hand movements can be executed over the full angular range only at rates below about 3 Hz. The faster the movements, the smaller the angles. In his own experiments, Freund (1983) found that the fastest alternating movements of the forefinger performed by 12 normal subjects were 7.5 ± 1.4 Hz (mean plus/minus SD) for the right and 6.8 ± 1.4 Hz for the left forefinger. These were minimum-amplitude oscillating movements that could only be executed for a few seconds. This amplitude–frequency relationship qualitatively resembles that found by Stiles (1976) for involuntary hand tremor and is probably at least partially due to mechanical restraints resulting from the contractile properties and activation patterns of the participating muscles (Freund 1983).

In the same vein, Rack (1981a) defies the common view that the limit-cycle oscillations possibly arising in the stretch reflex should be regarded as a defective operation of the system normally avoided by the animal. This need not be so. In activities such as running or hopping, the oscillatory propensity of the system may be employed to good advantage. "It is possible that the oscillatory tendencies of the stretch reflexes may be used in these situations, and the frequencies at which some of these repetitive movements usually occur are such that reflex activity could contribute to them" (Rack 1981 a; p. 244).

In this context, a suggestion by Alberts (1972) is of interest. He proposed that Parkinsonian hand tremor (at rest) results from a pathological release of (cortical) programes which normally generate rapid alternating movements at the rate of 3–5 Hz. (For a comparison of Parkinsonian postural hand tremor and large-amplitude physiological tremor see Stiles and Pozos 1976.)

It does not appear unreasonable to put forward the hypothesis that the different tremor states described above have functional equivalents in defined activity states of the nervous system (including skeletal muscle) that are used physiologically. Through various disturbances, these states may degenerate pathologically, and it would therefore not seem surprising to find pathological equivalents, too. Without going into detailed reasoning, a hypothesis is proposed here regarding the correspondence of physiological and pathological tremor states:

a) Medium-amplitude physiological tremor (Hagbarth and Young 1979); first region of large-amplitude tremor above low-ampl. region (Stiles 1976); "physiological clonus" (Agarwal and Gottlieb 1977)
6–8 Hz

Clonus (clonus is here interpreted as an oscillation not principally different from tremor)

6–8 Hz

b) Large-amplitude tremor (Stiles 1976)

Parkinsonian (postural) tremor (Stiles and Pozos 1976) cerebellar "intention" tremor (Vilis and Hore 1980)

3–5 Hz

3–5 Hz

This section may be concluded with some general considerations of Agarwal and Gottlieb (1977), which emphasise the possible role of meshed multiloop systems. They first point out that *conservative engineering design* tends to stress system stability and that this view is adopted by many physiologists when treating physiological regulating mechanisms. An alternative view of such mechanisms, particularly if they are composed of meshed loops, may be that some of their inner loops are inherently unstable. Homoeostasis may then be preserved by outer loops, which are called upon when some of the state variables, to be confined within set limits of some allowed "state space", approach these limits. The instability inherent in the inner loops is then only observed in pathological conditions when the safety mechanism provided by the outer loops is defective. These ideas look like a general comment on concepts outlined in Sect. 3.4.2. However, as pointed out in this section, the inherent instability may also be at the disposal of physiological functions, and it could be released to be used purposefully.

3.5 Plasticity in the Spinal Cord

As pointed out by Young and Hagbarth (1980), the transition from low-amplitude to "enhanced" physiological tremor may be promoted or even triggered by "plastic" changes in muscular or synaptic properties due to prolonged activity of the neuromuscular elements. Whereas changes in muscular properties might occur (e.g. Parmiggiani et al. 1982), their quality and strength are still controversial (Young and Hagbarth 1980; Gottlieb and Lippold 1983) and have to be studied more extensively in conjunction with changes in the responses of muscle receptors to altered motor unit contractions. In this connexion, we have recently obtained some evidence that motor unit fatigue, which reduces the gain of signal transmission from α-motoneurones to force output, is partially compensated by increased gain of signal transmission from force (or internal muscle length) to spindle afferents (Christakos and Windhorst 1986a) and homonymous α-motoneurones (Windhorst et al. 1986).

Changes in spinal synaptic efficacy on a time-scale compatible with the above transition might occur, but their mode of operation is presently unknown. One should keep in mind, however, that such processes may not only promote the transition from low- to larger-amplitude tremor, e.g. by increasing the gain of the monosynaptic Ia fibre-motoneurone linkage, but that, on the contrary, they could also delay this transition by temporarily stabilising the local stretch reflex loops dealt with in Sects. 3.3.1 and 3.3.2, or by some other mechanism in interneuronal circuits. It could be that a group of different processes acting at various time-scales contribute to such stabilisation. These processes might include plastic changes ranging from post-tetanic potentiation (time constant of several minutes; see Sect. 1.3.3) to short-term synaptic facilitation (time constant of the order of 50–100 ms). Short-term facilitation is present at central spinal synapses, such as those from recurrent motor axon collaterals to Renshaw cells (Windhorst et al. 1987b), and from cutaneous afferents to dorsal horn neurones (Kuipers et al. 1987).

The most likely kinds of synaptic change that could play a role are processes related to post-tetanic potentiation (PTP; recent review: Mendell 1984; see also Lloyd 1949; Lev-Tov et al. 1983; Lüscher et al. 1983; Davis et al. 1985, 1987). PTP is commonly regarded as taking place in the pathway that has been potentiated, and not in a convergent one (Davis et al. 1987). Lüscher et al. (1983) and Davis et al. (1985) showed that PTP also occurred in monosynaptic transmission from single Ia fibres to α-motoneurones. Lüscher et al. (1983) argued that PTP could best be explained by assuming that, in the nonpotentiated state, the conduction of action potentials at Ia fibre branching points into terminals was partially blocked (but cf. Davis et al. 1985). Hence, potentiation could result from at least two slightly different mechanisms: (1) relief of complete failure of transmission in projections to motoneurones, in which no EPSP had been elicited prior to tetanisation, and (2) relief of partial failure in projections to motoneurones which previously had some functioning connexions. This mechanism would also open an interesting possibility for the mutual interaction of several parallel presynaptic fibres, for, as suggested by Swadlow et al. (1980), activity in neighbouring neural structures may significantly alter the conduction properties and the probability of conduction failure at regions where conduction failure usually occurs (for further references see Lüscher et al. 1983; also Henneman et al. 1984; see Sect. 2.8.2.3). (Note that certain presynaptic correlation patterns might be of particular importance for this kind of interaction.) This would presumably require intricate anatomical interrelations and it stresses the importance of fine structural features for function, as has been emphasised several times before. A different explanation for PTP at the Ia fibre-motoneurone synapse was given by Davis et al. (1985). They argue that the probability of transmitter release is increased here as it is at neuromuscular junctions. Connexions generating small EPSPs can exhibit stronger potentiation than those generating large EPSPs, but the duration of potentiation appears to be smaller for small EPSPs. Davis et al. (1985) found it "reasonable to attribute the potentiation we have described to the 'facilitation' process. . ." (p. 1550), but the longer duration of potentiation of large EPSPs could justify its assignment to the augmentation process (see Sect. 1.3.3). Interestingly, these potentiations were obtained at intra-burst rates of 167 pps, lying well in the range of discharge rates of spindle Ia afferents during locomotion.

Mendell and co-workers (Nelson and Mendell 1979; Nelson et al. 1979; Cope et al. 1980) have also provided data indicating long- and short-term changes in the efficacy of synaptic transmission between single Ia fibres and α-motoneurones following spinal cord transection. In the present context, the most interesting results are those showing that the "projection frequency" of Ia fibres to homonymous motoneurones increases to nearly 100% immediately after more rostral spinal cord section (Nelson et al. 1979), whereas the increase of EPSP amplitudes takes some hours, indicating mediation by factors outside the neuraxis (Cope et al. 1980). Whilst in normal (anaesthetised) cats single Ia fibres project to about 80% of the homonymous α-motoneurones – two thirds (type X) projecting to most and one third (type Y) to about 65% of the motoneurones (Scott and Mendell 1976) – the projection frequency in spinalised animals is 99% irrespective of Ia fibre type. Nelson et al. (1979) suggest that some fibres have a low projection frequency (e.g. type Y fibres) because of the action of descending activity, which could restrict the access of the Ia fibres to all the motoneurones to which it sends terminals. The less widespread projection of some Ia fibres could therefore be the consequence of a physiological process rather than a more restricted system of terminals. However, these results could not be confirmed by Walmsley and Tracey (1983).

More generally, it is of interest that the (isolated) spinal cord is capable of exhibiting associative learning behaviour. Over the past decade, Durkovic and co-workers have extensively investigated classical conditioning of flexion reflexes in decerebrate spinal cats. Using flexor muscle responses to saphenous nerve (conditioned) stimulation and superficial peroneal nerve (unconditioned) stimulation as indicators, the above group showed (1) that the conditioned response (to saphenous stimulation) increases over the first repetitions of trials, and (2) that it reaches an asymptotic level, (3) that these response increases need temporal contiguity of conditioned and unconditioned stimuli, and (4) that response enhancement is of long duration (Durkovic 1983, 1985; Misulis and Durkovic 1984). Durkovic (1983) and Misulis and Durkovic (1984) showed that the conditioned flexion reflex requires the stimulation of Aδ fibres in the conditioned and unconditioned input, whence it was concluded that "the location of neural changes responsible for the conditioned facilitation is limited to specific interneuronal pools" (Durkovic 1983; p. 159). Durkovic (1985) excluded the possibility that this type of conditioning facilitation could be due to postsynaptic potentiation (PTP), which exhibits less persistence. He argued that, instead, the long-term effect more closely resembles the long-term potentiation in the hippocampus described by Levy and Steward (1983; Sect. 1.3.3). Yet, the neuronal mechanisms of such conditioning still have to be revealed.

The pertinent question remains whether any of the long- and/or short-term processes of synaptic modification and modulation suggested by von der Malsburg (Sect. 1.3.2) is at work in the spinal cord. These processes can be divided into two basic types: (1) long-term plastic changes responsible for the buildup (self-organisation) of particular, possibly topographically ordered connectivity patterns between neuronal (neuromuscular) elements; (2) short-term processes establishing functional neuronal assemblies on the basis of interneuronal correlations.

1) Schemes for the ontological development of topographic projections have been proposed for several sensory systems (see Sect. 1.3.3). The underlying processes of "synaptic learning" could also occur during the development of topographic connectivity patterns in the spinal cord, e.g. between Ia fibres and α-motoneurones (Windhorst 1978 b; see also Fawcett and O'Leary 1985). Remember that these processes may involve correlated activity between input (e.g. Ia fibre) and output (e.g. α-motoneurone) axons. Another dimension of plasticity and versatility would be gained if, in the spinal cord, correlations *between* input channels were able to modify synapses between these inputs and an output cell, as now seems demonstrated for supraspinal structures (see Sect. 1.3.3).

2) If such specific processes occur at all in the CNS, the question arises whether they are responsible only for the long-term buildup of connectivity, or whether, as von der Malsburg suggested (Sect. 1.3.2), they allow for the

rapidly alternating (short-term) construction and extinction of functional cell assemblies. If von der Malsburg's model is very speculative at the cortical level, its application to the spinal level would be even more so. But is seems worth looking for new models in view of our increasing knowledge of the true complexity of spinal neural networks which can show quite different activity patterns suggesting cooperative behaviour of varying neural assemblies. For example, it is conceivable that the low-amplitude tremor is temporally stable due to the transitory buildup of many parallel functional loops constituted, on the basis of some (topographical and "species-specific") anatomical order, by the short-term synchronised activity patterns of α- and γ-motoneurones, muscle receptor afferents and interneurones; the correlated activity patterns could cooperatively strengthen synapses so as to maintain the prevailing state for a while. This is very speculative but may initiate a new approach towards an alternative functional interpretation of short-term (and other types of) synchronisations that transcends the current restrictive attitude of using these correlations as a technical tool to investigate neuronal connectivity.

3.6 Summary

Besides the stretch reflex system (α-motoneurones and proprioceptive muscle afferent feedback), which may exist in various activity (tremor) states, the recurrent inhibitory system (α-motoneurones and Renshaw cells) may take on different states of activity in correspondence with those of the former system. That is, Renshaw cells might contribute to de-correlation of α-motoneurone discharges in low-amplitude tremor, and enhance and sustain rhythmic α-motoneurone correlation in larger-amplitude tremors. For such coordinated activity of both feedback systems to occur, their organisation and dynamic behaviour should be partially matched. This suggestion will be discussed in the next chapter.

The coupling of several feedback systems with different properties, such as gains and latencies, poses a difficult analytical problem, but also opens up new possibilities of stabilising the overall system. This is illustrated by the analytical and computer simulation studies of Oğuztöreli and Stein (1976). They concluded that "complex interactions are possible between the various reflex pathways, but under suitable conditions the presence of multiple pathways can be very effective in reducing the tendency for oscillation which will be inherent when any one pathway is present with a high gain" (Oğuztöreli and Stein 1976; p. 99).

A physiological example for the intricate balance of several feedback systems is provided by the studies of Vilis and Hore (1977, 1980) on cerebellar tremor elicited in monkeys by cooling the dentate region. This type of tremor might come about by disabling the predictive capacity of the cerebellum by which oscillatory tendencies in transcortical reflexes may normally be masked.

It is presumably short-sighted to conceive of tremor simply as a detrimental phenomenon. Small-amplitude tremor may also have beneficial effects in linearising the performance of muscle and possibly other structures. Large-amplitude tremor may result from processes that are also used physiologically in fast alternating movements. In these processes, the interaction of meshed loops with their inherent tendency to oscillate may play an important role.

In summary, Fig. 37 is an attempt at visualising the parallel-loop structure which is of importance for tremor mechanisms. Figure 37 A gives a schematic overview of the structures possibly involved in the genesis of cerebellar tremor as described by Vilis and Hore (1980). The reflex systems for the two antagonist muscles are coupled at three levels: peripherally through the muscle-load mechanics; at the spinal level (via Ia inhibitory interneurones denoted by "Ia IN" mediating reciprocal inhibition), and at a supraspinal level (via cerebello-thalamo-cortical connexions transmitting the cerebellar prediction). Also included at the spinal level are the recurrent inhibition via Renshaw cells of α-motoneurones and reciprocal Ia inhibitory interneurones (Ia IN). In Fig. 37 B the spinal (segmental) reflex system is further refined to make allowance for the different types of α-motoneurones and for muscle afferents from spindles and Golgi tendon organs. In Fig. 37 C the parallel structure of motor efferents, spindle Ia afferents and the recurrent inhibitory system is emphasised.

Processes of synaptic modifiability which could stabilise or change some of the motor phenomena are also known to occur in the spinal cord. Most of these processes have been studied at easily accessible synapses, such as the neuromuscular endplate or the Ia fibre-α-motoneurone synapse. Whether they are as elaborate and play physiologically similar roles as the corresponding processes at the cortical level (in the telencephalon or the cerebellum) remains to be determined. The physiological significance of spinal synaptic modifiability can probably only be properly assessed in conjunction with an elucidation of the function of interneuronal networks.

As sufficiently pointed out in Chap. 1, processes of synaptic modification probably play a major role in the self-organisation of complex neural systems. This almost certainly also holds true for the spinal cord.

4 Recurrent Inhibition and Proprioceptive Feedback

Correspondances

La Nature est un temple où de vivants piliers
Laissent parfois sortir de confuses paroles;
L'homme y passe à travers des forêts de symboles
Qui l'observent avec des regards familiers.

Comme de longs échos qui de loin se confondent
Dans une ténébreuse et profonde unité,
Vaste comme la nuit et comme la clarté,
Les parfums, les couleurs et les sons se répondent.

Ch. BAUDELAIRE [7]

As shown in the preceding discussion, correlations between the discharge patterns of neurones are ubiquitous and common at all levels of the nervous system. The correlative patterns encountered are rather similar varying from short-term to broad-peak synchronisation, negative and rhythmical cross-correlations. They are associated with different states of the activity of cell populations. Of particular interest are the short-term types of correlations and the related neural activity states which appear to reflect differentiated fine-grain information processing. Whereas these correlations, particularly synchronisations, between neural discharges have often been used to draw inferences about common inputs, the deeper and more important question is, of course, whether this feature of neural discharge is *nothing but* an epiphenomenon resulting from the particular neural connectivities, or whether it is used functionally to establish neural assemblies. There appears to be a greater reluctance to accept such a possibility for the spinal cord than for higher-order systems. The general feeling still seems to be that spinal cord function ought to be simpler and, perhaps, more "primitive" than cortical operation. Such an attitude is a carry-over from the old-fashioned reflex physiology, which regarded the spinal cord as a rapid throughway from sensory input to motor output (e.g. in flexion reflexes). This view is currently being revised and modified.

The question as to the code used is closely linked to the problem of function. In other words, any suggestion as to a specific code must be based on, or incorporated into, a more general theory of the function executed by a nervous structure. Such a theory is at present lacking for the spinal cord, even for subsystems, such as recurrent inhibition or the so-called stretch reflex, as will become apparent below.

---

[7] From Baudelaire C "Les fleurs du mal" (first two stanzas)

4.1 Fundamental Functions of the Spinal Cord

Superficially, the functions of the spinal cord can be summarised as follows:

1) Transmission or relaying of peripheral sensory information and of the states of spinal neuronal systems to higher perceptive and sensorimotor centres;

2) Conduction and relaying of motor commands and other descending signals from supraspinal centres to the spinal motor centres and neural systems controlling spinal cord input;

3) Integration of descending and afferent signals to generate a motor output;

4) Generation and maintenance of various activity patterns, such as locomotor rhythms.

(Note that the formulation of these functions is general enough to include those of the autonomic nervous system. In the following, we will only deal with the nonautonomic functions, and particularly with items 3 and 4.)

Described in this way, these functions appear to be straightforward. Viewed from the angle of motor output, the necessary signals used for information transfer seem to be defined along two major dimensions: recruitment (number) and firing rate of (moto)neurones. From this final shape of nervous information one might be led to extrapolate back to other neural – pre-motoneuronal – stages of signal processing where the same kind of information coding could be presumed to take place. Following these lines, this could apply not only to the spinal cord, but to the entire nervous system because motor output is the organism's way of talking to its environment, i.e. the nervous system can be viewed as a – very complicated – interface between sensory input and motor output (see Creutzfeldt 1983; Palm 1982). On this view, the functions listed above appear simple enough to be executed by the transmission and processing of signals whose magnitudes at any given time are defined as some product of the number of nerve cells, which – at each stage of processing – are active in parallel, and and their mean firing rates. Why, then, should there be a need of a more complicated code such as the pattern of correlation between neural elements?

It is certainly premature to answer this question positively. But it is increasingly recognised that the "eigen"-functions of the spinal cord (quite apart from those of the cortices; see Chap. 1) are more complex and more highly developed than previously believed. The spinal circuitry has turned out to be so complicated that earlier assertions as to its functions have been seriously challenged.

4.2 Specific Functions of Recurrent Inhibition and the Stretch Reflex

The difficulty of determining functions of neural networks becomes apparent by just listing proposals that have been made. We shall confront here suggestions related to the spinal subsystem of recurrent inhibition on the one hand, and to a more complex system (stretch reflex) on the other. This comparison will serve as a starting point for a more systematic comparison between recurrent inhibition and proprioceptive (muscle afferent) feedback in order to point out structural and functional correspondences which point to the need for an integrated model.

4.2.1 Functional Models of Renshaw Cell Inhibition

4.2.1.1 Self-Inhibition of α-Motoneurones via Renshaw Cells

The simplest scheme for a neuronal circuit involving Renshaw cells emphasises the recurrent excitatory connexions from α-motoneurones to Renshaw cells and the inhibitory connexions from the latter to the former (see Figs. 21 D and E, and 49). (Here and in the following, the term "α-motoneurone" includes β-motoneurones, for brevity.) This circuit is then essentially envisaged as a single-loop negative feedback system in which even single α-motoneurones might inhibit themselves (Cleveland and Ross 1985). The following first set of hypotheses concerning its physiological function is based on this concept.

1) "Stabilisation" of motor output: "Stabilization is understood as a reduction of sensitivity to changes in excitatory drive" (Haase et al. 1975; p. 115); and *"limitation"*: "Limitation expresses the fact that discharge frequency in response to a given input will be lower than that of the system without feedback" (Haase et al. 1975; p. 115). These two notions were proposed as "functions" of recurrent inhibition at an early stage and are propounded until today (see Pratt and Jordan 1987). But it has never been made clear why a complicated system such as the recurrent inhibitory one should have evolved just to limit α-motoneurone discharge, which could be achieved in much simpler ways, e.g. by limiting the input itself. To be sure, stabilisation and limitation are likely to be *consequences* of the negative feedback action exerted by Renshaw cells, but they are unlikely to be its (teleonomic) function.

2) Reduction of the effects of spurious inputs such as electrotonic coupling between α-motoneurones (W. K. Taylor 1965) or fluctuations in fusimotor bias to muscle spindles due to variations in the amount of post-tetanic potentiation (Granit 1970). More generally, the basic idea here is that systems subjected to a favourably tailored negative feedback are less sensitive to all kinds of perturbations arising internally or externally to the system (see Houk and Rymer 1981).

3) Speeding up of α-motoneurone responses to time-varying inputs (Wenstoep and Rudjord 1971; Cleveland 1977), i.e. contribution to the dynamic sensitivity of α-motoneurones (see Andreassen and Rosenfalck 1980). Here, too, there exist other means to achieve this end, e.g. intrinsic α-motoneurone properties (Andreassen and Rosenfalck 1980). The contribution of recurrent inhibition is again likely, therefore, to be a side-effect of the very existence of this circuit.

4) Suppression of the initial oscillations of α-motoneurone output in response to changing inputs (Cleveland 1977).

5) Contribution to the static firing characteristics of α-motoneurones whose current-frequency relationships typically show a division into a "primary" and a "secondary" range (Ross 1976; Pompeiano 1984; Cleveland and Ross 1985), and *generation of the higher sensitivity to current injection of large α-motoneurones than of small α-motoneurones.*

It is essential to recognise that all these "functions" require, albeit to different extents, that even a single α-motoneurone is able to effectively inhibit itself via its recurrent axon collaterals and a set of Renshaw cells. In general, however, the recurrent inhibitory circuit shows a pronounced convergence–divergence structure. This implies that even if a self-inhibitory circuit (one or a few α-motoneurones inhibiting themselves) is considered as the basic functional unit ("module"), these units are coupled among each other via cross-inhibitory links.

Recurrent inhibitory circuits are also known at supraspinal levels. An interesting interpretation of such a circuit in the olfactory bulb is given by Freeman (1981, 1988). Briefly summarising the circuitry in the bulbar superficial layers, he proposes that the negative feedback loop between the mitral and the granule cells establishes a neural oscillator receiving its input through each glomerulus. These oscillators are widely coupled by mutually excitatory axosomatic synapses and by mutually inhibitory interactions through cellular mechanisms not yet clearly identified (Freeman 1988). This organisation bears some resemblance to that among α-motoneurones and Renshaw cells. Its interpretation as coupled oscillators is consonant with the views presented above which give priority to "local" negative feedback circuits of single α-motoneurones and Renshaw cells, and with Elble and Randall's (1976) view of the involvement of recurrent inhibition in tremor genesis (see Sect. 3.4.1). The kind and geometry of the coupling between such oscillators are, not surprisingly, somewhat different from that in the olfactory bulb. Whereas the bulbar network forms a two-dimensional sheet, the α-motoneurone-Renshaw cell network (of each α-motoneurone pool) is essentially one-dimensional because motoneurones are aligned in rostro-caudal columns. (This view may be due to the lack of knowledge about fine features of organisation in the transverse plane.) As long as the spatially distributed cross-coupling between the local negative feedback circuits does not quantitatively predominate over the local feedback connexions, these local loops may be envisaged as modules; but the term "module" is not employed without ambiguity (Creutzfeldt 1983; Sect. 4.3.1).

4.2.1.2 Mutual Inhibition Between α-Motoneurones

More complicated models of the spinal recurrent system are established when further details of neuronal coupling are taken into account, e.g. the fact that Renshaw cells receive excitatory inputs from a collection of α-motoneurones and distribute their inhibitory output back to a slightly different set of α-motoneurones. The nonhomogeneous distribution of recurrent inhibition has two aspects: functional and spatial.

A) The *functional aspect* in turn has several facets:

1) *Inhomogeneous distribution of recurrent inhibition among different types of α-motoneurones* (see Sect. 4.3.2): Large α-motoneurones inhibit small α-motoneurones more strongly than they themselves are inhibited by the small ones, thus opening "the possibility that by this mechanism phasic extensor reflexes would tend to inhibit tonic ones" (Granit 1970; p. 157). It has been proposed that this would prevent the sluggish contraction of slow muscles from impairing the rapid contraction of fast muscles, but the latter assumption is not warranted (Hutton and Enoka 1986). Broman et al. (1985) discuss the possible involvement of the above mechanism in the suppression exerted by the recruitment of large motor units on the firing rates of units recruited at lower forces. The inhomogeneous recurrent inhibition might also compensate for the effect of the inhomogeneous distribution of certain input systems to α-motoneurones, e.g. of the homonymous monosynaptic Ia input (W. Koehler and Windhorst 1985; see also Pompeiano 1984).

2) *Exclusion of "fringe" α-motoneurones from firing*: "... the motoneurons firing at high rates effectively suppress those discharging at low rates while the inhibition of the latter on the former is of much smaller order" (Granit 1970, p. 155).

3) *De-correlation or de-synchronisation of α-motoneurone discharges* (Gelfand et al. 1963; Adam et al. 1978; Buahin and Rymer 1984). This idea has been explained in more detail in Sect. 3.4.1. Briefly, if the recurrent network interconnecting α-motoneurones is sufficiently random both in its detailed connectivity and its firing characteristics, it operates as a random signal generator injecting uncorrelated noise into the different motoneurones and thus de-correlating them. The prerequisite is that any larger group of Renshaw cells, which commonly project to a particular set of motoneurones, do not exhibit strong synchronised discharge, and this in turn requires that

any such group should not receive strong common input from the above (or another) set of motoneurones. That is, a substantial degree of irregularity or randomness should prevail within the convergence–divergence structure of the recurrent inhibitory network. From results obtained with a "lumped" model of recurrent inhibition, Stein and Oğuztöreli (1984) concluded that the loop via Renshaw cells reduces tendencies towards oscillations in the motor output.

B) The *spatial aspect* includes the *localisation of motor output* or the sharpening of motor contrast (Brooks and Wilson 1958, 1959; Windhorst 1979a; see Sect. 4.3.1). This view is based upon the idea of lateral inhibition serving as a contrast-enhancing mechanism.

4.2.1.3 More Complicated Models

Even more realistic models incorporate the recurrent inhibition of cells other than α-motoneurones and take account of inputs to Renshaw cells from sources other than α-motoneurones. In addition to α-motoneurones, Renshaw cells inhibit other spinal neurones and receive excitatory and inhibitory inputs from various supraspinal and segmental sources other than recurrent motor axon collaterals. Some recent models are:

1) Hultborn et al. (1979) proposed that recurrent inhibition serves as a *"variable gain regulator"* at the level of the motor "output stage" which is composed of α-motoneurones, γ-motoneurones and "reciprocal" Ia inhibitory interneurones mediating reciprocal inhibition. This idea may be refined when the spatial distribution of recurrent inhibition is considered (see Sect. 4.3.1). The above "motor output" cells linked by recurrent inhibition may again be interpreted as modules, but of higher complexity than those consisting only of self-inhibited α-motoneurones (see above).

2) It has been suggested that Renshaw cells play a *role in locomotion*. For example, Miller and Scott (1977) developed an intricate model including α-motoneurones, reciprocal Ia inhibitory interneurones with mutual inhibition and Renshaw cells with mutual inhibition (see Fig. 21) in order to show that with favourable parameter constellations, this model would be able to generate locomotor rhythms; this it did, not surprisingly. One main problem with this model is that α-motoneurones are not considered to belong to the locomotor generator proper (Grillner and Wallén 1985). Another suggestion was recently put forward by Pratt and Jordan (1987) who studied the discharges of α-motoneurones, reciprocal Ia inhibitory interneurones and Renshaw cells during fictive locomotion in decerebrate cats. They concluded that the latter two types of interneurones contribute to, but are not indispensable for the shaping of locomotor α-motoneurone rhythmicity. These interneurones, therefore, do not appear to belong to the spinal locomotor generator

(also see McCrea et al. 1980). However, it should be noted that generally the activity of neurones in these reduced preparations, particularly under such unnatural conditions as prevail in fictive locomotion, must be considered with caution (see Sect. 4.11).

A malevolent reader may be tempted to recommend that this list be included in the famous *Encyclopaedia of Ignorance* (Duncan and Weston-Smith 1977) because its sheer length and diversity impressively illustrate the problem we have in understanding neuronal networks at an intermediate level of complexity. However, on the other hand, it may well be that a network such as that containing Renshaw cells subserves more than one function (see Abeles' "multifunction" of synfire chains; Sect. 1.3.1).

4.2.2 Functional Models of the Stretch Reflex

The so-called "stretch reflex" (the term is a misnomer but will be employed in the following to comply with conventional usage) is an even more complex system than recurrent inhibition because the latter might be regarded as being nested within the former. In discussions of stretch reflex function, the structural implications, or the neuronal circuitry underlying the stretch reflex, are not always clearly specified. In other words, it is not necessarily spelled out, which subsystems are discussed to be involved. For simplicity, let us assume that it encompasses α-motoneurones, the muscle fibres they innervate and the feedback via muscle spindle and Golgi tendon organ afferents, thus excluding recurrent Renshaw cell feedback (other reviews: Rack 1981a; Houk and Rymer 1981). For comparison, the following brief overview is tentatively structured in about the same way as the corresponding Sect. 4.2.1 on recurrent inhibition, although this order appears sometimes somewhat strained and violates chronology.

4.2.2.1 Localised Feedback to α-Motoneurones

In parallel to recurrent inhibition via Renshaw cells (Sect. 4.2.1.1), the simplest scheme of the stretch reflex is based on the circuit consisting of α-motoneurones, the skeletal muscle fibres innervated by them and muscle spindles with their Ia afferents making monosynaptic connexions back to the homonymous α-motoneurones (the classical monosynaptic reflex loop).

Based on early work of Liddell and Sherrington (1924) and a suggestion of his teacher Fulton, L. A. Cohen (1953, 1954) presented evidence, in decerebrate cats, for the existence of "localised stretch reflexes". This idea has recently gained renewed interest and was formulated by Windhorst (1978b) (see also Windhorst et al. 1976) and Binder and Stuart (1980b). It is based on the *local* sensitivity of muscle spindle and Golgi tendon organ afferents to contractions of single motor units in conjunction with spatial patterns of Ia afferent connectivity (Windhorst 1978b, 1984; Binder and Stuart 1980b; see Sects. 3.3.1 and 3.3.2, and below). Binder and Stuart (1980b) suggest that the concept of the "final common path" (Sherrington 1904) should be expanded into one of a *fundamental functional unit*, which is composed peripherally of a muscle unit (the muscle fibres belonging to a motor unit) and the set of muscle receptors (spindles and tendon organs) it can influence, and centrally of the afferents of those particular receptors and the α-motoneurone innervating that muscle unit. Binder and Stuart (1980b) propose that, in this organisational scheme, the information carried by muscle afferents to α-motoneurones has a dual role. Each afferent provides somewhat "private" information to those α-motoneurones innervating muscle units whose contractions affect the afferent's discharge pattern, and each afferent simultaneously contributes to the "ensemble" afferent inflow. With regard to the large variety of movements in which individual muscles can participate, the significance of "discrete afferent lines" is probably variable. "Stereo-typed, repetitive movements, such as those associated with respiration and gaited locomotion in which the behavior of individual muscles is subordinated to the movement, are likely to be modulated largely by the ensemble signal (. . .), while discrete, localized afferent input is presumably more efficacious in the fine control of vernier movement, which require that the select muscles involved develop and sustain only a fraction of their maximum force output" (Binder and Stuart 1980b; p. 73).

This idea essentially defines a basic "module" ("fundamental functional unit"), in correspondence with local recurrent "self-inhibitory" feedback circuits (Sect. 4.2.1.1). The functions of this modular organisation are not yet clear. But they might be essentially the same as for recurrent inhibition:

1) *Limitation of α-motoneurone firing rates.* Recently Broman et al. (1985) showed that, upon recruitment of a motor unit, previously recruited motor units slowed down their firing rates (in human voluntary muscle contraction). A reason discussed by the authors in addition to recurrent inhibition (see above) could be some inhibitory proprioceptive feedback from the motor unit in question onto homonymous α-motoneurones. The authors suggested that this inhibitory mechanism could account for the relatively poor firing rate modulation of motor units in some human muscles and that this in turn could promote smooth muscle contraction.

2) *Reduction of the effects of disturbing inputs.* Disturbances to the system could originate internally or externally. For example, with regard to "small internal disturbances" sensed by Ia afferents, Binder and Stuart (1980b) suggest that the activation of individual motor units within a muscle represents a small, internal disturbance to which individual muscle receptors are exquisitely sensitive.

Similar proposals, though without explicit reference to localised feedback circuits, were made by Vallbo (1981) and Evarts (1981). They are based on the nonlinearity of primary spindle endings which are highly sensitive to small-amplitude length inputs and much less so to large amplitudes ("gain-compression nonlinearity"). The matter is complicated by the partial linearisation of spindle properties by fusimotor action, small movement sensitivity being reduced by both static (and to a lesser degree) dynamic γ-innervation, the extent of the gain-compression nonlinearity being diminished; Hulliger et al. 1977; for further references see Hulliger 1984). This fact led to the suggestion that the spindle feedback via Ia afferents could be used to compensate for small irregularities in tonic or slow-movement muscle contractions (Vallbo 1981). In the same vein, Evarts (1981), commenting on "Sherrington's concept of proprioception", argued that compensation for *large external disturbances* cannot be achieved by reflex feedback, but modulation of tonic α-motoneurone discharge in response to *small internal disturbances* can be. Such small internal disturbances arise in the muscle and/or the nervous system (for instance, by stochastic population properties of motor unit discharge; see Sect. 3.2) and create errors of movement even when external load changes are absent. The high dynamic sensitivity of muscle spindle Ia afferents for small length changes would allow the segmental reflex system to effectively compensate for such small disturbances. If signals from spindle afferents indeed are important in minimising the consequences of small internal disturbances, then postural instabilities should increase when inputs from spindle afferents are abolished. Evarts (1981) then refers to the work of Goodwin et al. (1978) who demonstrated such an effect in the control of voluntary jaw movements in the monkey. This study showed that surgical interruption of the reflex arc led to a considerable increase in the amplitude of spontaneous tremor during steady contraction. It may thus be suggested that muscle spindle afferents play an important role in "reducing errors of muscle length produced by fluctuating levels of motor discharge as well as those by external loads" (Goodwin et al. 1978; p. 83). Although it is not entirely clear what is meant by these internal fluctuations and how they come about, this hypothesis is reminiscent of similar proposals made for recurrent inhibition (Sects. 4.2.1.1. and 4.2.1.2) and of the role attributed to localised proprioceptive feedback in tremor suppression (Sect. 3.3).

3) *Speeding up of α-motoneurone response* to time-varying inputs. Andreassen and Rosenfalck (1980) pointed out a correspondence between summation of α-motoneurone afterhyperpolarisations, recurrent feedback and proprioceptive feedback via muscle stretch receptor afferents in regard to their potency of enhancing (single) α-motoneurone dynamic responsiveness. A similar correspondence, with the addition of presynaptic inhibition, was recently suggested by Calancie and Bawa (1985). More generally, consideration of dynamic properties in subsystems of the stretch reflex (Poppele and

Terzuolo 1968; Rosenthal et al. 1970; see also review of Stein 1974) would suggest that the high-pass properties of Ia afferents serve to compensate for the low-pass properties of skeletal muscle so that the frequency response of the overall system preserves a relatively constant gain over a broader frequency range. This would speed up muscle response to fast input transients (see Carli et al. 1967). In this sense, recurrent inhibition and proprioceptive feedback could subserve analogous functions, provided the respective feedback pathways are relatively private ("localised"; Sect. 4.3.1).

4.2.2.2 Convergence–Divergence Structure of Reflex Connectivity

A tendency towards "localised reflex" organisation should not distract from the possible functions of the wide divergence of connexions, in the muscle as well as in the spinal cord. That is, the local modules are interconnected, in the muscle by the fact that each motor unit affects the discharge of a smaller or larger set of spindles, and each spindle in turn is influenced by a set of motor units (see Fig. 12). In the cord each Ia afferent connects to virtually all homonymous α-motoneurones, and each α-motoneurone in turn receives input from nearly all homonymous Ia afferents. This organisational scheme is similar to that of recurrent inhibition (see Sect. 4.3.1). Hypotheses which explicitly take account of this organisation are rare for proprioceptive feedback. Nevertheless, the following aspects, which have been emphasised in such considerations, merit attention.

1) *Inhomogeneous distribution of Ia input to different α-motoneurone types.* This issue was briefly alluded to in Sect. 4.2.1.2 a1 and is dealt with again in Sect. 4.3.2.

2) *Improvement of signal transmission.* The multichannel, convergence–divergence structure of monosynaptic Ia connexions to α-motoneurones was shown, in theoretical and computer simulation studies, to enhance the fidelity of signal transmission from skeletal muscle to the α-motoneurones (e.g. Milgram and Inbar 1976; Feenstra et al. 1985).

3) *Partial decoupling of localised reflex loops.* An explicit explanation of the spinal "cross-connexions" between localised modules was attempted by Windhorst (1979 b) who argued that such cross-connexions might help decouple, to a certain extent, parallel reflex loops which are coupled mechanically in the muscle. He suggested that such structures could be relevant for tremor suppression, linearisation of neuromuscular properties, and minimisation of energy expenditure in muscle contraction. An important idea was to demonstrate that *localised stretch reflex units (modules) should not be independent, distinct entities.* For the partial decoupling to work in muscles with in-parallel structure (see Sect. 3.3.1), autogenetic Ia inhibition mediated by nonreciprocal Ia inhibitory interneurones (Sect. 2.9.1.1; Fig. 21 D) was

suggested to play a role (Windhorst 1979 a, b). However, this scheme was unnecessarily complicated for a number of reasons. A simpler and new scheme is presented in Sect. 5.1.2.

4.2.2.3 Other Models

Many models of stretch reflex function regard the efferent and afferent pathways as "lumped systems" without paying much attention to spatial differentiation within each limb. Several of these schemes are dealt with later in a different context (Sect. 4.7.2).

Thus, as with recurrent inhibition, the many functions assigned to the stretch reflex testify to the prevailing uncertainty (which is also well reflected in a recent article by Stein 1982, and commentaries on it by many specialists) and, more generally, to our problems in understanding complex systems.

4.2.3 Summary

There are two feedback systems onto α-motoneurones and other spinal neurones, recurrent inhibition via Renshaw cells and proprioceptive feedback via skeletal muscle and stretch receptor afferents, that have been extensively investigated but usually independently of each other. Based on the data obtained on each individual system, long lists of functions have been assigned to each, testifying to our difficulty in understanding even seemingly simple systems which are nested in others. Some functional proposals (and even properties) for the two systems are similar, suggesting that recurrent inhibition and proprioceptive feedback might well be considered together not least in order to gain new insights from an integrative approach.

4.3 Structural and Functional Analogies Between Spinal Recurrent Inhibition and Proprioceptive Feedback

Experimental and theoretical work on spinal recurrent inhibition via Renshaw cells and proprioceptive feedback via skeletal muscle and muscle receptors has mostly been concerned with one of the two systems in isolation, and only rarely have the two systems been treated together. One notable exception is the paper by Granit et al. (1957) who put forward the hypothesis that "the recurrent collaterals, in our view, are the natural efferent antagonists to the γ-driven tonic system" (p. 397), i.e. recurrent inhibition serves to counteract spindle excitation by the γ-fusimotor system. Haase et al. (1975) reviewed autogenetic inhibition via Golgi tendon organ afferents and recurrent inhibition via Renshaw cells in the same paper, but they have not integrated these two feedback systems into a common control scheme. However, such an integrated approach is not only worthwhile; one day or the other it will be indispensable if motor control is to be understood. In the following, a number of analogies between these systems are pointed out, sometimes in a somewhat forced way in order to paint a background against which differences may then be stressed and interpreted. Indeed these differences might give clues as to the function of the two systems. In many respects the following remains hypothetical due to the lack of adequate data.

Signals in the proprioceptive feedback pathway do not only reflect skeleto- and fusimotor activity, but also external inputs such as muscle length and its changes. However, in order not to complicate the comparison from the very outset, emphasis will be focussed on signal transmission around the stretch reflex loop with the external inputs constant (isometric conditions). Extensions to more general cases are discussed when appropriate. Yet, in principle the focus chosen reflects my conviction that what is signalled by proprioceptive feedback is one aspect of the performance and properties of peripheral motor devices, with external inputs providing modulatory signals.

4.3.1 Spatial Organisation of Recurrent and Proprioceptive Feedback

Each of the two feedback systems consists of two parts, the efferent path (α-motoneurones) common to both, and the afferent path, which consists of Renshaw cells in one case, and of muscle units and muscle afferents in the other. For certain muscles, at least, these paths are organised in a spatial pattern. This will be illustrated here with an emphasis on recurrent inhibition because principles of spatial organisation within the proprioceptive feedback system were extensively discussed above (Sects. 3.3.1 and 3.3.2).

1) Recurrent Inhibition
Three examples are discussed here.

a) Hindlimb muscles. Cullheim and Kellerth (1978a, b), by injecting horseradish peroxidase, have provided morphological data on the distribution of lumbosacral α-motoneurone axon collaterals in cats. (For the early postnatal development of these projections see Cullheim and Ulfhake 1985.) The "synaptic swellings" of the collaterals within the Renshaw cell area (ventromedial to the motoneurone area) are confined to a region, not exceeding 1 mm in rostro-caudal extent. The radial Renshaw cell dendrites extend up to several hundred μm (up to 1 mm according to van Keulen 1971; but see Jankowska and Smith 1973, and Lagerbäck and Kellerth 1985b), so that one motoneurone excites Renshaw cells within an area of limited rostro-caudal extent (1–2 mm). Thus the efferent path (from motoneurones to Ren-

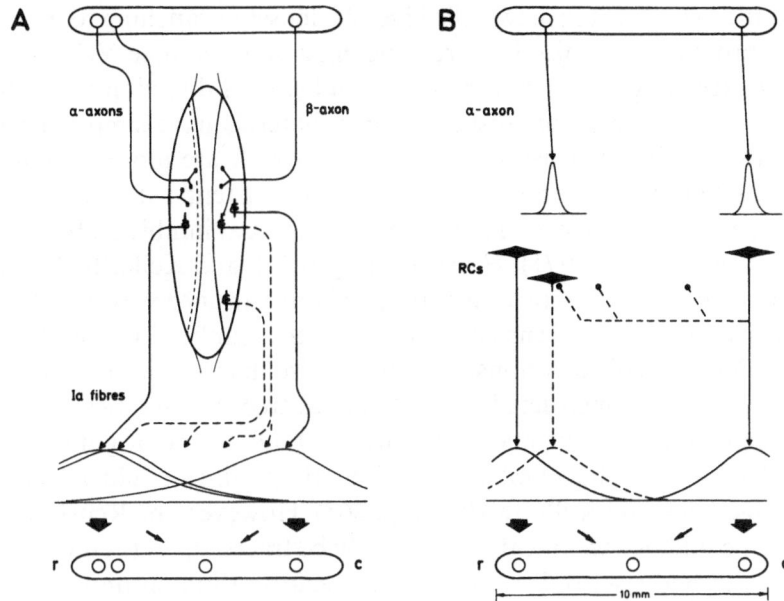

Fig. 38 A, B. Schemes contrasting the localised Ia feedback (**A**) and recurrent Renshaw cell feedback (**B**). **A** is reproduced from Fig. 35 for comparison; **B** The motoneurone pool, assumed to be of 10 mm rostro-caudal extent, is again drawn twice (at *top* and *bottom*). An α-motoneurone (upper pool) sends its recurrent axon collaterals and terminals with a slight caudal displacement to the Renshaw cells, the spatial (rostro-caudal) distribution of the synaptic boutons being fairly accurately represented by the *bell shape in the middle*. The Renshaw cells which can be contacted by the terminals, if only through their dendrites, are clustered within some mm. A Renshaw cell distributes its axon terminals to the α-motoneurones in a weighted fashion as indicated by the *broader bell-shaped curve*. In contrast, mutual inhibition among Renshaw cells and the resulting disinhibition of α-motoneurones have a more extended distribution, as indicated by *dashed lines*

shaw cells) is fairly well localised (see Fig. 38 B) in relation to the average lengths of lumbosacral motoneurone pools (6–12 mm, Romanes 1951).

The feedback path (from Renshaw cells to motoneurones) is much less localised. The output of Renshaw cells is distributed over considerable rostro-caudal distances (up to 12 mm: Jankowska and Smith 1973) since their axons partially travel in the ventromedial funiculus (Jankowska and Smith 1973; Ryall et al. 1971; van Keulen 1979; Lagerbäck and Kellerth 1985a). This distribution seems to depend on two major factors: *proximity of coupled motoneurones* (Eccles et al. 1961; Thomas and Wilson 1967) and on the *functional relation between these motoneurones* (Kuno 1959; Wilson et al. 1960; Thomas and Wilson 1967). Regarding the spatial factor, available electrophysiological evidence suggests that axonal branches of Renshaw cells projecting to motoneurone areas (Rexed's lamina IX) appear to extend up to ca. 4 mm longitudinally, with a frequency distribution declining with distance (Jankowska and Smith 1973). This agrees with anatomical findings of van

Keulen (1979). Renshaw cells give off densely branching axon collaterals to a "proximate" projection area encompassing laminae VII through IX and not exceeding ± 0.7 mm in rostral and caudal direction from the Renshaw cell somata. The more distant axon collaterals have simpler branching patterns with lower densities of terminals, which are more confined to medial areas (laminae VII and VIII). This is consistent with the notion that the longer-ranging axon collaterals mediate mutual inhibition between Renshaw cells (Ryall et al. 1971). However, Lagerbäck and Kellerth (1985a) provided some anatomical data which are partly at variance with van Keulen's findings. Although the Renshaw cell illustrated (number 5 in their Fig. 6) fits with van Keulen's observations, they emphasise that "... in the whole material of completely reconstructed axon collateral trees no significant correlation between the axonal distance from the cell body to the origin of the first-order collateral and the number of collateral swellings could be demonstrated" (Lagerbäck und Kellerth 1985a; p. 365). However, the Renshaw axon collaterals could, for technical reasons, only be traced up to (at most) 2.7 mm from the cell body. Also, it is known that recurrent inhibition is most prominent between homonymous and synergistic motoneurones so that if synergistic motoneurones are somewhat separated, Renshaw cell axons may span the distance (see Baldissera et al. 1981). Functional factors (see above) may therefore co-determine the distribution of Renshaw cell axon collaterals and synapses.

For the interconnexion between homonymous motoneurones, a topographically weighted distribution of Renshaw cell axons may still be assumed as illustrated by the idealised distribution in Fig. 38 B (solid lines). This is also supported by electrophysiological data. Hamm et al. (1987b) provided evidence that the recurrent action of single motor axon activation on single homonymous (medial gastrocnemius) α-motoneurones is localised. This evidence suggests that the amplitude of single-axon RIPSPs as well as their frequency of occurrence may be dependent upon topographic factors. Similarly, there are some unpublished results from Van Keulen's electrophysiological work in Kernell's laboratory which provided evidence that a single Renshaw cell affected a motoneurone only if the two cells were close to each other in rostro-caudal direction (unitary RIPSPs only seen at distance of less than 0.35 mm; Kernell, pers. comm.). However, in part these short distances may be due to the limited resolution of the spike-triggered averaging method employed. The Renshaw cell synaptic contacts on motoneurones are probably located near the soma (T. G. Smith et al. 1967; Burke et al. 1971), so that the distribution of the inhibitory influence from one motoneurone to another is determined by the size of the Renshaw axon collateral tree alone, whilst the size of the receiving motoneurone's dendritic tree would appear to be less important. Thus the distribution of inhibitory effects from a given motoneurone to rostral or caudal (homonymous) motoneurones would not exceed

5–6 mm, decreasing rather rapidly in strength within this range (see also Thomas and Wilson 1967; Windhorst 1979a). For hindlimb motoneurone pools, the probable situation is represented schematically in Fig. 38B, and is contrasted with the hypothesis of "localised" Ia feedback in a simple muscle in Fig. 38A (reproduced from Fig. 35).

Superimposed upon this localised recurrent inhibitory system (solid lines in Fig. 38B) is a disinhibitory one mediated by "mutual inhibition" of Renshaw cells (dashed lines in Fig. 38B). This system may be less, or not at all, localised. Mutual inhibition between Renshaw cells is almost certainly strongest between Renshaw cells associated with antagonist motoneurone pools (Ryall 1981). But probably there is also some mutual inhibition between Renshaw cells which receive their main excitatory input from axon collaterals of the same motoneurone pool and synergists. This disinhibitory connexion would have a much larger spatial extent than the recurrent inhibitory one because Renshaw cells are funicular neurones whose axons can project over long distances. Thus mutual inhibition has a particularly long reach (Ryall et al. 1971). This system would correspond to the excitatory spindle Ia feedback on the right side of Fig. 38A (dashed lines).

Except for disinhibitory influences between homonymous (and synergistic) α-motoneurones, there also appears to be a direct excitatory one. As shown by Cullheim et al. (1977), Lagerbäck et al. (1981a), and Cullheim et al. (1984), motoneurones send recurrent collaterals not only to the Renshaw cell area (lamina VII ventromedial to the motoneurone area), but also to the motoneurone area (lamina IX) in the spinal cord where they make monosynaptic (most probably excitatory) contacts with motoneurones. The rostro-caudal distribution of this projection is usually limited to less than 0.5 mm (Cullheim and Kellerth 1978b). The collateral terminals have been found localised in the motoneurone dendritic tree but only up to 0.7 mm or so away from the soma and are established with homonymous and synergistic motoneurones of various types, S to FF (Cullheim et al. 1984). Therefore, this excitatory cross-connexion between motoneurones is very localised. Khatib et al. (1986) adduced evidence for its existence and physiological effectiveness in phrenic motoneurones. However, what matters for functional considerations is that both disinhibitory and direct excitatory connexions are probably weak.

b) Respiratory muscles. Kirkwood et al. (1981) demonstrated the existence of recurrent inhibition in the thoracic respiratory (intercostal) motoneurone system (see also Hilaire et al. 1983). The spatial distribution of this inhibition via Renshaw cells has a similar form to, but is more extended than that in the lumbosacral cord. That is, recurrent inhibitory effects exerted by motor axons from one segment reach motoneurones up to three segments apart (ca. 20–30 mm; one thoracic segment is approximately 10 mm long) in rostral and caudal direction, but their strength declines steeply with distance.

The reasons for this decline are very probably the same as those given above for the lumbosacral cord (see Kirkwood et al. 1981).

c) Cervical muscles. Thomas and Wilson (1967) investigated the spatial distribution of recurrent inhibition between cat motoneurone pools in the lower cervical cord innervating forelimb muscles. The spread of inhibitory effects exerted by a given motoneurone pool on motoneurones of various species seldom extends more than half a segment farther rostrally or caudally from the source of inhibition. This is attributable to the principle of proximity, but functional factors act as modulators. More refined features as discussed above (item a) were not studied. A similar distribution also holds for the upper cervical cord where the splenius and biventer cervicis pools are situated. The structural prerequisites for recurrent inhibition among dorsal neck α-motoneurones, i.e. recurrent axon collaterals and Renshaw cells, exist (Keirstead et al. 1982; Brink and Suzuki 1987). As with lumbar Renshaw cells, upper cervical Renshaw cells receive input from motor axon collaterals that is spatially restricted. On the whole, however, recurrent inhibitory coupling between α-motoneurones spans a wider range than the excitatory input to Renshaw cells from motor axon collaterals. Recurrent inhibition originating from one segment decreases with segmental separation (Brink and Suzuki 1987), as is the case for the intercostal system (see above). But again functional factors are of importance, too (Brink and Suzuki 1987).

These patterns are merely topographical, i.e. related to rostro-caudal alignment of α-motoneurones connected via Renshaw cells. Superimposed upon them are almost certainly other patterns related to functional aspects of the motoneurones (see above) or perhaps spatial patterns of finer grain. What is not known so far is whether transverse coordinates within the motoneurone pool play a role in topographical connexions, and whether and to what extent species-specific connectivity occurs in the way demonstrated for monosynaptic Ia-motoneurone connexions (Sects. 2.8.2.3, 3.3.1 and 3.3.2). If they exist, such patterns would constitute hitherto unrevealed subtle features of organisation which might modulate the topographic pattern discussed above.

2) Proprioceptive Feedback

Some cases of localised proprioceptive feedback (particularly via Ia fibres) exist and have been extensively described in Sects. 2.8.2.3, 3.3.1, and 3.3.2. They are based upon two principles: topographical and species-specific connectivity. The studies on this kind of reflex organisation should be extended to more muscles in order to establish the generality of these principles. It would also be worthwhile, firstly, to relate the "local" stretch reflex units to possibly different functions of the muscle compartments, and, secondly, to investigate whether such local units combine smoothly across the borders of individual muscles, thus generating a spatially and functionally continuous mapping across synergists.

   The present context lends itself to a brief discussion of the module concept because the spatial principles of organisation present in the recurrent and proprioceptive feedback systems might be interpreted as establishing local, interconnected modules (see Sects. 4.2.1.1, 4.2.1.2, and 4.2.2). Unfortunately, the concept of modular organisation is fraught with diffuse notions. This point was clearly made by Creutzfeldt (1983) regarding the cortex. He writes that an essential shortcoming in the present model concepts of the "modular" or "columnar" organisation of the cortex is that two different and independent principles of organisation of the cortex are intermingled: (1) The topographical principle of the afferents which show discontinuities due to the coherence of fibre bundles of equal origin and which form the basis of "ocularity", "aurality", and possibly even of "orientation" columns (or better, stripes). (2) The modular cortical building blocks which are formed by the intracortical connexions and interactions. These intracortical modules cross not only the discontinuities between coherent sensory representations which are based upon afferent modules, but also borders between different representation fields, like those between areas 17 and 18, etc. Therefore, if we wish to speak of a modular or columnar organisation of the cortex, we must at least distinguish between these two mutually and independently overlapping modular principles of organisation which are based on the afferent and the intracortical connexions. Probably we will have to distinguish additional modular building blocks which might possess different geometric and functional parameters and superimpose independently and with certain local differences in different areas: Candidates are the modular systems of association fibres, of thalamo-cortical afferents, intracortical lateral excitation, intracortical lateral inhibition, of commissural fibres, etc. (see Creutzfeldt 1983; p. 414). These considerations also apply, mutatis mutandis, to the spinal cord. Therefore, it seems justified to speak of motoneurone-Renshaw cell modules or local stretch reflex modules as long as it is not forgotten that these are not isolated distinct elements. Furthermore, yet another type of module could well be defined, but now on functional rather than anatomical grounds. Whereas in the preceding anatomical definition a module appears as a rather static entity, a functional definition would emphasise dynamic aspects of cell assemblies and could be based upon neuronal discharge patterns and ensuing processes of synaptic modifiability (Chap. 1).

4.3.2 Feedback of Different Functional Types of α-Motoneurones

1) Recurrent Inhibition

It is now fairly well established that the input of Renshaw cells from α-motoneurones and their output to α-motoneurones is weighted according to motoneurone type. For a long time, the evidence had been indirect (Ryall et al. 1972; Hellweg et al. 1974; Pompeiano et al. 1975 a, b; see also Pompeiano 1984). Recently, two groups of workers provided more direct data.

a) Efferent path. Cullheim and Kellerth (1978 b) investigated the anatomy of recurrent collaterals given off by type-identified cat triceps surae motoneurones which were classified according to Burke (review: Burke 1981). On average, FF-type motoneurones produce more recurrent collaterals and "synaptic swellings" (92.9; this figure does not include the synaptic swellings found in the motoneurone area of the spinal cord) than do FR (45.3) and S motoneurones (32.2). This may be taken to indicate that the input to an average Renshaw cell (or the Renshaw cell pool) is largest from FF motoneurones and smallest from S motoneurones. This situation is symbolised by the different numbers of open circles representing synapses at the ends of the recurrent collaterals in Fig. 39.

b) Afferent path. Friedman et al. (1981) provided evidence that S-type motoneurones of the medial gastrocnemius motoneurone pool are more strongly inhibited by Renshaw cells than FR- and FF-type motoneurones. Recurrent IPSPs were generally larger in S-type than in FR-type, and larger in FR-type than in FF-type motoneurones. "Analysis of covariance confirmed that a significant relationship remained after adjusting for any effect of input resistance" (Friedman et al. 1981, p. 1349). This situation is symbolised by the differing numbers of inhibitory Renshaw cell synapses (filled black circles) on the various motoneurones in Fig. 39.

It thus appears that FF-type motoneurones exert the strongest recurrent inhibitory effect on S-type motoneurones and receive the weakest effect from the latter, FR-type motoneurones being intermediate (see Windhorst and Koehler 1983; for recurrent inhibition among neck α-motoneurones see Brink and Suzuki 1987).

2) Proprioceptive Feedback

The functional organisation (with respect to different motoneurone types) of the feedback path through skeletal muscle and Ia fibres is analogous to that of recurrent inhibition by Renshaw cells.

a) Efferent path. Signal transmission from skeletomotor fibres to Ia and spindle group II afferents is determined by several factors: (1) the number of muscle fibres per motor unit; (2) the size of the motor unit territory; (3) the force production per motor unit; (4) the distribution of muscle fibre and motor unit types within a muscle; (5) the distribution of muscle spindles within a muscle. Apparently, these factors add up in such a way that, on average, contraction of an FF motor unit influences the Ia and spindle group II afferent discharge from a muscle more strongly than an S-type motor unit does. This was demonstrated for the *statistical effects* of cat tibial posterior motor units activated at relatively low rates (Binder and Stuart 1980a; see Sect. 3.2.3). Munson et al. (1984) obtained similar results on cat medial gastrocnemius motor units activated at higher rates. These authors stimulat-

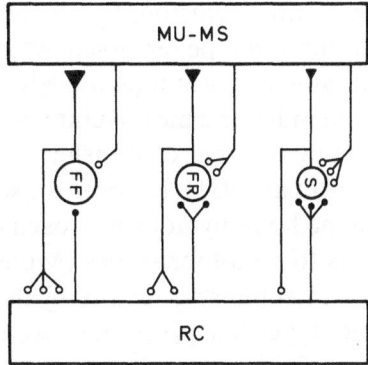

Fig. 39. Scheme of the inhomogeneous distribution of spinal recurrent and Ia feedback to different types of α-motoneurones. The *left block* labelled *MU-MS* represents the neuromechanical interface of muscle units and muscle spindles with motor efferents as inputs and spindle Ia afferents as outputs. The differently sized *filled triangles* at the efferent terminals symbolise the different strengths of signal transmission from skeletomotor axons to Ia afferents. The different numbers of synaptic terminals of the Ia fibres on the various motoneurone types represent the differential effects measured physiologically (see text). The *right block* labelled *RC* represents the Renshaw cell network, which receives more excitatory input from FF-type motoneurones than from S-type motoneurone, symbolised by the different numbers of (*open*) *circles*, and gives more inhibition to S-type than to FF-type motoneurones, as symbolised by the *filled circles* (for more details see text). Note the similarity of structure in the two feedback systems. The sign inversion in the peripheral proprioceptive system takes place in the muscle, that in the recurrent system at the motoneurones, by the inhibitory Renshaw cell synapses

ed type-identified motor units at tetanic rates of 40 pps (under isometric conditions), and quantified the effect on spindle afferents by expressing the change in *mean* afferent discharge rate as a ratio to the control rate (computed over 250 ms periods: "unloading index"). Again, the same trends were apparent: S-type motor units left more spindle afferents unaffected than FF units did, and the latter affected more afferents more strongly than the former. Moreover, Ia fibres were more often and more strongly affected than spindle group II afferents. Thus, both the *mean rate* and the *discharge modulation* around the mean of spindle afferents are more strongly influenced by FF- than by S-type motor units. This is symbolised by black triangles of different areas in Fig. 39.

 b) Afferent path. It is now well established that Ia fibres exert homonymous monosynaptic excitatory effects that are stronger on S-type than on FR and FF motoneurones (Fleshman et al. 1981; Harrison and Taylor 1981). Fleshman et al. (1981) showed that single-fibre EPSPs are larger in S-type than in FR-type, and larger in FR- than in FF-type motoneurones. Although EPSP size is correlated with input resistance across motoneurone types, input resistance does not appear to be an important factor for determining EPSP amplitude (see also Munson et al. 1982). An analysis of covariance revealed that, after adjusting for the effect of input resistance, a significant effect of motor unit type remained. Moreover, Ia fibres project to a higher proportion of fatigue-resistant (S and FR; 96%) than to FF (87%) motoneurones. These data indicate a motor-unit specific organisation of homonymous Ia input, in analogy to that of Renshaw cells to motoneurones of different type (Friedman et al. 1981). In contrast, such a motorunit specific organisation is absent in the homonymous monosynaptic connectivity from spindle group II afferents to α-motoneurones (Munson et al. 1982).

In summary, regarding the functional differentiation of motoneurones into types, the proprioceptive muscle-Ia fibre feedback system appears to be organised in a fashion comparable to that of the recurrent Renshaw cell feedback, as shown schematically in Fig. 39.

4.3.3 Static Input-Output Relations in the Pathways from Motor Axons to Renshaw Cells or Spindle Afferents

1) Recurrent Inhibition

Static and dynamic properties of the recurrent pathways via Renshaw cells have been intensively investigated by Cleveland and co-workers, and by ourselves.

If a group of motoneurone axons sending collaterals to a Renshaw cell is stimulated electrically (antidromically) at various fixed rates, the static discharge frequency f_{RC} of the Renshaw cell varies nonlinearly with input frequency f_{AD} (frequency variable), as illustrated in Fig. 40 (modified from Cleveland et al. 1981; see also Cleveland and Ross 1977). This relation can be described by a saturating function $y = cx/(k+x)$, where y and x are substituted for f_{RC} and f_{AD}, respectively, and c represents the saturation constant, i.e. the maximum Renshaw cell frequency attainable, and k the "semi-saturation" constant, i.e. the input frequency at which half the Renshaw cell saturation frequency is attained (see inset of Fig. 40). This static input-output relation thus has a slope declining with increasing input frequency. Whereas this decline of static Renshaw cell gain (being equivalent to the above slope) was well documented by the above authors, the same was not done for the second input variable, viz. recruitment of motoneurones. However, these two input variables (frequency and recruitment) seem to be equivalent, at least with respect to the dynamic Renshaw cell behaviour (Ross et al. 1982). This equivalence may also apply to the static nonlinearity, as suggested by a Renshaw cell membrane model put forward by Cleveland et al. (1981) to explain this nonlinearity. The static Renshaw cell nonlinearity is probably transmitted up to the motoneurone level (Cleveland et al. 1981).

Another important property becomes apparent when motor axons are stimulated not at various constant rates, but with a sinusoidally varying rate.

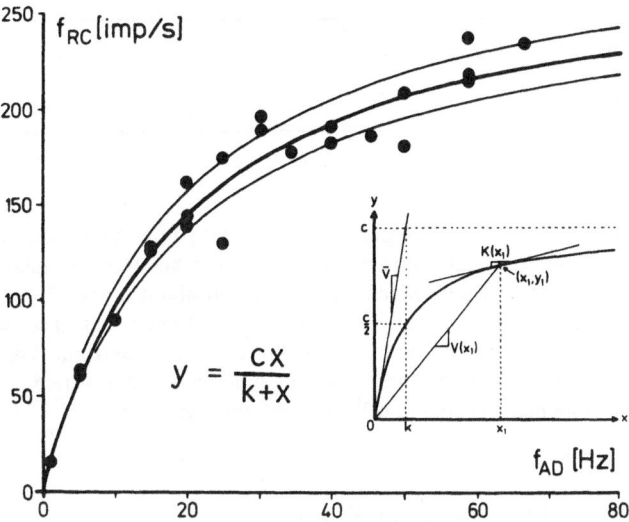

Fig. 40. Static input-output characteristic of a Renshaw cell. The adapted discharge rate f_{RC} (*ordinate*) is plotted against the antidromic stimulus rate f_{AD} (*abscissa*), the stimuli being applied to ventral root S_1 (350 mV). The *heavy line* is the rectangular function fitted to the points by the method of least squares. The *thin lines* indicate the range of standard deviation of the fitted function. The *inset* illustrates the definition of the parameters c and k. (With permission modified from Cleveland et al. 1981; their Fig. 1)

Figure 41 shows an example of a Renshaw cell (upper point display labelled RC) being activated by stimulating three different branches of the cat medial gastrocnemius muscle nerve (three lower point displays labelled S1 through S3) at rates varied sinusoidally between about 2 and 18 pps, the modulation frequency being 0.1 Hz. Figure 42 shows cycle histograms (left column; bin width 200 ms) and input-output relations (right column) for four Renshaw cells. In the left column, the thin curves represent the mean input rates (see legend) averaged over many cycles (associated ordinate at left). The thick curves are cycle-averaged output rates of the Renshaw cells (associated ordinates at right). In the right column, the time-related mean input (stimulus: abscissa) and output (Renshaw cell discharge: ordinate) rates from the corresponding left plots are plotted against each other, with the data from the rising phase of the sine input connected by a continuous line and those from the falling phase shown as individual points. The temporal sequence of successive points is indicated by arrows. The latter plots thus represent quasi-static input-output relations of mean rates. It appears that points and lines nearly fall on a common line indicating that, within this restricted range of input rates which closely resemble natural α-motoneurone firing rates, the input-output relations are fairly linear (see Cleveland and Ross 1977; their Fig. 3) and Renshaw cell firing shows no conspicuous hysteresis and only small phase advance at this low modulation frequency.

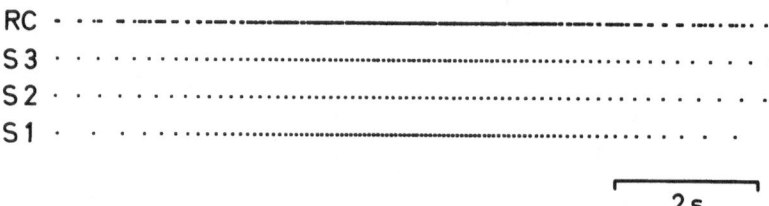

Fig. 41. Renshaw cell response to sinusoidally rate-modulated stimulation of three branches of the medial gastrocnemius nerve. These branches were stimulated with different stimulus sequences, the rates of which varied between about 2 and 18 pps at a low frequency of about 0.1 Hz. Stimulus strengths were submaximal such that each stimulus elicited one to a few discharges of the Renshaw cell. Dorsal roots L_6-S_1 were cut. The *lower three traces* show about one cycle (10 s) of the three different stimulus sequences used. The *upper trace* is the Renshaw cell response. (Rissing and Windhorst, unpublished)

2) Proprioceptive Feedback

Data on static gains of signal transmission from motor units to muscle spindles (or Golgi tendon organs) are available for the subsystems involved (motor units and muscle spindles or Golgi tendon organs), but are sparse for the overall system. Some data from our laboratory pertaining to this signal transmission *with the muscle held isometric* are presented here.

In the experiments underlying the following data, ventral root axons to motor units of the isometrically suspended cat medial gastrocnemius or soleus muscle were stimulated electrically at rates varied sinusoidally at a low frequency (ca. 0.1 Hz) between ca. 2 and 18 pulses per second (pps), as above for Renshaw cells. The responses of muscle afferents from the muscle under study were recorded in dorsal root filaments. An example of the data obtained is shown in Fig. 43, where the stimulus sequences are labelled S1 through S3, and spike trains from three spindle afferents are labelled MS A through MS C, and the muscle tension developed is shown at the top. Figure 44 shows cycle histograms (left column; bin width 200 ms) and input-output relations (right column) for four afferents from three experiments, in the same way as for Renshaw cells in Fig. 42. These specimens were selected to illustrate the range of patterns found. The input-output plots in the right column are quasi-static insofar as slow nonlinear dynamics enter in the form of hysteresis. The upper two rows show the behaviour of a medial gastrocnemius primary and secondary ending (A and B) recorded simultaneously in the experiment of Fig. 43. The mean firing rates of these two afferents were relatively unaffected until the input rate reached a certain threshold (roughly 10 pps), above which their firing rates first increased slightly (i.e. the afferents were excited) before decreasing moderately (i.e. the afferents were unloaded). Plots C show a (soleus) Ia fibre which reduced its rate approximately linearly with increasing input rate (continuous line in the right plot), whilst in the reverse direction (decreasing input rate) the same type of hysteresis became

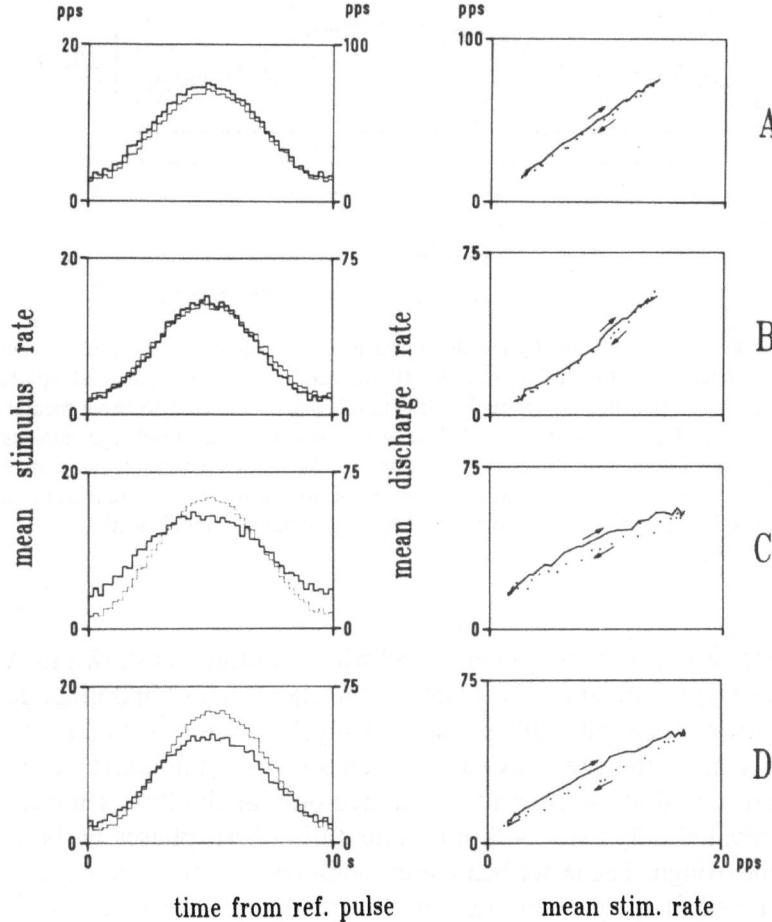

time from ref. pulse   mean stim. rate

Fig. 42 A–D. Four examples of input-output relations between sinusoidally rate-modulated stimulus sequences applied to motor axons and Renshaw cell discharges. *Left column*: cycle histograms computed for input stimuli (*thin curves*) and output discharge (*thick curves*); the *curves* represent the mean pulse densities averaged with respect to cycle triggers (between 49 and 65; bin width 200 ms). For the *upper two rows* (**A** and **B**) the *thin* (input) *curves* are also averages of the respective cycle histograms belonging to each of the two or three stimulus patterns used to stimulate different portions of motor axons. For example, for **A**, three branches of the medial gastrocnemius nerve were stimulated with different sinusoidally rate-modulated stimulus sequences (the rate was varied between about 2 and 18 pps at a low frequency of about 0.1 Hz in all cases; see Fig. 41). One cycle of this data is shown in the preceding figure. For **B**, the posterior biceps and anterior semitendinosus nerves were stimulated with two independent patterns of the same kind. In **C**, the ventral root S_1, and in **D**, the posterior biceps nerve were stimulated, each with a single sinuisoidal pattern. Stimulus strengths were submaximal yielding one to a few spikes from the recorded Renshaw cell. *Right column:* Input-output relations in terms of phase plots; bin contents of the left input histograms (*abscissa*) and time-related bin contents of the left output histograms (*ordinate*) are plotted against each other. Points on the upstroke of the input cycle are connected by *solid lines*; the time sequence is indicated by *arrows*. As apparent from the left cycle histograms and the right Lissajou plots, the RC response is nearly sinusoidal. The relation between stimulus sequence and Renshaw cell discharge is approximately linear and exhibits a slight tendency for the Renshaw cell response to phase-advance the input. (Rissing and Windhorst, unpublished)

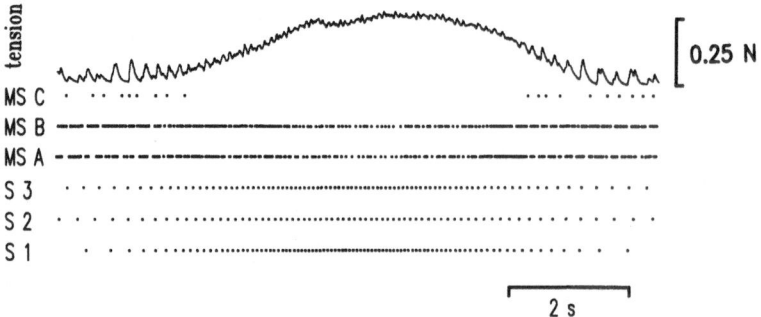

Fig. 43. Muscle spindle afferent discharge and muscle tension in response to sinusoidally rate-modulated stimulation of motor units. 10 s (about 1 cycle) of sampled and replotted data. *Lower three traces* (labelled *S1* through *S3*): Stimuli (*dots*) delivered to three medial gastrocnemius motor units. *Next three traces* (labelled *MS A, MS B, MS C*): discharge patterns of one Ia fibre (*MS A*) and two group II spindle afferents (*MS B* and *C*) from the medial gastrocnemius. *Upper trace*: muscle tension fluctuations in response to the motor unit activations (quasi-isometric recording). (Schwarz, Windhorst and Meyer-Lohmann, unpublished)

manifest as was also encountered with the afferents shown in A and B. The (soleus) Ia afferent in D (from another experiment) had unloaded even more showing a considerable pause in firing; its hysteresis had a course opposite to that in the previous cases. Even stronger unloading occurred in other cases, so that an afferent was active only at the lowest input rates or was activated only twice within a cycle during brief phases on both sides of the sine trough. The latter behaviour again resulted from activation of an otherwise silent receptor by the unfused motor unit contractions.

In some cases we also stimulated a medial gastrocnemius motor unit at constant rates varied from 3 to 30 pps in steps of 3 pps with intermediate pauses. The static input-output relations for four afferents recorded simultaneously again differed considerably, depending obviously on the mechanical coupling between motor units and afferents. (It is noted in passing that the *passive* dynamic and static input-output relations of spindle discharge to cat ankle joint angle also show hysteresis; see Burgess et al. 1982.)

Judged by the data of Cleveland and Ross (1977) and Cleveland (1980) and by those illustrated in Fig. 42, hysteresis is apparently absent in the pathway from motoneurone axon collaterals to Renshaw cells. In contrast, it is very prominent in the pathway from α-motoneurone axons to spindle afferents, even without interference of external muscle length changes (Fig. 44). This is an important difference between recurrent inhibition and proprioceptive feedback via skeletal muscle and spindle afferents, which could be relevant in comparing their physiological significance. In order to appreciate this difference, it is important to enquire a bit into the possible sources of hysteresis in the proprioceptive pathway. In theory, the hysteresis

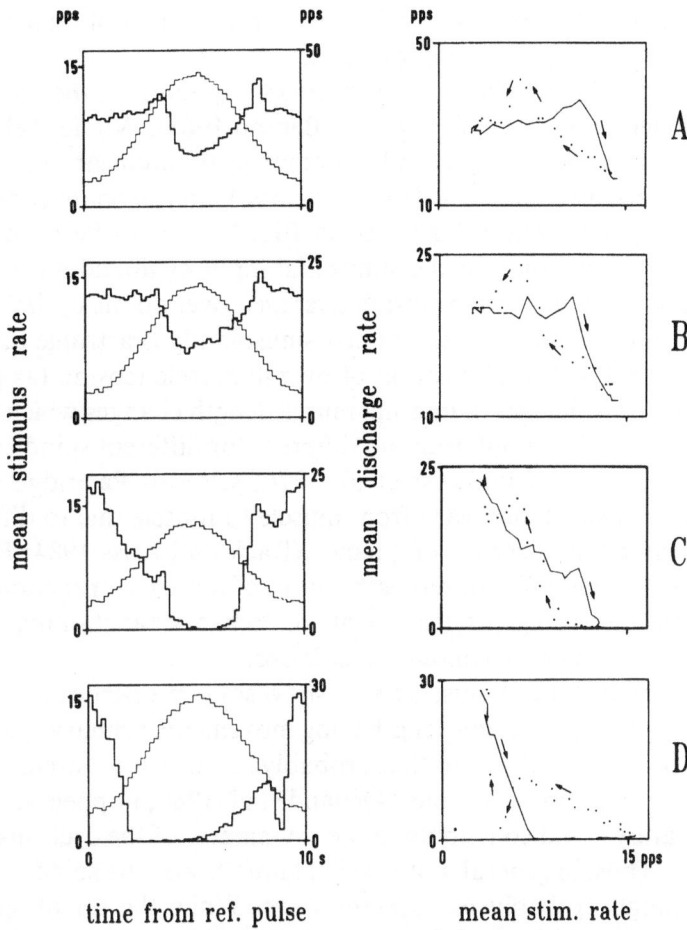

time from ref. pulse   mean stim. rate

Fig. 44 A–D. Four examples of input-output relations between motor unit stimulus and receptor discharge patterns. *Left column:* cycle histograms computed for input stimuli (*thin curves*) and output discharge (*thick curves*); the *curves* represent the mean pulse densities averaged with respect to cycle triggers (between 53 and 84; bin width 200 ms). The *thin* (input) *curves* in addition are averages of the respective cycle histograms belonging to each of the two or three stimulus patterns used to stimulate different motor units (see Fig. 43, *traces S1* through *S3*); this averaging was performed in order to be able to refer to a common measure of the two or three inputs. *Right column:* Input-output relations in terms of phase plots; bin contents of the left input histograms (*abscissa*) and time-related bin contents of the left output histograms (*ordinate*) are plotted against each other. Points on the upstroke of the input cycle are connected by *solid lines*; the time sequence is indicated by *arrows*. The *upper two rows* (labelled **A** and **B**) refer to the spindles labelled *MS A* and *MS B* in Fig. 43 and show moderate modulation of mean discharge rates of the receptors in response to activation of the three medial gastrocnemius motor units. The *third row* (**C**) shows the response of a Ia fibre to activation of two filaments each containing two to three motor unit axons to the soleus muscle. The *lower row* (**D**) illustrates the response of a soleus Ia fibre. (Schwarz, Windhorst and Meyer-Lohmann, unpublished)

illustrated in Fig. 44 could arise from a number of separate processes as well as from a combination of these.

a) The first factor contributing to hysteresis is the nonlinear behaviour of skeletal muscle (Partridge and Benton 1981). Whole skeletal muscles as well as single motor units exhibit a number of nonlinear properties in response to various input patterns (e.g. Burke 1981; Parmiggiani et al. 1982; Sect. 3.2.2). The phenomenon illustrated in Fig. 43, where the isometric tension developed in response to the sinusoidal input exhibited a distortion from a pure sine wave (as did the tension averaged over all the cycles), has also been seen with entire muscles stimulated sinusoidally (Partridge and Benton 1981).

b) The transformation of overall muscle tension (as discussed in a) into the local changes in internal muscle length changes which provide the spindle input is often nonlinear and different for different spindles (Meyer-Lohmann et al. 1974; Windhorst et al. 1976; see also Partridge and Benton 1981). Moreover, it may vary from muscle to muscle due to differences in internal anatomy and series compliance (Rack and Ross 1984; Rack and Westbury 1984). The different representations of muscle contraction in individual spindles' output signals again relate to "sensory partitioning" that might be used in establishing localised reflex loops.

c) Intrafusal muscle fibres show some properties which might contribute to hysteresis. In slow lengthening movements primary spindle endings exhibit a nonlinear behaviour that probably results from "stretch activation" of bag_1 intrafusal fibres (Emonet-Dénand et al. 1980; Poppele and Quick 1981). This feature could contribute to the asymmetry of the cycle histograms in Fig. 44.

Thus, in general, both skeletal muscle and the spindle behave nonlinearly, though probably in different ways. While the set of spindles in a muscle present a picture of the distributed mechanical events that take place in the muscular tissue, their output alone is a poor representation of the motor effects actually exerted unless the motor input signal to the muscle is known. This can be represented in a relatively undistorted way in the Renshaw cell discharge. Thus, comparing proprioceptive feedback signals and recurrent feedback signals would give a closer approximation to the mechanical events taking place in the muscle. This could be a function for recurrent feedback via Renshaw cells, but this issue cannot be discussed without an account of the other inputs to Renshaw cells and fusimotor neurones.

4.3.4 Dynamics of Signal Transmission from Motor Axons to Renshaw Cells or Spindle Afferents

1) Recurrent Inhibition

The dynamic behaviour of the recurrent inhibitory pathway is that of two subsystems, one system being represented by the synapses of motoneurone

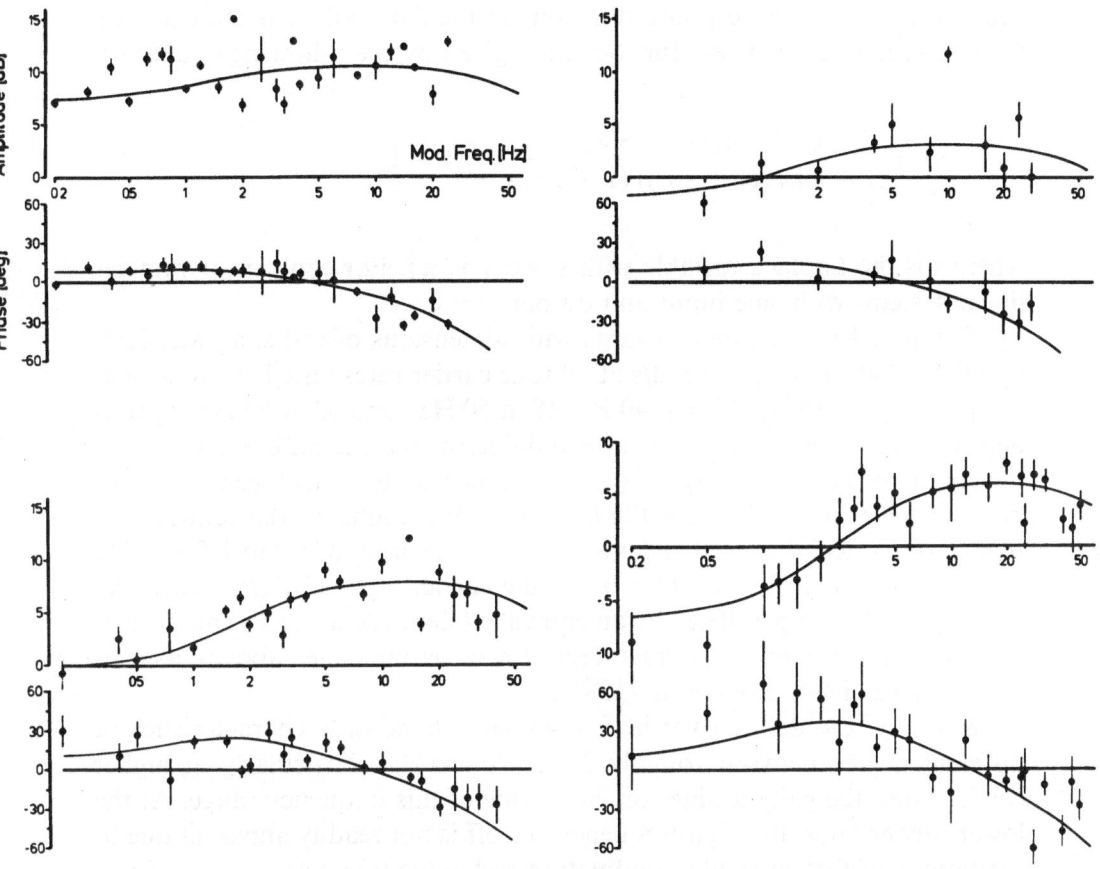

Fig. 45. Averaged frequency response (Bode) plots for all the Renshaw cells studied. The *solid points* are mean values (plus/minus standard errors) of the Renshaw cell discharge modulation amplitude and phase shift relative to the amplitude and phase of the stimulus pulse train. Centre frequency, modulation depth of stimulus train and maximum number of cells included in average are as follows: 30 Hz, 23.5%, 11 (*upper left plots*); 40 Hz, 17.6%, 4 (*upper right plots*); 50 Hz, 14.2%, 11 (*lower left plots*); 60 Hz, 11.8%, 3 (*lower right plots*). The *continuous curves* represent transfer functions fitted to points. (Courtesy of S. Cleveland 1980; his Fig. 20)

axon collaterals onto Renshaw cells, and the other by the Renshaw cell synapses on motoneurones. The first system was investigated by Cleveland and Ross (1977), Cleveland (1980) and by us, using different linear systems analysis methods. The former authors stimulated ventral root portions. Their results are summarised in Fig. 45, which shows four Bode plots of averaged Renshaw cell responses to antidromic stimulation, the rate of which was sinusoidally modulated at frequencies plotted on the abscissae; the mean ("carrier") antidromic rates were different in the four plots (from Cleveland 1980; his Fig. 20; see also Cleveland and Ross 1977; their Fig. 9). The authors fitted frequency response functions to the data points as indicated by the continuous lines. These functions are given by the following equation:

$$R(s) = \frac{(1 + s/a_1)(1 + s/a_2)}{(1 + s/b_1)(1 + s/b_2)(1 + s/b_3)} \exp(-Ls), \tag{11}$$

where s is the Laplace variable with $s = c + j2\pi f$. For constant-parameter linear systems with sine input and output, $c = 0$.

The a_i and b_j are rate constants with dimensions of rad/s. a_1 was 0.36, a_2 7.0, b_1 0.48, and b_3 120 rad/s at all four carrier rates tested; b_2 was 10 at a carrier rate of 30 Hz, 12.5 at 40 Hz, 18 at 50 Hz, and 32 at 60 Hz; b_2 thus depended upon the mean input rate and determined the different shapes of the continuous curves in Fig. 45 (this is another nonlinearity; see also Windhorst and Koehler 1983). L is the latency from stimulus to the ventral root filaments and Renshaw cell response and is of the order of 1.0 to 1.5 ms. The dependence of the rate constant b_2 on the carrier input rate (rate variable) would probably be paralleled by an equivalent dependence on the number of activated motor axon collaterals (recruitment variable; see above), as suggested by results of Ross et al. (1982).

Basically, the gain curves in Fig. 45 show band-pass characteristics of different degree between roughly 2 and 30–50 Hz. Particularly at higher carrier rates, the gain declines on either side of this frequency range. At the lower carrier rates, the high-frequency cut-off is not readily apparent due to limitations of the sinusoidal modulation technique (see below).

Essentially similar results were obtained by us using random motor axon stimulation. This stimulus form has the advantage of extending the frequency range over which Renshaw cell responses can be studied, whereas Cleveland and Ross' sinusoidal modulation is limited to modulation frequencies *below* the carrier (mean stimulus) rate. We also extended their findings by stimulating, instead of ventral roots, muscle nerves whose axons had effects of different excitatory strength on the Renshaw cell discharge. Figure 46 shows examples of the dependency of Renshaw cell dynamic behaviour on the strength of synaptic input. This cell was responsive to stimulation of two different muscle nerves (lateral gastrocnemius-soleus and medial gastrocnemius) but this to a very different degree. The two nerves were subsequently stimulated supramaximally with the same random pattern at a mean rate of about 22 pps. The average responses of the cell to the two inputs are shown as PSTHs in the upper row of Fig. 46. It is obvious that stimulation of the lateral gastrocnemius-soleus (upper panel in A) elicited a much stronger response than medial gastrocnemius stimulation as seen in the mean Renshaw cell discharge rates (see PSTH baseline levels) as well as in the peaks after reference pulse occurrence ($\tau = 0$ ms).

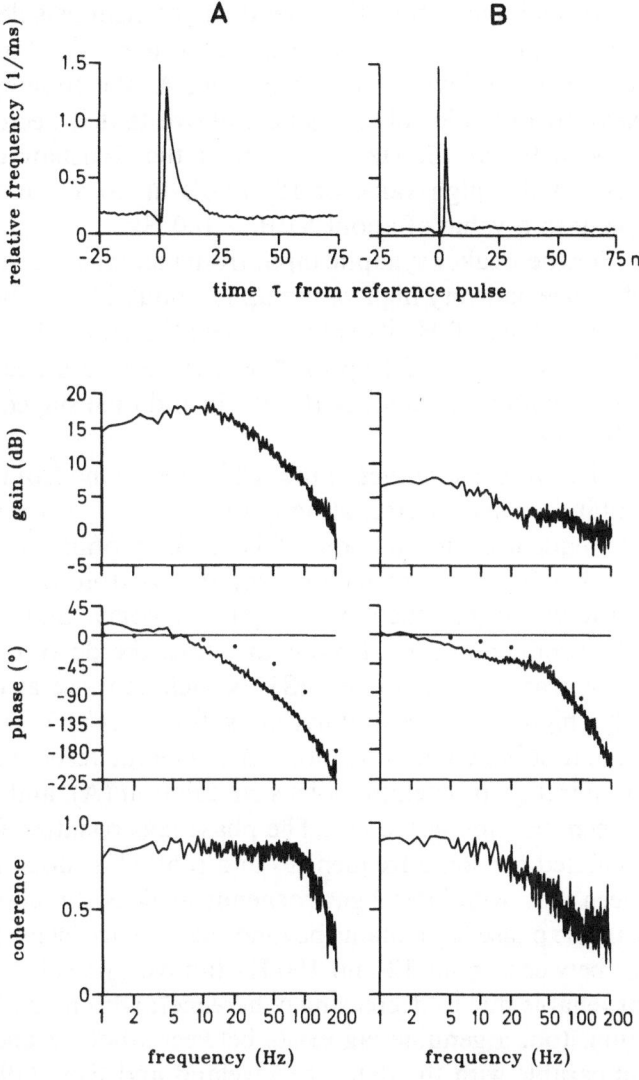

Fig. 46 A, B. Examples of PSTHs and spectral measures for a Renshaw cell excited by gastro-cnemius nerve stimulation with quasi-Poisson stimulus patterns. **A** (*left column*) Effects of stimulation of the lateral gastrocnemius-soleus (LGS) with 200 mV. Strong coupling as indicated by the early peak after $\tau = 0$ ms; **B** (*right column*) Effects of stimulation of the medial gastrocnemius (MG) nerve with 600 mV. Coupling of medium strength. Mean stimulus rate about 22 pps in both cases. The *dots* in the phase curves show phase lags expected from Renshaw cell response latencies calculated from PSTHs at high resolution (bin width of 0.1 ms). The PSTHs *on top* were calculated with a bin width of 1 ms, so that a relative frequency value (on the *ordinate*) of 1 indicates the occurrence of, on average, 1 spike per bin. Number of reference events: **A**, 2654; **B**, 2659. (Christakos, Windhorst, Rissing and Meyer-Lohmann)

With stimulation of the lateral gastrocnemius-soleus (Fig. 46 A), the linear range of operation of the Renshaw cell extended from 1 to about 100 Hz. As seen in the lower panel in column A, the coherence was about 0.8 for nearly two decades of frequency; above that it decreased rapidly, but still exceeded 0.5 at 150 Hz. The gain of the Renshaw cell (second row from above) had a high value of 15–17 dB up to about 10–12 Hz, then it fell rapidly to a value of about 5 dB at 100 Hz.

For the weaker synaptic input from medial gastrocnemius (Fig. 46 B), the coherence had very high values up to about 20 Hz, above which it decreased rapidly; above 50 Hz its value was less than 0.6. The gain was relatively low with values of 5–7 dB up to 20 Hz. It showed a decline by a factor of 1.5 (3.5 db) within the band of 10–30 Hz and then stayed relatively constant up to 100 Hz.

The two phase curves of the cell in the second from bottom row of Fig. 46 exhibit a similar course, although the left curve is more phase-advanced at low frequencies (up to about 15 Hz). The prominent feature is the phase lag at higher frequencies which in part results from the response latency of the Renshaw cell (conduction and synaptic delays and cell utilisation time for spike production). To demonstrate this, the delay of the Renshaw cell response was estimated from PSTHs such as those at the top of Fig. 46, but with a higher time resolution (bin width of 0.1 ms). The delay from $\tau = 0$ ms to the first bin containing more than average response spikes was ca. 2.5 ms for lateral gastrocnemius-soleus stimulation (A), and 2.8 ms for medial gastrocnemius stimulation (B). The phase lags resulting from these delays were calculated for some frequencies and plotted as dots into the phase plots. It is seen that, with lateral gastrocnemius-soleus stimulation (A), a genuine cell response phase lag remains beyond that due to delays (see above) at frequencies between about 12 and 100 Hz (above this value phase calculations are not reliable due to decreasing coherence); with medial gastrocnemius stimulation, too, a genuine lag exists between about 2 and 30 Hz. This is again comparable with the data of Cleveland and Ross (1977; Fig. 45, upper left plot).

Essentially similar changes in Renshaw cell dynamics to those occurring with a decrease in synaptic input (Fig. 46 A to Fig. 46 B) occurred with an increase of the mean stimulus rate from about 10 to 45 pps (not shown).

It may be argued that the synchronous activation of many (or at least several) motor axons as employed in the above experiments to excite the Renshaw cells is unnatural and therefore does not give valid results. However, it should be recalled that Renshaw cells can be excited as strongly by single motor axons as by groups of axons (Ross et al. 1975; van Keulen 1981).

In summary, both Cleveland and Ross (1977) and we found that Renshaw cells commonly show band-pass characteristics in a frequency range from ca. 2 to 15 Hz, depending on input parameters (mean rate and synaptic input strength). The band-pass properties were more pronounced in Cleveland and Ross' (1977) study than in ours in that the gain declined more strongly towards lower frequencies. Thus, Renshaw cells can follow particularly sensitively inputs in a frequency range encompassing the steady firing rates of many α-motoneurones.

2) Proprioceptive Feedback

The dynamics of signal transmission from motoneurones to muscle spindles are determined by the respective dynamics of two subsystems, those of the motor units and those of the spindles. The dynamic behaviour of motor units depends on motor unit type. In a linear approximation, the dynamic response of motor units tested under isometric conditions can be represented by second-order (low-pass) frequency response functions. FF-type motor units with their fast contraction and relaxation would display low-pass characteristics with higher cut-offs than S-type motor units with their slow twitch waveforms. The dynamic response of spindle afferents to small-amplitude muscle length changes imposed *externally* by stretches can be modelled by a linear high-pass with a cut-off at about $1-2$ Hz, with an additional cut-off for primaries at a somewhat higher frequency (Matthews and Stein 1969; Poppele and Bowman 1970; Poppele 1981), and with a final cut-off (leading to a high-frequency roll-off) around 100 Hz (Goodwin et al. 1975; Cussons et al. 1977). From these results obtained with separate approaches, one might predict that the overall gain characteristic of the system between motor axons and spindle afferents is relatively flat because the high-pass properties of spindles could compensate for the low-pass properties of motor units. However, both subsystems coupled in series may show nonlinearities as illustrated in Sects. 3.2.2 and 3.2.3. Before quantifying the signal transmission from motor unit axons to spindle afferents in terms of linear transfer functions (frequency responses), one should therefore check linearity by means of coherence functions. This was done in Fig. 47 in the same way as for Renshaw cells in Fig. 46. (Note the different scales: they are linear in Fig. 47.) In the underlying experiment a small set of motor units of the isometric cat medial gastrocnemius muscle was stimulated with a Poisson input train at a mean rate of ca. 7.5 pps, and the discharge of a Ia fibre affected by at least one of the motor units was recorded. Figure 47 A shows the estimated coherence between input and output pulse trains. It is relatively high in a low-frequency range up to about 35 Hz and then declines. That is, the output is accounted for linearly by the input within the low-frequency range, and only in this range is it reasonable to compute the frequency response (gain in B and phase in C). The (linear) gain factor declines gradually from a low-frequency maximum, whereas the phase lag (lag upwards) of the output with respect to the input increases with increasing frequency.

In general, the range of high coherence values depended upon two factors. Firstly, the mechanical coupling between motor units and spindle afferents

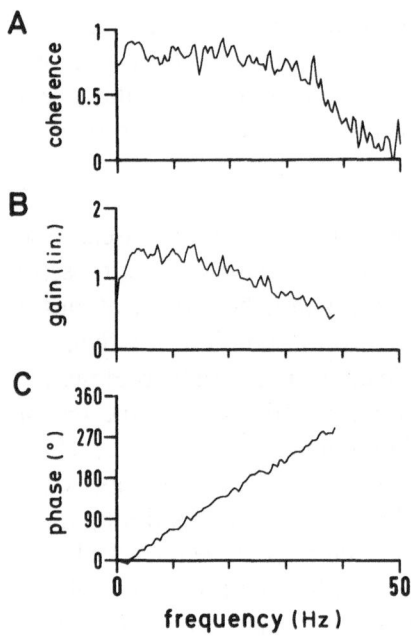

Fig. 47 A–C. Frequency response characterisation of the system from motor unit axons to a Ia afferent, estimated by using random stimulation at a mean rate of about 7.5 pps. Note the linear frequency scale. **A** Estimated coherence between a Ia fibre spike train and the stimulus sequence applied to a small set of motor units of the cat medial gastrocnemius muscle under quasi-isometric conditions; **B** Gain factor (linear scale) for the system linking the stimulus and the afferent trains; **C** Phase curve (in degrees) for the same system, phase *lag* being plotted upwards. (With permission from Christakos et al. 1984; their Fig. 4)

had to be strong for the coherence to attain high values at all. Secondly, the width of the low-frequency range with high coherence was determined by the contractile properties of the motor units: It was generally larger for fast motor units than for slow ones; for the latter it could extend up to 10 Hz and decline above it (Christakos et al. 1985). Within the low-frequency range of high coherence, the gain was usually relatively flat (Christakos et al. 1985; Christakos and Windhorst 1986a).

4.3.5 Responses of α-Motoneurones to Activation of Single Homonymous Motor Axons Mediated Through Renshaw Cells or Motor Units and Muscle Afferents

1) Recurrent Inhibition

The effects of spinal recurrent inhibition have usually been studied by stimulating *sets of motor axons*. Sometimes close to single motor axons have been stimulated in order to study their effects on Renshaw cells (e.g. Ross

1976; Cleveland 1980; Bergmann-Erb 1985). However, the recurrent effects on membrane potentials of α-motoneurones elicited by stimulating single homonymous motor axons have only recently been investigated (Hamm et al. 1987 a). Some representative examples are now displayed.

In anaesthetised (urethane-chloralose 600/60 mg/kg initially) or ischaemically decapitate cats, medial gastrocnemius α-motoneurones were recorded intracellularly, and single motor axons in continuity to the spinal cord were stimulated electrically at a low rate (6–7 pps) either by intra-axonal penetration of the stimulating electrode or by intramuscular stimulation of axon branches. (Dorsal roots were cut.) Noise was removed by averaging the motoneurone membrane potential over usually at least 1000 stimuli. Figure 48 presents some typical examples of recurrent IPSPs. The profiles of responses presumed to be recurrent IPSPs were characterised by a long-lasting hyperpolarisation that was clearly distinguishable from the control (prestimulus) portion of the average. They are also different from the profile of extracellular control averages, which were taken after intracellular averaging from 7 cells which exhibited recurrent IPSPs (see Fig. 48 D, middle trace). An initial depolarisation that could come about through direct excitatory crossconnexions between α-motoneurones has not yet be observed, possibly due to their narrow distribution or weakness (see Figs. 21).

In anaesthetised animals, the mean amplitudes of the presumed recurrent IPSPs were 12.0 ± 6.9 μV (mean ± SD), the half-width 24.1 ± 12.4 ms. In decapitate cats the corresponding values were 46.2 ± 44.7 μV and 18.5 ± 1.0 ms. The difference in mean amplitudes may be due to the inhibitory effect of chloralose on Renshaw cells (Haase and van der Meulen 1961 b; Biscoe and Krnjevic 1963).

2) Proprioceptive Feedback

As described in Sect. 3.2.3.3, stimulation of single motor units via their ventral root axons leads, via the muscle afferent feedback, to average membrane potential changes in homonymous α-motoneurones (see Fig. 33). Although the precise time course of these changes may be more complicated than that of the membrane potential trajectories elicited via the recurrent pathway (Fig. 48), the durations and amplitudes of the potential trajectories are of the same order of magnitude.

It may be argued that the comparison of the two types of feedback effects is complicated by the fact that fast motor units may fatigue, whereby the evoked average membrane potential changes should decrease in amplitude and possibly change their time-course. Whereas this is partially true (W. Koehler et al. 1984 a), the loss of gain in the motor unit system is in part compensated by an increase in gain of the subsequent receptor system transforming force changes into afferent discharge variations so that the overall gain is not diminished as strongly as would be expected from motor unit fatigue (Windhorst et al. 1986). This effect could result to an appreciable

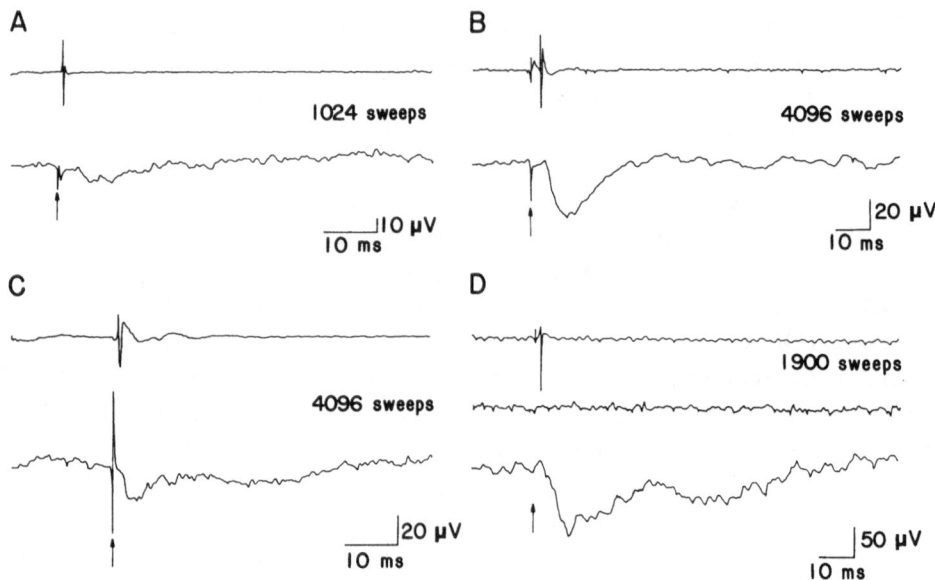

Fig. 48 A–D. Examples of averaged recurrent inhibitory postsynaptic potentials (RIPSPs) elicited in single cat medial gastrocnemius α-motoneurones by stimulating single homonymous
motor axons at rates between ca. 6 and 7 pps. *Upper traces* are extracellular ventral-root
recordings and *lower traces* are intracellular motoneurone recordings. *Arrows* indicate stimulus
onset. **A–D** Averages based on 1024, 4096, 4096, and 1900 stimuli (sweeps), respectively. *Middle
trace* in **D** is an extracellular recording, taken from just outside the motoneurone. This control
average suggests that the RIPSP was a true transmembrane hyperpolarisation. **A, C** Records
from chloralose-urethane preparations; **B, D** Responses from ischemic-decapitate animals. Two
negative-going peaks are observable in the profile of the **C–D** RIPSPs, as seen in 10 of the 76
RIPSP averages. Note different voltage calibrations and time scales. (With permission from
Hamm et al. 1987a; their Fig. 2)

extent from a similar behaviour of muscle spindle afferents (Christakos and
Windhorst 1986a). This could therefore be a mechanism to preserve a high
quality of afferent information about motor unit contractions.

4.3.6 Differentiated Feedback?

1) Two Types of Recurrent Feedback?

Cullheim and Kellerth (1981) have recently adduced pharmacological
evidence for the existence of two types of Renshaw cells, one with glycinergic
synapses and the other with GABA-ergic synapses on motoneurones (see also
Polc and Haefely 1982). The first type exerts shorter, the latter longer-lasting
synaptic effects, i.e. the time courses of the recurrent IPSPs differ (see van
Keulen 1981). Cullheim and Kellerth (1981) found that "... no organization-
al pattern based on motor pool category or motor unit type has been detect-

ed, and both inhibitions seem to be mediated via both short and fairly long (at least 2 mm) intramedullary routes" (p. 222). Indeed Lagerbäck (1983) found no discrete distributions of most parameters characterising the cell architecture of presumed Renshaw cells (soma diameter and shape, number of dendrites, dendritic diameter, synaptology) except for one feature. Three of his cells had a lower somatic and dendritic synaptic packing density of F-type (flattened vesicles) and S-type (spherical vesicles) and a lower F/S ratio than the remainder (12 cells). However, this finding, which might indicate the existence of separate Renshaw cell populations, will have to be confirmed and correlated with pharmacological characteristics.

Regarding the possible functional significance of the two Renshaw cell types they propose that it need not be primarily coupled with the recurrent effects of α-motoneurones, but might be connected with the recurrent inhibition of other spinal neurones (γ-motoneurones, Ia inhibitory interneurones, other Renshaw cells, VSCT cells). And there is still another possible explanation for having two types of recurrent inhibition. The different types of Renshaw cells could receive different kinds of specific synaptic inputs from other sources than the α-motor axon collaterals. "In this way, one or the other of the recurrent systems could be selectively used in situations with special demands" (p. 222).

2) Two Types of Muscle Spindle Feedback

Note that the idea of Cullheim and Kellerth (1981) quoted at the end of the last paragraph also applies to the two types of spindle afferent which underlie different types of fusimotor control, the different fusimotor neurones possibly being influenced by different input systems. It would be interesting to compare the inputs to the two types of Renshaw cells and muscle spindle afferents.

4.3.7 Presynaptic Inhibition

1) Recurrent Inhibition

There is sparse data pertaining to the possibility that motor axon collaterals receive presynaptic inhibition. Lagerbäck and Ronnevi (1982) performed an ultrastructural study of synapses on presumed Renshaw cells and distinguished two main types: S-type synapses (with spherical vesicles) and F-type synapses (with flattened vesicles). Into the first category fall the synapses from motor axon collaterals. And among the S-type synapses are also some (called S-type P, found in 8 of 17 cells), which are presynaptically contacted by one to three synaptic terminals with irregular vesicles. For technical reasons, their frequency of occurrence may have been underestimated. This synapse arrangement is usually considered to be the morphological substrate of presynaptic inhibition. Lagerbäck and Ronnevi (1982) discuss the possibility that the S-type P synapses stem from dorsal root afferents; it is unlikely that they originate from motor axon collaterals because Lagerbäck et al. (1981 b) did not see axo-axonic synapses on motor axon collaterals. But since such synapses appear to be difficult to find, the matter deserves further study. To my knowledge, nothing is known about the possible presynaptic inhibition of Renshaw cell axons on motoneurones (and other spinal neurones). It is this site that corresponds to the Ia afferent terminals on α-motoneurones.

2) Proprioceptive Feedback

It is known that Ia and other afferents are subject to presynaptic inhibition from a variety of sources (Schmidt 1971; Baldissera et al. 1981). Autogenetic presynaptic inhibition of Ia afferents from group I afferents of homonymous and synergistic muscles appears to be common in man (e.g. Iles and Roberts 1986), but weaker in cats (Barnes and Pompeiano 1970), although it could be enhanced with convergent input from other sources (Brink et al. 1984). Presynaptic inhibition from other, even antagonistic, muscle nerves is widespread and shows a complicated pattern (Iles and Roberts 1986; Brink et al. 1984). The number of interneuronal systems mediating presynaptic inhibition from and to various group I afferents may lie between two and six, but within each system a further fractionation might be caused by differential patterns of input from other segmental and descending fibre systems (Brink et al. 1984; also Rudomín et al. 1983). Interestingly, monosynaptic EPSPs in a triceps surae α-motoneurone elicited by two different homonymous or synergistic Ia fibres may be differentially reduced by presynaptic inhibition evoked by stimulation of the posterior biceps-semitendinosus nerve, and the EPSPs evoked by one Ia fibre in two different α-motoneurones may also be subject to different amounts of presynaptic inhibition from the same source (Clement et al. 1987). This argues for a fine-grain organisation of presynaptic inhibition.

Thus, whereas a substantial amount is known about the complicated structure of presynaptic inhibition of Ia afferents, little if anything is known about presynaptic inhibition of the recurrent inhibitory pathway.

4.3.8 Recurrent Inhibitory vs Proprioceptive Feedback on γ-Motoneurones

To appreciate the structural similarities, to be summarised now, between spinal recurrent effects via Renshaw cells and proprioceptive feedback onto neurones *other than α-motoneurones*, one should recall the net effects of an increase in excitation of an agonist motoneurone pool *before considering parallel effects of descending command signals on Renshaw cells and γ-motoneurones*. This increase causes an increased inhibition via the "homonymous" Renshaw cells, a reduced firing of spindle afferents (a disfacilitation) and an augmented firing of Golgi tendon organ afferents due to muscle contraction.

1) Recurrent Inhibition of γ-Motoneurones

It is generally agreed that γ-motoneurones receive recurrent inhibition via Renshaw cells predominantly from α-motoneurones (Ellaway 1971; Ellaway and Murphy 1980b; Appelberg et al. 1983b; see Fig. 49C). But there are contradictory reports on their contribution to recurrent inhibition. Kato and Fukushima (1974) found some Renshaw cells that were excited from γ-motor axons. Westbury (1982) described γ-motoneurones with recurrent collaterals (but cf. Cullheim and Ulfhake 1979), but found no evidence for α-motoneurones being inhibited from γ-axons (Westbury 1980), in agreement with Ellaway and Murphy (1980b). Westbury (1980) suggested that γ-motoneurones might inhibit themselves via Renshaw cells, and Appelberg et al. (1983b) provided some evidence for this, although the effects were weak. The latter authors suggested that there are γ-motoneurones which receive recurrent inhibition only from γ-motoneurones, but not from α-motoneurones. "Thus there seem to be differences between the recurrent control of α- and γ-motoneurones which add to the arguments against the concept of stereotyped α–γ linkage" (p. 302).

2) Proprioceptive Feedback on γ-Motoneurones

As briefly mentioned in Sect. 2.9.1.2, muscle stretch receptors exert oligo- and polysynaptic actions on γ-motoneurones that are similar to but more complicated than those on α-motoneurones.

4.3.9 Recurrent Inhibition vs Ia Excitation of Reciprocal Ia Inhibitory Interneurones

1) Recurrent Inhibition of Reciprocal Ia Inhibitory Interneurones

It is well established, and has been excellently reviewed several times (Hultborn 1972, 1976; Lindström 1973; Haase et al. 1975), that Renshaw cells also inhibit "reciprocal" Ia inhibitory interneurones mediating reciprocal inhibition between antagonist muscles (see Figs. 21E and 49C). The pattern of distribution is generally such that Renshaw cells which receive excitatory input from an agonist (say extensor) motoneurone pool inhibit reciprocal Ia inhibitory interneurones which receive homonymous or synergistic Ia excitatory input and inhibit antagonist motoneurones. (This applies at least to motoneurone pools acting around the same joints.) This distribution pattern is one reason for the recurrent facilitation which was early described by Renshaw (1941) and which was shown to work predominantly between antagonists (see review by Haase et al. 1975).

2) Proprioceptive Feedback to Reciprocal Ia Inhibitory Interneurones

As evident from the preceding paragraph, the reciprocal Ia inhibitory interneurones subjected to recurrent inhibition from one motoneurone pool receive homonymous or synergistic Ia excitation (Hultborn 1972; Baldissera et al. 1981).

4.3.10 Mutual Inhibition Between Renshaw Cells vs Reciprocal Inhibition

1) Mutual Inhibition between Renshaw Cells

Renshaw cells are known to inhibit each other (Ryall et al. 1971; Ryall 1970, 1981). The main pattern of this "mutual inhibition" appears to be such that Renshaw cells excited by agonists (say extensors) most strongly inhibit those excited by antagonists (flexors; Ryall 1981; see Fig. 21 E), and vice versa. Thus, excitation of "agonist" Renshaw cells entails disinhibition of antagonist α-motoneurones, and vice versa. It seems unlikely that the "antagonist" Renshaw cells inhibited by "agonist" Renshaw cells form a separate subgroup of "antagonist" cells, with a different distribution of inputs and outputs, although nothing is known about this issue. If so, this would be an example of Lundberg's economy principle (Lundberg 1979). The disinhibition of the antagonist motoneurones would then be weighted by the latter's α-motor output in addition to descending and segmental influences.

2) Reciprocal Inhibition

The disinhibition of antagonists via mutual inhibition of Renshaw cells parallels the reciprocal inhibition of antagonists by agonist Ia afferents and facilitation of the antagonists by agonist Ib afferents.

Why then is there a "mutual inhibition" among reciprocal Ia interneurones (see Baldissera et al. 1981)? The answer to this question can only be speculative. These interneurones inhibit each other in such a way that interneurones excited by agonist (say extensor) Ia afferents inhibit those excited by antagonist (flexor) Ia afferents, and vice versa (see Fig. 21 E). This pattern is similar to that prevailing in the mutual inhibition between Renshaw cells. This suggests that the mutual inhibition between interneurones which either serve, or are served from, antagonist muscles is a general feature of spinal cord organisation.

4.3.11 Supraspinal Inputs to γ-Motoneurones and Renshaw Cells

The muscle stretch receptors are endowed with their own efferent control system mediated via specialised γ-motoneurones, and they are affected in addition by β-motoneurones, which innervate both skeletal and intrafusal muscle fibres. Since the fusimotor effects can be functionally divided into static and dynamic ones, there are in principle four types of fusimotor innervation: static and dynamic γ, and static and dynamic β innervation (the dynamic β-motor units being of S-type and the static β-motor units of F-type). Whether this diversity is exploited to the fullest extent under physiological conditions is unknown. In particular, the selective supraspinal control of the various fusimotor neurones is still at issue. The supraspinal control of the Renshaw cell feedback has been less well studied, though a number of studies are available (see Haase et al. 1975). It cannot therefore be decided at present whether this control is as diversified as the fusimotor control of spindles.

The effects exerted on Renshaw cells or muscle spindles by supraspinal motor command signals, which *bypass the α-motoneurones*, can be considered under three aspects: (a) a general consideration of the excitatory or inhibitory effects fed back to α-motoneurones; (b) change in static sensitivity; (c) change in dynamic sensitivity.

1) Recurrent Inhibition

Renshaw cells are also subject to supraspinal influences that, at least in part, bypass the α-motoneurones (early work is reviewed in Haase et al. 1975; see also W. Koehler et al. 1978; Hultborn and Pierrot-Deseilligny 1979; Henatsch and Windhorst 1981; Henatsch 1983; Henatsch et al. 1986; Pompeiano 1984). The loci from which definite and conjectured modulating effects upon Renshaw cells can be exerted are briefly summarised below.

a) Any excitatory or inhibitory influence on Renshaw cells is mediated by them with opposite sign to motoneurones. Hultborn and Pierrot-Deseilligny (1979) maintain that during voluntary tonic soleus contraction of increasing strength or during ramp contraction (in man), the Renshaw cell recurrent inhibition is progressively diminished (from an initially facilitated level), probably by supraspinal descending influences bypassing the motoneurones. The inhibitory influence exerted on Renshaw cells from supraspinal sources appears to dominate and to offset the excitatory input derived from motoneurone excitation.

b) The influence of descending motor control signals on the static sensitivity of Renshaw cells has not yet been studied experimentally in the same quantitative terms as employed by Cleveland et al. (1981) (see Sect. 4.3.3). In their membrane model designed to account for the nonlinear static Renshaw cell behaviour, the saturation constant c depends upon resting membrane potential E_r and conductance G_r in such a way that inhibitory inputs would decrease c. This would amount to a scaling down of the Renshaw cell output and, hence, to a proportional change of the slopes or gains (at any given input frequency) of the nonlinear static relationship. The "semi-saturation" constant k also determines these slopes and could be changed by inhibitory inputs through variation of G_r, but the influence of G_r depends on other unknown membrane characteristics and cannot be easily predicted. Thus it remains to be shown that inhibition of Renshaw cells from supraspinal sources can not only proportionally scale down the Renshaw cell response to the same excitatory input but can also flatten (linearise) the static input-output relation by augmenting k (and thereby shifting a large part of the curvature out of the physiological input range).

c) The influence of supraspinal motor commands on Renshaw cell frequency responses, as investigated by Cleveland and Ross (1977) and Cleveland (1980), has also not yet been studied experimentally. If the dynamic Renshaw cell gains are reduced proportionally at all modulating input frequencies, i.e. exactly as the static gain (at zero frequency), then a scaling down of the Renshaw cell responses due to inhibitory input as discussed above (b) would also reduce the dynamic Renshaw cell sensitivity.

Motor cortex and corticospinal tract: Henatsch et al. (1961), using conditioning cortex stimulation, found excitatory and inhibitory influences on recurrent inhibition independent of effects on α-motoneurones (see also Henatsch 1983). MacLean and Leffman (1967) observed inhibitory effects from the pericruciate cortex (mediated via the pyramids) on antidromic Renshaw cell responses to ventral root stimulation. W. Koehler et al. (1978) found that Renshaw cell responses to antidromic stimulation of ventral roots or muscle nerves were inhibited in 74% of the cases and facilitated in only one, when conditioned by stimulation of the contralateral capsula interna. In 84% of the cases in which antidromic Renshaw cell responses as well as monosynaptic reflexes were tested, Renshaw cell depression concurred with reflex facilitation.

Cerebellum: Granit et al. (1960), stimulating the anterior cerebellar lobe, described inhibitory and facilitatory influences on recurrent inhibition of extensor α-motoneurones. Haase and Vogel (1971) showed that, upon stimulation of the ipsilateral nucleus interpositus, gastrocnemius α-motoneurones may be suppressed whilst Renshaw cells coupled to them may be excited simultaneously. Benecke et al. (1976), upon stimulation of the interpositus nucleus, often found facilitation of monosynaptic extensor and flexor reflexes concomitant with depression of related orthodromic Renshaw cell responses, particularly when group II afferents were recruited. These findings indicate a flexible coupling of Renshaw cells to their input-giving α-motoneurones.

Extrapyramidal nuclei (striatum, globus pallidus, thalamus, and substantia nigra): MacLean and Leffman (1967) observed inhibitory effects of stimulation of the ventral thalamus (field of Forel, zona incerta), the anterior thalamus and of the globus pallidus on antidromic Renshaw cell responses to ventral root stimulation. Benecke et al. (1975) found only slight influences of substantia nigra stimulation on Renshaw cell spontaneous activity or antidromic early responses, these effects not being dissociated from alterations of motoneurone activity.

Vestibular nuclei: Ross and Thewissen (1987), upon stimulating semicircular canal afferents, found a depression of Renshaw cell responses to antidromic stimulation of extensor and flexor muscle nerves. The conditioning effect could last as long as 600 ms. In the vicinity of the inhibited Renshaw cells neurones were found that were excited by the supraspinal stimulus with the appropriate discharge pattern to be candidates for mediating the inhibition on the Renshaw cells. Ross and Wittrock (1987) reported changes in lumbar Renshaw cell activity correlated with vertical whole-body movements, indicating that macular afferents exert an inhibitory influence on Renshaw cells. Pompeiano et al. (1985a) studied the effects of head and body rotation on the activity of Renshaw cells coupled to gastrocnemius α-motoneurones (next section).

Nucleus ruber: Henatsch et al. (1986), upon conditioning stimulation of the contralateral nucleus ruber, found predominant facilitation of various (extensor and flexor) monosynaptic reflexes associated with depression of antidromic and orthodromic Renshaw cell discharges, again indicating a flexible coupling of the latter to their input-giving α-motoneurones.

Reticular formation: MacLean and Leffman (1967) found inhibitory effects of stimulation of the medio-lateral mesencephalic and of the bulbar reticular formation on antidromic Renshaw cell responses to ventral root stimulation. Haase and Vogel (1971) found that stimulation of the ipsilateral (and also contralateral) nucleus reticularis ventralis facilitated monosynaptic gastrocnemius reflexes and background discharges whilst simultaneously depressing orthodromic Renshaw cell responses to the motoneurone actvitity (see also Haase and van der Meulen 1961a).

Locus coeruleus: Fung et al. (1987) found a suppression of recurrent inhibition of lumbar flexor and extensor motoneurones upon stimulation of the locus coeruleus, which usually increased the excitability of such motoneurones (Fung and Barnes 1987). A differential effect on flexor or extensor motoneurones could not yet be distinguished.

2) Proprioceptive Feedback

The descending influences on spindle output bypassing the extrafusal muscle are very complex (recent review: Hulliger 1984). These influences first act on four kinds of fusimotor neurones: static and dynamic γ-motoneurones and static and dynamic β-motoneurones. These act in turn on (at least) three types of intrafusal muscle fibre: dynamic and static bag fibres (bag$_1$ and bag$_2$ fibres, respectively) and chain fibres. From sensory endings on these intrafusal muscle fibres, the signals on group Ia and II spindle afferents are composed in various proportions. This complex signal processing cannot be dealt with here in any detail (for reviews and overviews see, e.g. A. Taylor and Prochazka 1981; Laporte et al. 1981; Boyd 1981; Boyd and Smith 1984; Hulliger 1984). Just a brief summary is given here.

a) Excitatory and inhibitory descending influences onto fusimotor (γ- and β-)motoneurones would be transmitted with the same sign to spindle afferents, although the efficacy of the transmitting channels, that is dynamic fusimotor effects through bag$_1$ fibres to Ia fibres and static fusimotor effects

Table 2. Effects of static and dynamic γ-efferents on the static and dynamic sensitivity of primary and secondary muscle spindle endings

		dyn. fusimot.	static fusimotor	
		bag$_1$	bag$_2$	chain
primary	basic activity	$\uparrow$	$\uparrow\uparrow$	driving[a])
	stat. sens.	$0-\uparrow$	$0-\downarrow$[b])	
	dyn. sens.	$\uparrow\uparrow$[c])	$\downarrow-\downarrow\downarrow$	
secondary	basic activity	$0-(\uparrow)$	$0-\uparrow$	$\uparrow\uparrow$
	stat. sens.	$(\uparrow)$	$0-(\uparrow)$	$(\uparrow)-\uparrow\uparrow$
	dyn. sens.	$(\uparrow)$	$0-\downarrow$	$\downarrow$

The *arrows* indicate increase (head up) or decrease (head down) of the activity or sensitivity. *Single arrows* indicate small effects, *double arrows* strong effects. *Brackets* indicate that the effects may also be absent.
[a]) Driving 1:1 up to 60 pps; length signal disrupted. [b]) Driving very rare. [c]) According to Boyd (1981) dynamic fusimotor activity increases the length-sensitivity of primary endings during the dynamic phase of (ramp) stretching, this effect being velocity-dependent.

through bag$_2$ and chain fibres to Ia and group II fibres, varies appreciably (e.g. Boyd 1981; see Table 2). The supraspinal control of static and dynamic γ-motoneurones seems to be partially independent of that of α-motoneurones (see below).

b) The static sensitivity of spindle afferents to length changes and its dependence on fusimotor action has only been examined by applying external length changes, i.e. by stretching the parent muscle or the spindle itself. Thus the sensitivity to length changes effected by skeletal α- or β-motor input (which is the input of interest in the present comparison) has not been investigated. Fusimotor effects on the static and dynamic sensitivity of primary and secondary spindle endings to large-amplitude ramp stretches have been studied repeatedly. A recent study differentiating the effects according to the involvement of the three types of intrafusal muscle fibre was presented by Boyd (1981) and is summarised in Table 2 (for further discussion and earlier literature, see that paper).

c) The fusimotor effects on the dynamic sensitivity of spindle endings, particularly primaries, should be differentiated into effects on responses to small- and large-amplitude inputs. For the latter see Table 2. Fusistatic stimulation "... has been shown to decrease the sensitivity of muscle spindles to small stretches (Crowe and Matthews 1964; Chen and Poppele 1973; Goodwin et al. 1975), and to decrease or even remove the dependence of sensitivity on muscle length (Goodwin et al. 1975; Fig. 6) ... the results obtained with fusistatic stimulation imply that an activation of the contractile machinery reduces muscle stiffness to small stretches and, further, reduces the dependency of muscle stiffness on muscle length" (Poppele et al. 1979; p. 410). Fusidynamic stimulation paradoxically decreases the sensitivity of Ia fibres to small, especially low-frequency, length changes around a constant mean muscle length (Goodwin et al. 1975; Hulliger and Sonnenberg 1985). However, if the small-amplitude length changes are superimposed upon a large-amplitude triangular movement, dynamic γ-fibres tend to uniformly enhance Ia sensitivity (Hulliger and Sonnenberg 1985).

In the following a brief summary is given of supraspinal loci from which fusimotor effects can be elicited by electrical stimulation. For more details and references see Hulliger's (1984) review.

Motor cortex and corticospinal tract: In the baboon a monosynaptic co-activation of finger (and hindlimb) flexor α- and γ-motoneurones occurs with a preference for static compared with dynamic γ's. In the cat there is a variable balance of the effects on static and dynamic γ's and only partial co-activation (flexors) or co-inhibition (extensors) of α- and γ-motoneurones.

Cerebellum: Often reverse effects from supraspinal and deep loci on α- and γ-motoneurones ("decoupling" or "α-γ independence") are seen. Mostly tonic effects are exerted on static γ-motoneurones.

Extrapyramidal nuclei (striatum, globus pallidus, thalamus, and substantia nigra: Stimulation of the caudate excites static extensor γ's and (predominantly) dynamic flexor γ's. Excitation of static γ's occurs from the substantia nigra.

Vestibular nuclei: Monosynaptic co-activation of extensor α- and mostly static γ-motoneurones is exerted from the lateral vestibular nucleus (see below).

Nucleus ruber: Predominantly a co-activation of extensor and a co-inhibition of flexor α- and γ-motoneurones is found. A nearby area, the *MesADC* (mesencephalic area for dynamic control), appears to excite dynamic γ-motoneurones exclusively.

Reticular formation: From the medial longitudinal fascicle, flexor α- and predominantly static γ-motoneurones are monosynaptically co-activated. From other parts of the pontine and from the bulbar reticular formation effects on dynamic γ's prevail.

Very generally, then, effects on Renshaw cells can be elicited by stimulation of nearly the same supraspinal structures that also influence fusimotor neurones. Unfortunately, however, it is not yet clear in many cases how the supraspinal effects on Renshaw cells are related to those on the α- (and/or γ-) motoneurones to which the Renshaw cells may be predominantly related. It would be worth while to study such conditioning effects in parallel on fusimotor neurones and on Renshaw cells with a clear identification of both neurone types as regards their allocation to motor pools.

In view of the division of fusimotor spindle control into dynamic and static types and their possible separate supraspinal control it would be interesting to know whether a similar division exists in the Renshaw cell recurrent feedback. This possibility seems to exist, and a separate control of two types of Renshaw cells has indeed been suggested (see Sect. 4.3.6).

4.3.12 A Phylogenetic Comparison

The foregoing discussion suggests that proprioceptive feedback via muscle spindles and recurrent inhibition via Renshaw cells can be controlled from the same supraspinal centres and may thus act as parallel systems. Additional support for this notion of parallel organisation comes from the concomitant phylogenetic development of γ-motoneurones and Renshaw cells.

a) In amphibians, the system subserving spinal motor control is much simpler than in mammals (see Fig. 49 A and B). Only one type of motoneurone (β-motoneurone) innervates both extra- and intrafusal muscle fibres (see Ottoson 1976). Thus skeletomotor and fusimotor innervation are truly "linked" (see Sect. 4.11) in that they cannot be controlled separately. The message from muscle spindles is carried to the CNS via spindle afferents whose monosynaptic reflex connexions to motoneurones are relatively weak in frogs (see Simpson 1976, for review). The proprioceptive loop, whose efferent limb is made up of β-motoneurone axons, can formally be envisaged to be split functionally (although not necessarily structurally) into the intrafusal or "inner" loop and the extrafusal or "outer" loop (see Fig. 49 A). The former transmits the "β-signals" positively, the latter inverts their sign (on average). The inner loop is nested within the outer loop, which implies that the extrafusal muscle contraction would also affect the transmission of signals via the inner loop. But there still may be a difference in the origin and time course of the signals (therefore *two* afferents were drawn to represent the two loops). For example, a strong near-synchronous motor volley may first excite the spindle afferents (by way of early discharges) and then inhibit them (by spindle unloading). Murthy (1978) points out that the most prominent peripheral effect of a β-fusimotor action is the pronounced initial burst of spindle discharge at the onset of a tetanic contraction followed by an unloading due to extrafusal muscle shortening. For less compliant loads, the spindle discharge may even persist during the plateau of muscle contraction (Murthy 1978).

This scheme is not yet complete. In amphibians, there are no γ-motoneurones and no Renshaw cells either (Fig. 49 A). But there are recurrent effects. Activation of one motoneurone has strong excitatory effects on neighbouring motoneurones. It is not yet quite clear whether these effects are exerted through recurrent excitatory axon collaterals or solely through electrotonic coupling (dendritic gap junctions) between motoneurones (Simpson 1976; Schwindt 1976). At any rate, there is a positive feedback as indicated by the

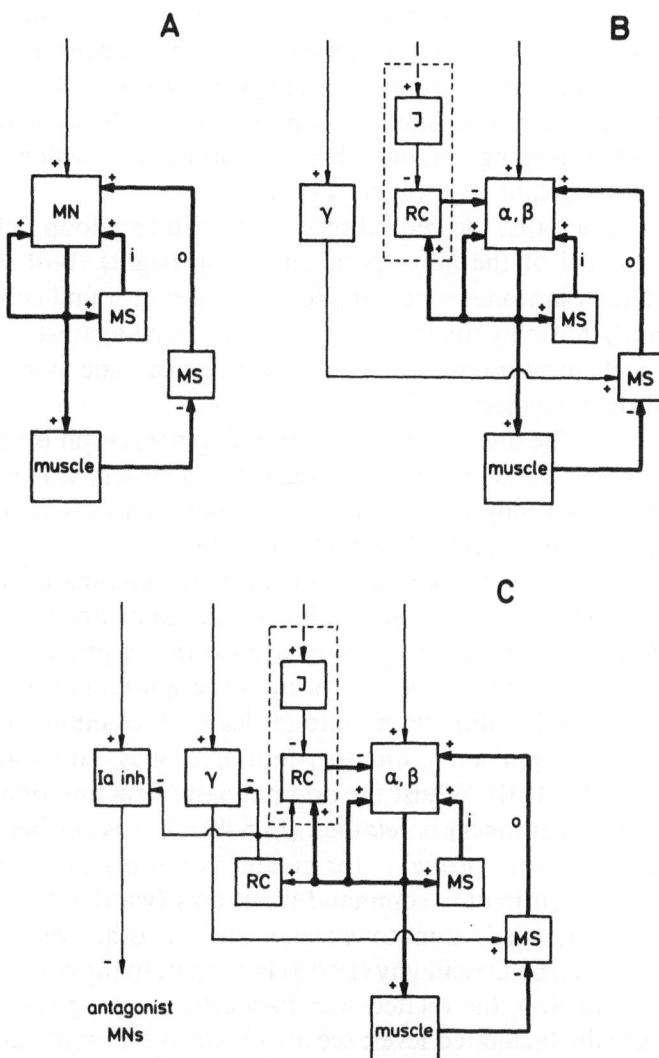

Fig. 49 A–C. Three simplified schemes of some motor control circuits for amphibians (**A**) and mammals (**B** and **C**). The *blocks* represent pools of neurones or muscle fibres. In **A** there is a direct recurrent excitatory loop from motoneurones onto themselves (*upper left loop*), an "inner" peripheral loop (labelled *i*) established by motoneurone axon collaterals to spindles and from their afferents back to motoneurones, and an "outer" loop (labelled *o*) established by motoneurone axons, skeleletal muscle fibres, muscle spindles and their afferents. The inner loop is nested within the outer loop because a mechanical influence is of course exerted by skeletal muscle on the spindle integrated in the inner loop; **B** reproduces the scheme of **A** and adds γ-motoneurones and the basic Renshaw cell circuit as two further motor input systems; the *block* labelled *J* represents the interneuronal system inhibiting Renshaw cells. For simplicity, only the γ-motoneurone input that influences the muscle spindles in the outer loop has been drawn; **C** in turn reproduces the scheme in **B** and adds the recurrent inhibition of reciprocal Ia inhibitory interneurones, γ-motoneurones and Renshaw cells ("mutual inhibition"). For more details see text

loop on the left in Fig. 49 A. Simpson (1976) argues that the short-latency interaction does not so much initiate motoneurone discharge as facilitate discharge and bring about the synchronisation of as yet unknown populations of motoneurones, perhaps to improve the dynamics of fast movements, such as leaping. So, here there is a situation in which correlations could play an important physiological role.

Formally, the central recurrent excitatory loop (left) may be regarded as a model of the inner peripheral loop (right). Both loops signal "events", namely motoneurone firing (central) or motor unit contraction (inner peripheral), but only the latter events are checked against the prevailing peripheral mechanical conditions. Thus, we have a cascade of models, each representing different aspects.

b) The discussion in the preceding paragraph is not quite adequate insofar as it outlines ideas with regard to a system which probably cannot (and need not) fully stand up to the demands. This discussion should therefore be considered a prelude to the following.

In the amphibian system (Fig. 49 A), commands are sent directly to the β-motoneurones via descending tracts; there are no γ-motoneurones and no Renshaw cells. It will now be argued that a prominent novel design feature of the mammalian motor system is the splitting of the command input to the peripheral motor system into (at least) three inputs: one to α-motoneurones (as in amphibians), one to γ-motoneurones, and another to Renshaw cells (see Fig. 49 B). Whilst the original basic structure (including a proportion of β-motoneurones) is retained (Fig. 49 A), it is further supplemented by new features, very probably for greater versatility and flexibility.

The fusimotor command input acts (via the "γ-loop") on homonymous and synergistic α-motoneurones and on other central neurones. Hultborn and Pierrot-Deseilligny (1979) claim that, in the course of augmenting muscle contraction, the related Renshaw cells are progressively inhibited (from an initially facilitated level; see also Sects. 4.3.11 and 4.3.12). This would imply that both the Renshaw cells and the fusimotor pathway would exert an influence of the same sign (by disinhibition) on α-motoneurones and on other central neurones (system enclosed by dashed lines in Fig. 49 B). However, in various other aspects the two command inputs to Renshaw cells and to fusimotor neurones differ both from each other and from those to α-motoneurones (Sect. 4.10).

The important point, then, is that from amphibians to mammals both the γ-fusimotor and the Renshaw cell system have evolved in parallel and thus appear to be linked functionally. It will later be proposed that they represent "designed" muscle contractions which are checked against prevailing circumstances at two different levels: centrally by Renshaw cells against the actual spinal α-motor output, and peripherally against the mechanical conditions in muscle (contractile state). These notions are taken from the theory of adaptive model-reference control systems and have already been applied to the physiology of motor control (Sects. 4.6 and 4.7).

4.3.13 Summary

There are a number of structural and functional analogies between recurrent inhibition via Renshaw cells and proprioceptive feedback via skeletal muscle and muscle receptor afferents onto α-motoneurones and other spinal nerve cells. These analogies concern the spatial organisation in the two feedback paths, the inhomogeneous distribution of feedback from and onto different types of α-motoneurones, the response characteristics of the feedback paths, supraspinal control of these paths, and some further similarities of input to various spinal neurones. However, the differences between the two systems are as important to note because they may turn out to yield clues to the functions of the two systems. Thus, the static and dynamic input-output relations of the two feedback paths are not exactly the same; the supraspinal and segmental afferent control might show interesting differences that have not yet been studied in any detail; the distribution to various motoneurone pools is more widespread for recurrent inhibition than for monosynaptic Ia input; and the relative importance of recurrent inhibition and proprioceptive feedback may vary from proximal to distal muscles. Be that as it may, the comparison of these two systems should incite new research, experimental as well as theoretical, oriented towards developing an integrated concept of motor control. An attempt in this direction is made in the following sections.

4.4 The Reafference Principle

In view of the remarkable analogies between the organisation of recurrent inhibition and proprioceptive feedback discussed above, it is tempting to envisage the former as a central model of the latter. A related idea is that recurrent inhibition could provide the "efference copy" of the motor signal leaving the central nervous system. To clarify these notions let us take a look at von Holst and Mittelstaedt's "reafference principle" (see also Evarts 1972; and MacKay 1966, for an alternative).

In 1950 von Holst and Mittelstaedt published their now famous and widely discussed "reafference principle", by which they wished to propose one among several possible solutions to a ubiquitous problem (von Holst and Mittelstaedt 1950). The general problem can be stated as follows (see Mittelstaedt 1971): The organism usually responds in a different way to sensory messages that originate from changes in its environment than it does to those which originate from movements caused by the organism itself. How does this come about? Von Holst and Mittelstaedt (1950) gave a solution in very general terms which led to some misunderstanding. In 1971 Mittelstaedt therefore felt compelled to scrutinise the original notions. His reassessment provided a great deal of clarification. The following description closely follows his 1971 article.

Those sensory messages that are caused exclusively by the environment were called "exafference", those resulting from movements of the organism "reafference". The problem then is the mechanism by which the organism can distinguish between these two "afferences". (Note that "afference" here means a whole complex of single sensory messages implying that it is a *signal vector* rather than a scalar quantity.) Now the proposal goes that this distinction crucially depends on the "state of the nervous system". This suggestion reversed the perspective commonly adopted. Mittelstaedt (1971) makes this point very clearly. The very successful method used most often before was to present to the sensory organs certain, if possible, quantifiable stimuli and to measure the reaction of the effector organs. In this approach, the influences of all the variables not measured – including the spontaneous actions of the central nervous system – are legitimately regarded as *disturbances*, which are eliminated either in advance by an experimental procedure or afterwards by statistical methods. The method tends to mislead us by regarding the system as a kind of complex automaton transforming input data into output data. The perspective of the physiologist proceeded from the sensory organs through the CNS to the effector organs, i.e. followed a – more or less complex – reflex pathway. Instead, von Holst and Mittelstaedt chose to start at the central end, the CNS, which, according to von Holst's previous results, is active also when no stimuli are present to affect it. Imagine an active CNS which issues commands to the effector organs and receives messages from its sensory organs. All the messages coming in regularly when nothing happens in the environment evidently result from the own activity, are reafferences. All the messages coming in when no commands are issued are exafferences, whose meaning is change – caused by external forces – in the environment or in the latter's relation to the organism.

The general reafference principle can be described as follows (see also Fig. 50 A). The command k initiates a motor act by means of the motor system (M). The resulting movement changes the spatial relation (y) between the animal and its environment. The same can occur as a consequence of external disturbances (z). Both superimpose and generate the actual spatial situation (x) of the animal. The variable x is transformed by the sensory system (S) into the "total afference", which is thus a mixture of reafference and exafference. Total afference and command are now compared at a level labelled Z_1. The result of this comparison generates a "message" that is sent to a higher CNS level (labelled Z_i) for further evaluation, be it for "recogni-

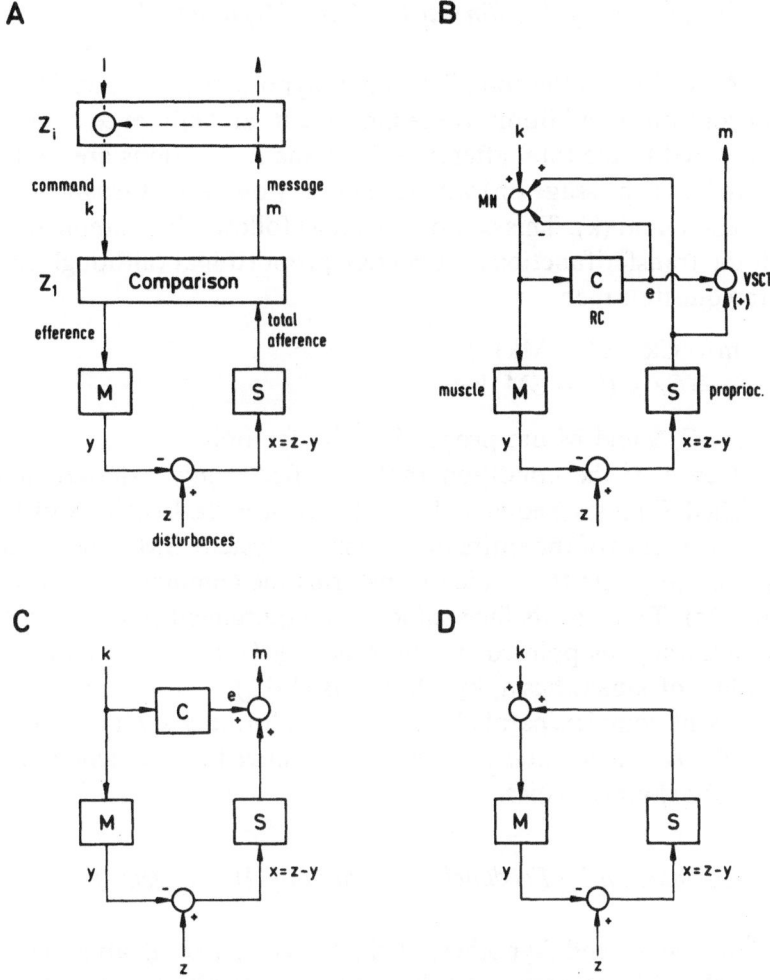

Fig. 50 A–D. Different schemes involving the reafference principle. **A** General scheme used by Mittelstaedt (1971; his Fig. 1a) and completed by details from von Holst and Mittelstaedt (1950; their Fig. 4); **C** One specific ("copy") hypothesis realising the reafference principle (modified from Mittelstaedt 1971; his Fig. 1b); **D** Another ("regulating") hypothesis realising the reafference principle (modified from Mittelstaedt 1971; his Fig. 1c); **B** Specific neurophysiological hypothesis showing one possible combination of the schemes in **C** and **D** and still compatible with the reafference principle. *MN* motoneurones (represented as a summing point); *RC* Renshaw cells; *VSCT* ventral spinocerebellar tract cells. For fuller explanation see text

tion" (in perception) or for a compensatory reaction (which may be unconscious). The general principle is thus that the comparison functions in such a way that the message reflects only external disturbance inputs (z), and *not* the endogenous activity resulting from the command (k). Any compensatory reaction would then counteract only the exogenous influences.

Let us now consider some specific hypotheses.

a) *Comparison by Feedforward ("Copy Hypothesis")*

According to the so-called "copy hypothesis" (Fig. 50 C), the command (k) controls, in addition to the movement, an "efference copy" (e), which is compared to the total afference. If all the interactions are calibrated appropriately, the message (m) will depend only on the disturbances (z) and not on the command (k). This can be shown as follows. For simplicity, assume that all the transfer functions are merely proportional (although this is no strict prerequisite); then:

$$\begin{aligned} m &= Ck + S(z - Mk) \\ &= Sz + (C - SM)k, \end{aligned} \tag{12}$$

where C, S and M are proportionality factors.

Hence, if the condition that m reflects only external inputs is to be satisfied, C must be equal to SM, whence m = Sz. In other words, the transfer characteristics of the entire motorsensory system, SM, must be equal to those of the subsystem (C), which transforms the command (k) into the efference copy (e). Taken at its face value, this requirement puts a very high demand on accuracy, as pointed out by MacKay (1966) and also discussed, in the context of kinaesthesia, by Matthews (1982). One reason is that SM and C represent complex parallel systems transforming signal vectors; thus many *details and fine features* of these systems have to be matched in structure and transfer characteristics.

b) *Comparison by Feedback ("Regulating Hypothesis")*

In the second hypothesis (Fig. 50 D), the total afference is compared directly with the command (k) itself. This results in a simple negative feedback loop. The following relations hold (with the same proportionality assumptions as above):

$$\begin{aligned} x &= z - y \\ y &= M(k + Sx) \end{aligned} \tag{13}$$

Solving for x yields:

$$x = \frac{z - Mk}{1 + SM} = \frac{z}{1 + SM} - \frac{k}{1/M + S} \tag{14}$$

Hence, for the influence of z (external disturbances) to become small, the total loop gain (SM) must be high. For the internal influence (command k) to become great, M must be large compared to S.

According to Mittelstaedt (1971) the above two hypotheses both comply with the reafference principle, but they are not functionally equivalent. The feedback circuit yields a compensatory reaction but no "message" to higher centres. The "efference copy" hypothesis yields such a message but no compensatory reaction (only a higher centre could generate such a reaction from the message). Therefore, the two hypotheses are not alternatives, but rather complementary to each other. Indeed they can be combined.

4.5 The Renshaw Cell System as an Efference Copy?

a) With respect to the first formulation of the reafference principle (Fig. 50 C), it is tempting to equate the Renshaw cell feedback with the efference copy, and the sensory feedback with the proprioceptive feedback carrying information on internal (y) and external (z) variables. However, this simple equation does not work out easily. The signs do not tally. At the summing point (Fig. 50 C), where the message is generated from a superposition of total afference and efference copy, both act with positive signs (because the sign inversion occurs in the periphery, here in skeletal muscle, between motor signal and proprioceptive feedback). But Renshaw cells are inhibitory. (The sign problem would be alleviated in MacKay's 1966 more general scheme where an "evaluator" replaces the simple summing point. This scheme was designed for recognition processes at the cortical level, and Matthews 1982 resorted to it to reconcile some difficulties with corollary discharges and muscle spindle afferent feedback in the context of kinaesthesia. Such an evaluator would probably consist of an entire, possibly complicated, neural network; this concept is therefore not apt to solve the sign problem with the Renshaw cells.)

b) The strongest recurrent feedback is on motoneurones, i.e. directly on the motor output. Thus, a scheme such as that in Fig. 50 D would be more appropriate for description. One would then be left with a negative feedback loop, or, more correctly, with two nested loops, since Fig. 50 D originally represents the proprioceptive feedback (discussed by von Holst and Mittelstaedt 1950). What then does the Renshaw cell feedback do? Furthermore, what role is to be ascribed to the descending influences upon Renshaw cells and muscle spindles (via the fusimotor system)?

c) As pointed out by Mittelstaedt (1971), there are many possible combinations of the schemes depicted in Fig. 50 C and D. This is demonstrated here for one particular combination (Fig. 50 B) designed specifically with the recurrent and proprioceptive feedback in mind. Renshaw cells inhibit some ventral spinocerebellar tract (VSCT) cells (Lindström and Schomburg 1973), which also receive a complex proprioceptive input. (There are four types of VSCT cells: those excited by Ia afferents, a second type excited by Ib fibres, a third type excited by both Ia and Ib fibres, and those excited by neither of these two afferents; but VSCT cells also receive inhibition from Ia fibres; for more details see Ito 1984). The VSCT cells recurrently inhibited are those, which also receive monosynaptic Ia input. They could hence generate a message to higher motor centres. But the effects of Renshaw cells on VSCT cells are again inhibitory (Kim et al. 1982 provided evidence for direct excitatory input to VSCT cells from motor axon collaterals; however, neither VSCT cells nor motoneurones were identified); and the Renshaw cell inhibition of motoneurones must also be incorporated.

Proceeding in the same way as for Fig. 50 C (Sect. 4.4), the message (m) is derived as:

$$m = Sz - (C + SM)(k + Sx)/(1 + C) \tag{15}$$

Hence, if m is to depend solely on z, C must be equal to -SM, which is not the case. Only if Renshaw cells *and* proprioceptive feedback (Sx) both had negative (inhibitory) effects on VSCT cells (this does not appear to be totally impossible) would the condition change to $C = SM$, as before (Sect. 4.4). The message sent up would have a negative sign, but this would not necessarily be a disadvantage. Remember that the VSCT axons enter the cerebellum as mossy fibres and ultimately, after being processed (with positive sign) by the granule cells, *excite* Purkinje cells which in turn *inhibit* nuclear cells. The sign would therefore be inverted again.

In the latter case, i.e. if $C = SM$, the variable x would obey the relation:

$$x = \frac{(1 + C)z - Mk}{1 + C + SM} = \frac{(1 + SM)z - Mk}{1 + 2SM} \tag{16}$$

In 1971, Lundberg put forward the hypothesis that VSCT cells signal the state of activity of central spinal systems. Arshavsky et al. (1983) agree with this view, because in their studies of locomotion and scratch reflexes in decerebrate cats VSCT and spinoreticulo-cerebellar cells were rhythmically modulated with locomotor activity, this rhythmical pattern depending predominantly on spinal networks and persisting after deafferentation or curarisation (see also Mori 1987). The rhythmicity of these neurones could be derived from diverse central sources, two among them being Renshaw cells (whose activity is also rhythmically modulated in locomotion, see McCrea et al. 1980; Pratt and Jordan 1987) and recurrent motor axon collaterals (Kim et al. 1982). Apparently, however, the VSCT pool is fractionated by its diversified inputs (see above) as other interneurone pools appear to be (see Sect. 2.9.1).

4.6 Model-Reference Control Systems

Mittelstaedt's (1971) interpretation of the reafference principle endows it with a high versatility. The possible integration of negative feedback schemes opens the whole field of modern control theory. The main drawback of the specific hypotheses concerning recurrent inhibition and proprioceptive feedback (Sect. 4.5) is that supraspinal and segmental inputs to fusimotor neurones and Renshaw cells have not been taken into account. To do so, it is worth while to discuss some notions adopted from control systems engineering which have been repeatedly used in the description of motor control schemes designed to explain the role of *muscle spindles* in motor acts. These schemes do not generally take account of recurrent inhibition (but see Sect. 4.7.2). We shall first deal with model-reference control.

4.6.1 Basic Notions

The basic idea of model-reference systems is outlined in Fig. 51 A in terms of control engineering. The controlled *system* (or plant) is duplicated by a *model*. The command signal from an *input system* is directed to both plant and model, whose respective outputs, d (t) and c (t), are compared at a summing point generating a reference error (or mismatch) signal, e (t), whenever the outputs differ. (It is worthwhile to note that, up to this point, the scheme in Fig. 51 A resembles that in Fig. 50 C, the model corresponding to the efference copy and the error signal to the message; only the use made of the error signal differs). In fact, the error signal can be used in more than one way, as indicated by the numbered dashed lines. It can be fed back to the controlled system (2) or to the model (1). Physiological interpretations of the first possibility are presented in the next section.

In physiological terms, the second possibility implies feeding back the error signal, e (t), to fusi-motoneurones (Fig. 51 C: dashed path). Houk and Rymer (1981) point out the theoretical advantage of this configuration by stating that such a positive-feedback loop, if appropriately tuned, can actually result in improved stability along with greater attenuation of errors. Errors can be reduced to zero if the gain of the positive-feedback loop is precisely 1. Such a system is called a zero-sensitivity system, since it is insensitive to disturbances. Note that this possibility is structurally realised for β-innervation, but that there is also weak spindle feedback to γ-motoneurones (Sect. 2.9.1.2). [Rymer and Grill 1985 estimated the gain of the positive feedback loop via β-motoneurones and muscle spindles (the β-path) for soleus and medial gastrocnemius muscle in decerebrate cats: It is close to zero for the former and averages 0.46 for the latter. They think these values could be underestimated due to the influence of muscle series compliance.]

4.6.2 Hypothetical Motor Control Schemes

A physiological interpretation of the abstract scheme in Fig. 51 A can be given regarding β-innervation (Fig. 51 B), which prevails in amphibians and

A

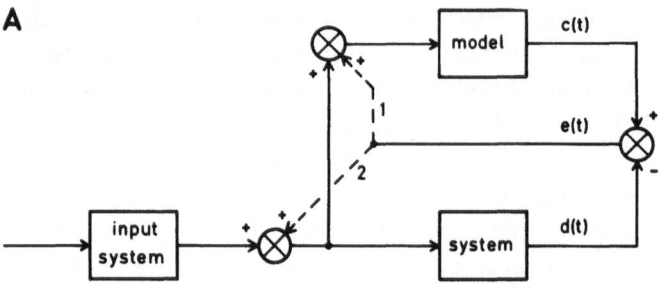

B

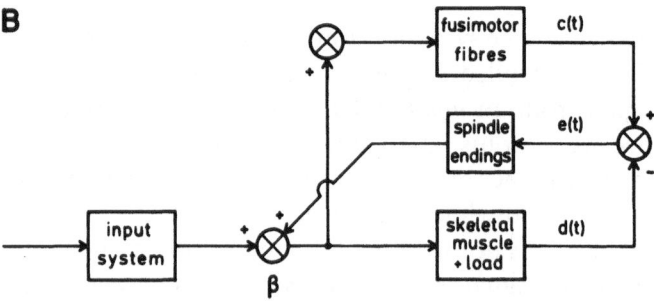

C

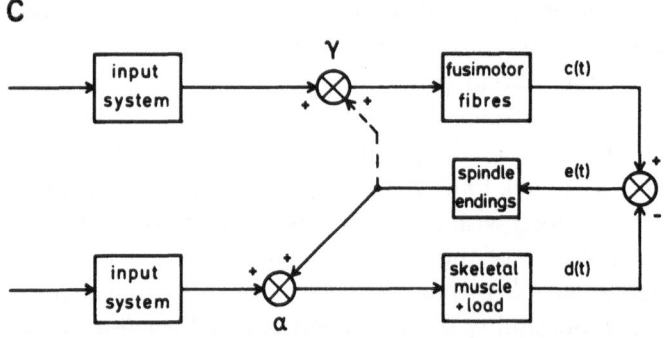

Fig. 51 A–C. A model-reference control scheme and two applications to motor control. **A** The *system* (or plant) is "duplicated" by a *model* (with similar properties). Both subsystems get the same input from an *input system*, and thereupon generate outputs $c(t)$ and $d(t)$, which are compared at a summing point. If different, an error signal $e(t)$ is generated. This signal may be fed back to the model (*1*) or to the system (*2*); **B** Skeleto-fusimotor (β-) control as in amphibia and reptilia; **C** Parallel skeleto- (α-) and fusimotor (γ-) control as in mammalia. For more details see text

reptilia, but is also found in mammals. The plant is the skeletal muscle including its load (Houk and Rymer 1981), the model is the intrafusal muscle of the spindle. The input to both is delivered by skeleto-fusi-motoneurones, and the two outputs, d (t) and c (t), interpreted as muscle and spindle lengths, respectively, are compared by the spindle endings. (Note that length is a one-dimensional variable.) The connexion of spindle afferents back to β-motoneurones generate a "conditional feedback system". This term implies that the action of the afferent spindle signals comes into play upon the condition that the desired muscle length output, as represented in the muscle spindle length, deviates from the actual muscle length. If muscle length actually does not deviate from its desired value, the conditional feedback vanishes. As pointed out by Houk and Rymer (1981), what is really meant by a vanishing conditional feedback is that spindle afferent discharge stays constant at some nonzero level, rather than actually dropping to zero pulses/s. The discharge is thus biased to an elevated level, which is important because it permits the detection of bidirectional reference errors. The function could be to correct for deviations of actual muscle length changes from intended ones. If, for instance, a movement is impeded by an unusually large load, skeletal muscle would shorten less than required for the muscle spindle to maintain its reference rate of discharge: The spindle would increase its rate, and vice versa. Provided the feedback gain is high enough, disturbances such as unexpected loads and changes in system properties (e.g. muscle fatigue) can be effectively compensated for. In principle, the same reservation applies to this control scheme as that raised by MacKay (1966) against the reafference copy (Sect. 4.4): The demand on accuracy is very high. And it is not met (see Sect. 4.11).

The fact that, in mammals, fusimotor innervation (input to the model) is not exclusively provided by branches from β-axons, but also by γ-motoneurones (see Fig. 51 C), opens new possibilities for motor control.

a) *Merton's "follow-up length servo" hypothesis* (1951, 1953) explicitly took account of an additional γ-fusimotor spindle input by endowing it with the priority in the initiation of motor acts (at least in slow precise movements). The basic idea was taken from control theory and envisaged the loop made up of α-motoneurones, skeletal muscle, muscle spindles and Ia afferents as a lumped negative feedback system used as a servo. The motor command from supraspinal centres would be directed to the γ-motoneurones and then relayed around the reflex loop to α-motoneurones. The problem with this hypothesis is that it requires a high loop gain and that it is fairly slow.

b) Motor commands are normally directed to both α- and γ-motoneurones (Fig. 51 C). In a strict sense, this physiological configuration behaves like a model-reference control system (Fig. 51 A) *only* if the activation of α- and γ-motoneurones is "linked" (Granit 1970) or if both types of motoneurone are co-activated (Phillips 1969). However, there are some problems with these assumptions, which will be detailed in Sect. 4.11. This conditional model-reference scheme also includes Matthews' "servo-assistance" hypothesis (1972), which took account of the two fusimotor types, static and dynamic, and of the low gain around the loop. It was assumed that static γ-motoneurones modulate their discharge in accord with α-motoneurones so as to prevent spindle silence during muscle shortening, and that dynamic γ-motoneurones are α-independent, possibly tonic in discharge, in order to maintain a high sensitivity of Ia afferents to small disturbances during the movement.

One problem with some of the above schemes is that, if the spinal stretch reflex is regarded as a servo system for the control of limb position (or muscle length), it has a gain which is "disappointingly low" (Rack 1981 b), i.e. rather ineffective in load compensation (see also Houk and Rymer 1981; Loeb 1984). Vallbo (1981) summarised a number of studies by stating "that spindle afferent activity does not provide a major or indispensable excitatory drive to the skeletomotor neurones" (p. 266). Although this view was recently challenged by Hagbarth et al. (1986) who argued that the γ-loop did indeed contribute non-negligible power to maximal voluntary muscle contractions in man, and although the gain can almost certainly be different for different muscles and under different conditions (e.g. levels of contraction), one would certainly be well advised to look for other possible functions of the muscle afferent information to the CNS. In general, the above schemes are far too simple to explain the behaviour of neural elements in animals awake during natural movements and the complexity of spinal circuits (Sect. 4.11 and Chap. 5; see also Loeb 1987).

4.7 Adaptive Control Systems

4.7.1 Basic Notions

In addition to the idea of model-reference control introduced above, the notion of adaptive control systems is also of importance. Interestingly, this notion originated from the study of biological control systems (see Inbar 1972), rather than from engineering control theory, and Inbar (1972) in fact was one of the first to apply it to problems of motor control. His approach is briefly outlined here to give a general impression. Inbar (1972) defines "self-adaptive" systems as having the means of monitoring their own performance relative to a certain criterion, and as modifying their own controlling parameters in a closed loop action, in order to approach a desired output. This necessitates the ability to perform the following functions:

1) Definition of the performance criterion, i.e. of the desired performance.

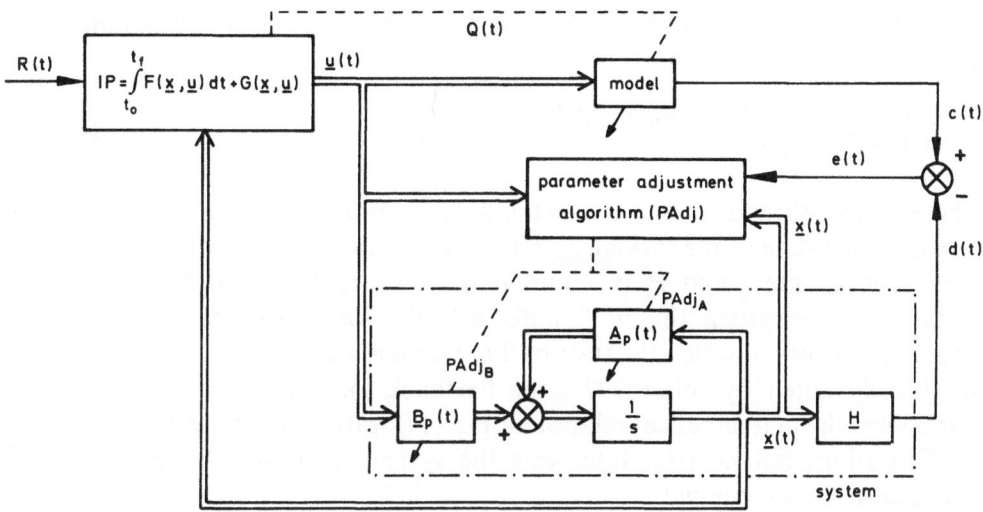

Fig. 52. General scheme of a model-reference *adaptive* system. The *system* (plant) to be param-
eter-adapted is *at the bottom* (enclosed by a *dashed line*). It is represented in standard state
variable form so that the block denoted $\underline{A}_p(t)$ does not signify a real feedback path. The *model*
is again on top. As in Fig. 51 A, the system output, $d(t)$, is compared with the model output, $c(t)$,
yielding the error signal, $e(t)$, which in this case, however, is not directly fed back to the system
or model input, but to a *parameter adjustment system* (PAd_j). This also receives the common
input, $\underline{u}(t)$, and information on the system state $\underline{x}(t)$. On the basis of the information received,
this algorithm adjusts the parameters of the plant (indicated by *two dashed arrows*) according to
a performance criterion (*left upper block*) which receives the command input $R(t)$ and state
feedback, $\underline{x}(t)$, and produces the common input, $\underline{u}(t)$, and a signal, $Q(t)$, possibly changing
model behaviour. *Double lines* signify signal vectors. (With permission slightly modified from
Inbar 1972)

2) Measurement of its performance, i.e. identification of system dynamics.

3) Comparison of actual performance with the desired performance and
achievement of a decision on how to realise the desired performance.

4) Adjustment of the system parameters to produce the necessary compensating action.

Figure 52 (modified from Inbar 1972) shows a very general scheme of a
system that would be able to perform the required functions. It should be
noticed that double lines represent signal *vectors*, which consist of several
signals ("several" ranging from a few to rather large numbers). Thus, multi-
input, multi-output aspects are emphasised. The *system* (plant) whose
parameters (gains, dynamics) are to be adapted is shown at the bottom
(enclosed by the dotted line). It is presented in a form corresponding to the
standard state variable form (DeRusso et al. 1965). Assuming linearity for
simplicity of description (though this is no indispensable requirement), sys-

tem behaviour can be described by the following linear vector differential equations:

$$\dot{x}(t) = \underline{A}_p(t) \cdot \underline{x}(t) + \underline{B}_p(t) \cdot \underline{u}(t),$$
$$d(t) = \underline{H} \cdot \underline{x}(t), \qquad (17)$$

where $\underline{u}(t)$ is the input vector, $\underline{x}(t)$ is the state vector representing the momentary dynamic state of the system, $\underline{A}_p(t)$ and $\underline{B}_p(t)$ are matrices of the parameters to be adapted, and $\underline{H}$ is a unit matrix. $\underline{A}_p(t)$ describes the inherent dynamics of the system, that is, its unforced behaviour when left alone. $\underline{B}_p(t)$ describes the way in which the state of the system is changed by the input. [As specifically noted by Inbar 1972, the feedback loop in the system block symbolises the mathematical relation in Eq. (17), rather than a real feedback configuration; but he later interprets the system in terms of such a real configuration; see below.]

The (reference) *model* at the top gets the same input $\underline{u}(t)$ as the system below. In the model-reference scheme, it is reasonable to assume that the model behaviour can be described by the same type of equations as the system behaviour:

$$\dot{y}(t) = \underline{A}(t) \cdot \underline{y}(t) + \underline{B}(t) \cdot \underline{u}(t)$$
$$c(t) = \underline{H} \cdot \underline{y}(t). \qquad (18)$$

The essence then consists in adapting the variable parameters in Eqs. (17), so that the dynamic behaviour of the first set of Eqs. (17) becomes equal to that of the second set (18). This is done by a parameter adjustment subsystem on the basis of information from three sources: (1) the outputs of the model and of the system, $c(t)$ and $d(t)$, which are compared (here at a summing point) yielding a reference error signal, $e(t)$; (2) the input signal, $\underline{u}(t)$, and (3) feedback from the system state, $\underline{x}(t)$. The parameter adjustment subsystem thereby receives all the information necessary to adapt the system parameters comprised in $\underline{B}_p(t)$ and $\underline{A}_p(t)$ to the prevailing conditions and demands. This adjustment is carried out according to a criterion (index of performance: IP) which may very generally be given by

$$IP = \int_{t_0}^{t_f} F(\underline{x}, \underline{u}) \, dt + G(\underline{x}, \underline{u}), \qquad (19)$$

where $t_f - t_0$ is the control interval, and $F(\underline{x}, \underline{u})$ and $G(\underline{x}, \underline{u})$ are respectively accumulative (integrated) and instantaneous "penalty" or "pay-off" functions which have to be minimised. The index of performance may be chosen very differently according to the tasks performed, e.g. as energy expenditure, deviation from the desired course, time required for a motor task etc. (Recently Hogan 1984 used minimisation of jerk to model a simple single-joint movement. Hasan 1986 adopted the product of joint stiffness and change of equilibrium position as a criterion.) Finally, the model may also be adapted according to the desired performance through the function $Q(t)$. Thus, in general, the model itself may be parameter-adjusted, this possibility providing for a high flexibility.

4.7.2 Interpretations

a) Inbar's (1972) tentative interpretation of his model in physiological terms is not very detailed and not always quite clear. It must be admitted, however, that a precise allotment of system components (Fig. 52) to physiological subsystems is difficult. In brief, Inbar's interpretation is as follows. The actual system output, $d(t)$, is monitored by joint receptors if $d(t)$ corresponds to limb position (this is unlikely by current knowledge; see Burgess et al. 1982; Matthews 1982), or by Golgi tendon organs if $d(t)$ corresponds to tension. This is an interesting idea: Obviously depending on the task, the actual system output, $d(t)$, could be position or tension (or something else). Anyway, on Inbar's view, it is not the muscle spindles' task to provide the signal $d(t)$. The desired output, $c(t)$, as well as the index of performance are determined at the supraspinal level (cortex, subcortical nuclei, cerebellum), using state feedback information, $\underline{x}(t)$. Hence, the command signal, $\underline{u}(t)$, descends from supraspinal structures and adapts spinal and peripheral parameters. This adjustment concerns, firstly, parameters collected in matrix $\underline{B}_p(t)$, which represent motoneurone bias or *Renshaw cell feedback properties*, and, secondly, parameters collected in matrix $\underline{A}_p(t)$, which represent *muscle spindle properties* to be changed by fusimotor input. It should be noticed that, in this interpretation, the system enclosed by a dashed line in Fig. 52 consists of the monosynaptic spindle-motoneurone reflex loop plus, perhaps, the Renshaw cell feedback. Muscle spindles, therefore, do *not* serve as models of skeletal muscle and its load (Fig. 51 B), but are integrative components of the system to be parameter-adjusted. They do not identify system dynamics (see Inbar's item 2 above), but contribute to them. This scheme is conventional insofar as it preserves the prevailing view that descending signals change the parameters of spinal circuits (e.g., gain of recurrent inhibition: see Hultborn et al. 1979; or gain of the spindle afferent pathways: see Hulliger 1984, for review). (Note that even if only the *gain* of the muscle spindle, and not its phase, is changed, the dynamic properties of the whole loop are altered.) The truly new feature (in 1972) was the recognition and incorporation of the fact that the monosynaptic reflex loop has too small a gain to serve as an efficient servo (Inbar 1972; Rack 1981 a, b), the gain being great enough, though, to enable sufficient change of system dynamics. Models are not really necessary in Inbar's scheme, it concentrates on parameter adaptation. This then clearly demonstrates the distinct nature of model-reference and adaptive systems.

b) An interpretation of spinal stretch reflex mechanisms, particularly of the role of muscle spindles in motor control, which tends towards an adaptive control scheme, was also roughly outlined by Gottlieb and Agarwal (1980). First they emphasise that muscle spindles provide the CNS with timely and precise information about current selected state variables of the motor system, including initial conditions etc. Information about unexpected perturbations to the movement can be utilised in various ways; but a particular consequence, namely the tendon jerk and other short-latency stretch-evoked responses *per se* are – to every neurologist's dismay – interpreted as epiphenomena of little physiological importance. Instead, these short-latency mechanisms are envisaged as *modulating* the more potent but delayed responses mediated via long-latency routes (see below). Gottlieb and Agarwal (1980) emphasise that this interpretation turns conventional servo thinking on its head. In such conventional thinking, supraspinal centres are usually assumed to modulate the gain of segmental reflex pathways.

Gottlieb and Agarwal's (1980) interpretation is interesting and nonconventional for two reasons. Firstly, in contrast to Inbar (see above), they change the relative roles of descending and afferent segmental influences. Secondly, these afferent influences are conceived of as "modulating" (adapting) the signal flow from descending tracts, thus placing adaptation at a spinal level.

This has also some bearing upon the relation of spinal and "long-latency" stretch reflexes (overview: Desmedt 1978). Stretch reflex function is often investigated by applying external perturbations, e.g. various forms of muscle stretch (as also done by Gottlieb and Agarwal 1980; for a discussion of the limitations of this approach see Evarts 1981). The contracting muscle (and its synergists), which are stretched, usually respond in three or four phases: with an immediate inherent response (due to muscle stiffness) and with two or three reflex responses, the first of which is of segmental origin, the following ones being "long-latency" responses. Whereas the latter are usually rather effective in compensating for the change in load, the first spinal reflex, commonly accepted as being mediated by monosynaptic Ia fibre-motoneurone connexions, is too weak in many limb muscles to compensate for the perturbation to any considerable extent (but see Calancie and Bawa 1985). The pathway mediating the longer-latency muscle responses to fast disturbance transients is still heavily debated. The early suggestion (1) of supraspinal long-loop reflex pathways possibly traversing the cortex (see Sect. 3.4.2.2) has been challenged by the proposals that (2) segmentation of reflex responses is due to muscle oscillations ("resonance hypothesis"; Eklund et al. 1982), (3) the second reflex component is initiated by spindle group II afferents (Matthews 1984), and (4) that this second component results from activation of cutaneous receptors (Darton et al. 1985). Such quibbles aside, however, it is clear that the part played within these multi-loop systems by short-latency segmental reflex pathways, with their low gains and their concomitant inability to effectively compensate for external disturbances, will have to be reconsidered.

4.8 An Integrated Model Including Proprioceptive Feedback and Recurrent Inhibition

Discussing the possible involvement of spindle afferent signals and corollary discharges in kinaesthesia, Matthews (1982) noted that "the fusimotor system still occupies a paradoxical position in relation to our present elementary ideas about efference copy" (p. 213). This is so because, on the one hand, the fusimotor system requires corollary discharges in the CNS so that the spindle signals can be decoded, but, on the other hand, it has some of the characteristics of an efference copy in relation to skeletomotor discharge. Perhaps part of the paradox can be removed by integrating recurrent inhibition into the scheme. This will be attempted in the following. So far, the scheme to be developed is qualitative. Quantitative formulations in terms of a mathematical description or computer simulation remain a task of the future. But at this stage, the proposed scheme can at least serve to initiate further research (Sect. 4.13).

In Fig. 53 is depicted a general scheme comprising the two feedback systems. This scheme is held to be general enough to enable an interpretation in terms of whole groups of muscles involved in a movement or position-holding task (including also antagonists). That is why vector notation has been used stressing multiple-input, multiple-output characteristics which *encompass spatial aspects of organisation.*

The controlled *system* (or plant) is again given, as in Fig. 51 C, by skeletal muscle(s) and its (their) load. The *model* is split into two parts, one represented by Renshaw cells (the central model) and the other by spindles (the peripheral model). It follows that there are two model outputs, $\underline{e}_{RC}(t)$ and $\underline{e}_{MS}(t)$, the Renshaw cell output and the muscle spindle output. The first signal, $\underline{e}_{RC}(t)$, represents a comparison between the spinal skeletomotor output signal, $\underline{d}_{MN}(t)$, and the descending and other inputs to the Renshaw cell model, $\underline{u}_1(t)$. The second signal, $\underline{e}_{MS}(t)$, correspondingly represents a comparison between the peripheral motor output (muscle) signal, $\underline{d}_{MU}(t)$, and descending and other inputs to the fusi-motoneurones, $\underline{u}_2(t)$.

The *combination of the two motor output signals,* $\underline{d}_{MU}(t)$ and $\underline{d}_{MN}$ (encircled), contains information on the *properties of the plant that is not contained in either of them alone.* It bears emphasis that this is a very important novel feature of the present scheme. Since the comparisons performed by the two models use the two different versions of motor output (central and peripheral), it appears intuitively reasonable that they should also receive somewhat

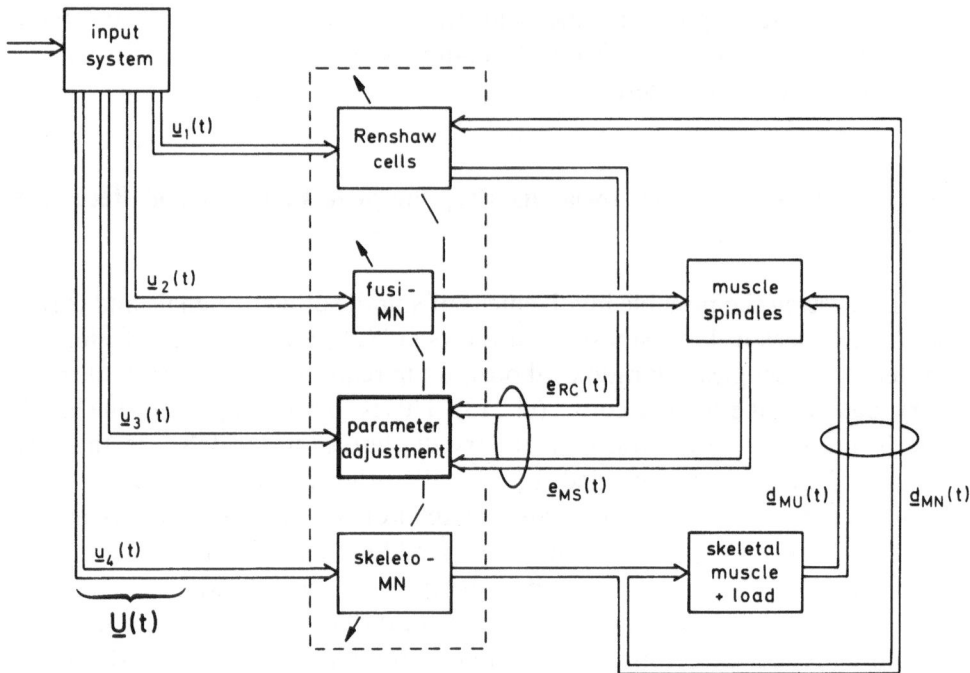

Fig. 53. General scheme of a model-reference adaptive motor control system incorporating recurrent inhibition. All signal paths are represented by *double lines* and *arrows* in order to indicate the multi-input, multi-output character. This scheme is so general as to encorporate control of groups of muscles including antagonists. The *system* (or plant) is represented, as in Fig. 51 C, by the skeletal muscle and its load. System output is designated $\underline{d}_{MU}(t)$ and is sensed by muscle spindles. Additionally, plant input or motoneurone output is designated $\underline{d}_{MN}$ and sensed by Renshaw cells. These two signals, $\underline{d}_{MU}(t)$ and $\underline{d}_{MN}$ (*encircled*), contain information on the properties and performance of the plant. Renshaw cells and muscle spindles get two components of the command input, $\underline{u}_1(t)$ and $\underline{u}_2(t)$, respectively. Their outputs, $\underline{e}_{RC}(t)$ and $\underline{e}_{MS}(t)$ (*encircled*), are fed back to spinal neuronal networks, including skeleto- and fusi-motoneurones as well as Renshaw cells, for parameter adjustment. The parameter-adjustment subsystem is not sharply delineated from the latter neuronal systems, as symbolised by the *dashed box*. Yet it may contain additional interneuronal systems, such as reciprocal Ia inhibitory interneurones. It also may receive an input signal component, $[u_3(t)]$ different from those to Renshaw cells, fusi-motoneurones and skeleto-motoneurones $[u_4(t)]$. The entire input vector is designated $\underline{U}(t)$

different inputs. Two further input components are directed to the parameter adjustment system ($\underline{u}_3(t)$) and to skeleto-motoneurones ($\underline{u}_4(t)$). The entire input vector is designated $\underline{U}(t)$.

The combined outputs from the two models, $\underline{e}_{RC}(t)$ and $\underline{e}_{MS}(t)$ (encircled), adapt the parameters (spatio-temporal excitability distributions, parameters of input-output relations, i.e. thresholds and gains, etc.) in some spinal neuronal networks, including skeleto-motoneurones (encompassing the known categories FF, FR, and S) and fusi-motoneurones (encompassing static and dynamic categories) and including even Renshaw cells. The parameter-adjustment subsystem should not be conceived of as being sharply delineated from skeleto- and fusi-motoneurones because these receive recurrent inhibition and proprioceptive (spindle) feedback; this is symbolised by the dashed box. The former nonetheless contains additional interneuronal systems, such as reciprocal Ia inhibitory interneurones.

The parameter adjustment is done in accordance with the varying functional requirements of the tasks to be performed (see Sect. 4.11). This task-dependence is partially co-determined by the state of the peripheral system. That is why Renshaw cells and fusimotor neurones get all kinds of segmental inputs from various peripheral receptor systems (this influence is not represented in Fig. 53), albeit in somewhat different patterns than with the descending inputs. The state feedback to the models once again increases the complexity and flexibility of the entire system above and beyond those attained by letting the performance criterion influence the model by the function $Q(t)$ in Fig. 52. The optimum performance criterion is generated either supraspinally (for "voluntary" motor acts) or spinally (e.g. in locomotion).

Why parameter adjustment? One reason is the same as that put forward by Inbar (1972): low gain of the stretch reflex and probably also of the recurrent inhibitory feedback loop (W. Koehler and Windhorst 1983, 1985; Windhorst and Koehler 1983, 1986). Another is that spindle afferents and Renshaw cell axons diverge to different neuronal systems that do not all form simple single-loop feedback loops with them. The concept of regulation and servo is more difficult to apply in this case. The more general concept would be that of parameter adjustment. This by no means excludes feedback mechanisms, but on this view a feedback subsystem is treated as a special case rather than as a general principle.

In this context, it is important to emphasise that feedback need not only be used to regulate the signalled variable by a servo. It can also be used to change the response of a system to other inputs according to peripheral conditions. This distinction is particularly pertinent in nonlinear systems, where linear superpositions may be supplanted by multiplications, threshold effects and other nonlinear operations. In biological systems (such as the neuromuscular system with its complicated proprioceptive feedback), the primary "purpose" of feedback is often not easily recognisable. The *observable* effects of activating certain feedback pathways (for instance, by stretching muscles) may be secondary by-products, misleading the observer to draw premature conclusions. One should thus be cautioned before generalising from results obtained by using very special (and sometimes artificial) experimental paradigms such as those often employed in studying "stretch reflex function".

4.9 Remarks on the Function of the Integrated System

It is suggested that the dual model of muscle spindles and Renshaw cells serves to detect properties of the plant between spinal motor output and spindle input. Among the plant properties that can be recognised by use of *input and output* signals to and from muscle are nonlinearities, spatial inhomogeneities and so forth (see Sect. 5.1.2). Skeletal muscles vary widely in structure, properties (and their spatial distribution), and "inherent" loading resulting from the muscle's location within the complex of the body or limb (see Partridge and Benton 1981; and Chap. 5). As argued below (Sect. 4.11), it cannot generally be stated exactly which plant properties are detected by the two models, but a simple exemplifying proposal for restricted conditions is presented in Sect. 5.1.2. Some general remarks are nevertheless indicated at this stage.

New insights may often be gained from following up the deviations that any model exhibits from the reality to be modelled. For example, some functional schemes of motor control assume that muscle spindles essentially are models of skeletal muscle and its load (see above). However, for various reasons, muscle spindles cannot be perfect models. One reason is that they can, at best, represent the mechanical state of the muscle fibres in their vicinity, not of the whole muscle, nor of the load (see also next chapter). Moreover, the mechanical behaviour of spindle and skeletal muscle may be different. Furthermore, skeletal muscle is a three-dimensional structure with inhomogeneously distributed properties. Spindles are dispersed (not always homogeneously) throughout this volume and sample some distributed parameters, but do not necessarily, by their afferent synaptic connexions, represent an ensemble average. Thus, the CNS, upon issuing a skeletomotor neural command, cannot *know a priori* how it will be translated into a motor act, partly due to the nonlinearities inherent in the muscular response, which may arise from various sources (see Demiéville and Partridge 1980; Partridge and Benton 1981; Partridge 1982; Sects. 3.2.2 and 4.3.3). But, from the spindle afferent signals alone, it also cannot *learn a posteriori* what has happened to its motor command, because the nonlinearities are not reflected fully or faithfully in spindle afferent discharge variations (Sect. 3.2.3). In order to estimate the quality and amount of the nonlinearity arising in skeletal muscle, the spindle afferent signals should be compared with the skeletomotor command itself which should be represented in a central model. This can be done by the Renshaw cells that monitor motor output fairly faithfully (Cleveland 1980).

It is therefore the *differences between the feedback delivered by Renshaw cells and by muscle spindle afferents* that may be of physiological importance. These differences can only be evaluated against the background of similar functional structures for recurrent inhibition and proprioceptive feedback. Both features have been summarised in Sect. 4.3.

There are two issues that deserve further attention in the context of the present hypothesis, namely some differences in the distributions of recurrent inhibition and proprioceptive feedback.

The distribution of recurrent inhibition between different motoneurone pools is such that pools connected in Ia synergism (monosynaptic Ia connexions) are also connected via recurrent inhibition, this being a similarity; however, the distribution of recurrent inhibition is wider (Baldissera et al. 1981), that is, it interconnects more α-motoneurone pools than Ia monosynaptic connexions do.

Secondly, there are muscles (in cat) lacking recurrent inhibition and proprioceptive feedback (via Ia fibres), such as many cranial motor pools. The diaphragm contains only a few muscle spindles, and phrenic motoneurones are subjected to little recurrent inhibition (Hilaire et al. 1983; Lipski et al. 1985). However, there are small plantar muscles that appear to lack recurrent inhibition (Cullheim and Kellerth 1978a; Cullheim and Ulfhake 1985; see also Egger et al. 1980), but are supplied with spindles. This may reflect the tendency for recurrent inhibition to be more prominent in proximal than in distal muscles (Baldissera et al. 1981), which is not the case for spindle feedback.

How could these differences be explained or at least made plausible in the context of the motor control scheme outlined above? Regarding the first issue, it may be assumed that a second model (recurrent inhibition) is the more necessary, the less adequately the peripheral model (muscle spindles) represents the skeletal muscle and its load. Now, progressing centrally from limb periphery, extrafusal muscles are increasingly loaded by the mass of the limb peripheral to their insertion. Relative inertial load (apart from changing with limb position, owing to gravity) increases from peripheral to proximal muscles. Furthermore, proximal muscles have more widespread mechanical actions than more distal muscles do. For instance, ankle muscles directly control the foot dynamics, whereas knee muscles control the "intersegmental" dynamics between shank and foot (J. L. Smith and Zernicke 1987; see Sect. 5.2). If muscle spindles were to model skeletal muscle inclusive of its load, they would have to be increasingly complemented by a central model such as the recurrent inhibitory system.

Concerning the second issue, it should be kept in mind that even spindle feedback consists of two afferent systems (disregarding Golgi tendon organ feedback for the moment). It is thus not fair to compare the distribution of recurrent inhibition among various motor nuclei exclusively with the distribution of monosynaptic Ia feedback. That of group II spindle afferents should also be taken into account. Indeed, the latter distribution appears to be very wide and complicated (Schomburg and Steffens 1985; Lundberg et al 1987). Cavallari et al. (1986) described a group of interneurones in segments L3/L4, which excited or inhibited triceps surae or hamstring α-motoneurones and received monosynaptic excitation from group I and/or group II afferents from various sources. Thus, group I and II spindle afferents might also converge on common interneurones intercalated in disynaptic excitatory or inhibitory pathways. In general this indicates that the recurrent inhibitory and Ia feedback, to which the above comparison applies, represent just one functional layer of connectivity which is certainly accompanied by further layers whose functional importance, however, is not yet known. But it may be interesting to find out whether the postulated two categories of Renshaw cells, if they indeed exist (Sect. 4.3.6), have different distributions to various motor nuclei, as Ia and spindle group II afferents are likely to have.

Golgi tendon organs have so far been mentioned only briefly in the context of the new concept. The same negligence had occurred to them also in previous motor schemes. Indeed even Nichols and Houk's (1976) stiffness regulation hypothesis which was to endow Golgi tendon organ afferents with the role of force feedback, turned out to be able to dispense with them (Houk and Rymer 1981). So what is the role of Golgi tendon organs?

This question is not easy to answer, particularly because the central connectivity of Golgi tendon organ afferents is widespread and probably less well known than that of Ia afferents. Many motor schemes (except for Inbar's, Sect. 4.7.2) rely on muscle spindles as models of skeletal muscle and its load. None of the other mechano-afferents are explicitly incorporated into them, but they remain in the background. For the time being, therefore, they might perhaps be conceived of as forming the ground upon which muscle spindle and spinal recurrent feedback play their roles as leading actors. This background would co-determine the significance of spindle feedback, so that, in certain conditions, Golgi tendon organ afferents may assume a more prominent and specific role (see Sect. 5.3). For their role, it is probably of importance that Golgi tendon organs are active significantly only during muscle contraction. They might thus "gate" pathways from descending and afferent sources that share common interneurones to motoneurones and other central neurones, or which are subject to presynaptic inhibition from Ib afferents (Jankowska 1984). This gating is akin to that attributed by Appelberg et al. (1983c) to muscle spindles. It is also of significance that signals from Golgi tendon organs, too, bear a local sign (Binder and Osborn 1985), i.e. react to local motor activity.

In general, it is not surprising that proprioceptive feedback and its central action are more complicated than recurrent inhibition and its action because, as compared to spinal motor output, the mechanical periphery on which proprioceptive feedback reports is *richer*, showing more diverse peculiarities and subjected to more modulating (external mechanical) influences. One reason is simply that the instruments used by the nervous system to execute its mechanical goals are highly complicated nonlinear devices whose properties have to be taken into account by the CNS (see Sects. 5.1 and 5.8).

4.10 Are the Inputs to γ-Motoneurones and Renshaw Cells Matched to Each Other?

The foregoing discussion suggests that the inputs to γ-motoneurones and Renshaw cells should be functionally matched to each other, at least to a certain extent. This is indeed exemplified by recent results of Pompeiano and co-workers (Boyle and Pompeiano 1984; Pompeiano et al. 1985a, b; see also Pompeiano 1984; see Fig. 54), who investigated the effects onto Renshaw cells and γ-motoneurones exerted by the macular labyrinths. They used sinusoidal head rotation in decerebrate cats with neck afferents cut to study the responses of gastrocnemius α-motoneurones, Renshaw cells linked to them, and spindle afferents from the gastrocnemius muscles. From the results they postulated the following system organisation. On each side of the animal, two supraspinal nuclei appear to be involved (let us call the left side ipsilateral): the lateral vestibular (Deiters) nucleus (denoted LVN in Fig. 54), and the medullary inhibitory area of Magoun and Rhines (denoted RF). Vestibulospinal neurones excited by ipsilateral macular labyrinth receptors increase their firing rates on ipsilateral head-down movements and decrease it on upwards movement ("α-response"). They in turn excite the ipsilateral extensor (e.g. gastrocnemius) α- and γ-motoneurones mono- and polysynaptically, whereby the motor response is of type α, too. (It is the static γ-motoneurone species that is monosynaptically excited by the vestibulospinal tract; see Hulliger 1984.) The other supraspinal system, the ipsilateral medullary reticulospinal system (nucleus: RF, and tract: RST) shows a "β-response" (not to be confused with the β-fusimotor system; instead simply 180° out of phase with the α-response) to head-down movement; this response is facilitated by a pontine cholinergic system. This β-response results from excitatory input to the RF neurones from the *contralateral* labyrinth, mediated by the contralateral vestibulospinal and the crossed spinoreticular (denoted cSR in Fig. 54) tracts. The RF system is postulated by Pompeiano and co-workers to excite the ipsilateral Renshaw cells which are coupled to the ipsilateral extensor (gastrocnemius) α-motoneurones and to show suppressed activity during ipsilateral head-down movement, at least with appropriate facilitation of the RF system. Therefore, during ipsilateral head-down movement, extensor α- and γ-motoneurones are excited from the vestibulospinal tract, and at the same time the α-motoneurones are disinhibited by reduced Renshaw firing. Of course, the inputs to the motoneurones and Renshaw cells are not identi-

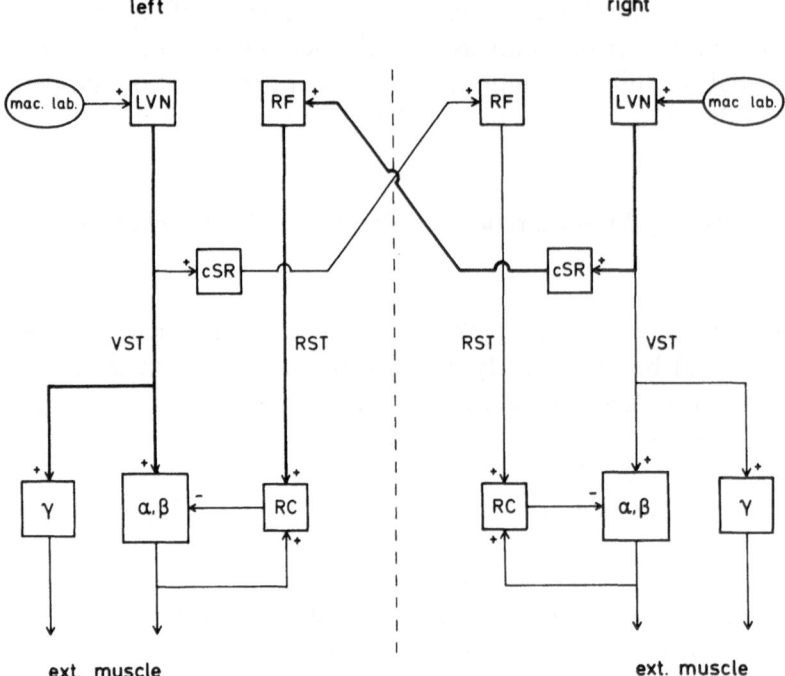

Fig. 54. Scheme of the influences exerted by the macular labyrinths (*mac. lab.*) on extensor fusi- and skeleto-motoneurones, and associated Renshaw cells. The lateral vestibular nucleus (*LVN*) excites ipsilateral fusi- and skeleto-motoneurones (γ, α and β) via the vestibulospinal tract (*VST*) and, via the crossed spinoreticular connexion (*cSR*), excites also the contralateral reticular formation (*RF*). Each *RF* nucleus in turn excites ipsilateral Renshaw cells (*RC*) coupled to extensor skeleto-motoneurones. The influences exerted by the *two* labyrinths on the *left* motoneurone-Renshaw cell system are symbolised by *thick lines*. For more details see text. (After Pompeiano et al. 1985 b; their Fig. 9 B)

cal because they originate from different supraspinal systems with different dynamics (Pompeiano et al. 1985 a). (The influences on the left motoneurones and Renshaw cells are traced by thick lines.) But the whole organisation is compatible with the triple command scheme. It is interesting that, in the present case, the inhibitory input to Renshaw cells acting in conjunction with an excitatory input to α- and γ-motoneurones derives from excitatory labyrinth input, but from the contralateral side. Ross and Thewissen (1987) showed that Renshaw cells are also inhibited from bilateral semicircular canal afferents, and they adduced evidence that the interneurones mediating the inhibition are found in the vicinity of the Renshaw cells (see Fig. 49 B).

Pompeiano et al. (1985 b) also propose that the reticulospinal pathway exerts its effects on Renshaw cells, either directly or through interposed interneurones, through excitatory synaptic contacts located predominantly on the cell body and/or the proximal dendrites of the Renshaw cells. In

contrast, the recurrent axon collaterals of the small gastrocnemius motoneurones probably make synaptic contacts more distally in the dendritic tree. If
this also holds true for collaterals from larger α-motoneurones, it might be
interpreted as a priority given to the inputs from the reticulospinal (RS) tract,
whereas those from motoneurone collaterals then would mainly play a modulatory role.

This type of organisation is also carried over to the reciprocal effects on
flexor motoneurones. The lateral vestibulospinal tract inhibits ipsilateral
flexor α-motoneurones disynaptically via the Ia inhibitory interneurones that
mediate reciprocal inhibition. These interneurones are also inhibited by the
same Renshaw cells which inhibit the extensor α-motoneurones.

Another set of data suggesting the same input pattern to Renshaw cells,
and α- and γ-motoneurones as that obtained by Pompeiano and co-workers
in vestibulospinal reflexes relates to fictive locomotion in decerebrate cats.
Renshaw cell firing patterns in these preparations comply with the expectation that they are excited in the phase of the step cycle when the α-motoneurones from which they receive their strongest excitatory input are active
(McCrea et al. 1980). These results were confirmed, in high spinal cats, by
W. J. Koehler et al. (1984), who in addition found that the excitability of
Renshaw cells was slightly depressed during the active phase in which these
cells and the associated α-motoneurones were active. The authors concluded
that, during fictive locomotion, the Renshaw cells are inhibited during the
active phase by a nonmotoneuronal synaptic input. In conclusion, together
with the finding of Sjöström and Zangger (1976) of a tight α-γ linkage during
fictive locomotion in spinal cats, these results would extend Pompeiano's
input scheme to Renshaw cells, α- and γ-motoneurones to another situation.

Thus, these two examples of the organisation of input to Renshaw cells,
α- and γ-motoneurones are consistent with the scheme outlined in Fig. 49 B.
However, there are two reservations concerning the above results, which
might well render generalisations premature. One concerns the (technically
unavoidable) low frequency of head rotation, which leaves uninvestigated a
considerable physiological range of behaviour. The other concerns the preparations (decerebrate and/or spinal paralysed cats), which may have shown
response patterns different from alert animals (see Sect. 4.11).

Nevertheless, in the present context further examples of matched inputs
to γ-motoneurones and Renshaw cells may also be mentioned.

1) Both types of neurone (γ-motoneurones and Renshaw cells) lack a
strong, in particular a monosynaptic Ia input from homonymous (and synergistic) muscle nerves (Haase et al. 1975; Appelberg et al. 1983a; see
Sect. 2.9.1.2).

2) Group II muscle spindle afferent input to γ-motoneurones follows a
very intricate pattern (see Appelberg et al. 1983b). What appears as an
outstanding feature, however, is the predominance of excitation particularly

of dynamic extensor (e.g. gastrocnemius) γ-motoneurones by stimulation of group II afferents from homonymous (gastrocnemius), but also from the biceps-semitendinosus muscle (Noth and Thilmann 1980; Appelberg et al. 1983 b). (The latter muscle is partly a flexor, partly an extensor; see, e.g. Loeb 1984). Fromm et al. (1977) suggested that recurrent inhibition of extensor α-motoneurones could be suppressed by stimulation of the gastrocnemius nerves in the group II range. Thus, the extensor α-motoneurones might be disinhibited via the Renshaw cell loop by activity in homonymous spindle group II afferents.

3) If the above hypothesis of a parallelism between γ-motoneurones and Renshaw cells indeed is general, it would be required that, if γ-motoneurones are recurrently inhibited by Renshaw cells, Renshaw cells themselves should also be inhibited in this way (see Fig. 49 C). This is indeed the case, as discussed in Sect. 4.3.10. One might argue that this twofold inhibition violates the sign requirement, i.e. the rule of opposite signs for inputs to γ-motoneurones vs Renshaw cells. However, remember that the main part of "mutual inhibition" of Renshaw cells comes from antagonistic α-motoneurones (Ryall 1981).

4.11 Relation of Fusi- and Skeletomotor Activation Patterns: Task Dependence

What information is contained in the combined signals from muscle spindles and Renshaw cells, e_{MS} (t) and e_{RC} (t) in Fig. 53? This question cannot be answered easily. The reason is that the nature of the measured variables and their central evaluation also depends on the nature of the descending (and segmental) inputs to fusimotor motoneurones and Renshaw cells in relation to skeletomotor activation during natural motor tasks. Very little is currently known about the descending input patterns to Renshaw cells in normal movements (see Hultborn and Pierrot-Deseilligny 1979; see also above). But even those to γ-motoneurones and their relation to skeletomotor activity in motor acts are so far confusing rather than simple, as required, for example, by conditional model-reference control schemes (Sect. 4.6). In these systems, the error signal e (t) should usually be zero unless a disturbance makes the model output deviate from the actual plant output. To obtain such a precise match requires a precise match of the inputs to model and plant. This does not appear to be the case generally. The matter is complicated by the fact that the fusimotor spindle innervation is divided into static and dynamic systems. Thus, a long list of proposed patterns of activation of α-, static γ- and dynamic γ-motoneurones has been reviewed (Hulliger 1984; Prochazka 1985, 1986). Only a brief overview is given here (see Fig. 55).

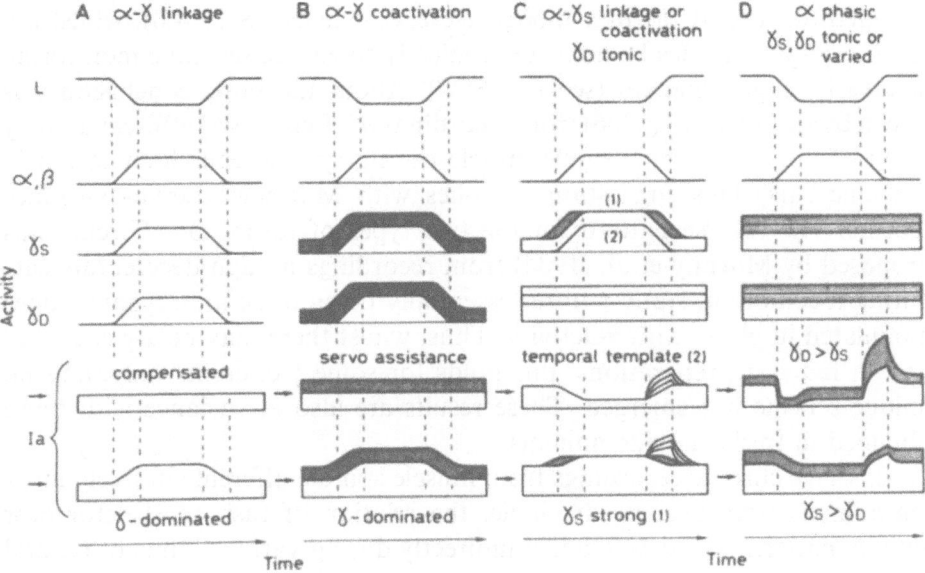

Fig. 55 A–D. Schematic display of α-, β-, γ-efferent and spindle Ia afferent firing rates during idealised voluntary movements, for different strategies of fusimotor control. *Top to bottom:* muscle length (*L*) shortening downwards; α-, β-, γ_S-, γ_D-firing rates; *Ia* firing rates (*upper* assuming weak γ-action, *lower* assuming strong γ-action). **A** α-γ-Linkage (extreme case: strict proportionality between α- and γ-firing rates); **B** α-γ-Coactivation (*shaded areas* indicate the range of possibilities); **C** α-γ_S-Coactivation, γ_D-tonic (*1* γ_S-coactivation strong; *2* γ_S-modulation a "temporal template" of the length changes). Note increasing levels of tonic γ_D-firing rates, reflected in increasing Ia responses during muscle lengthening; **D** Independence of both γ_S- and γ_D-activation from α- and β-activation (special case: constant γ_S- and γ_D-firing rates, phasic α- and β-activation). (With permission from Hulliger 1984; his Fig. 18)

The experimental evidence on fusimotor-skeletomotor activation patterns rests on three bodies of data: (1) recordings of α- and γ-activity or spindle afferent activity in "reduced" animal preparations (mostly anaesthetised, decerebrate and/or spinalised cats), (2) chronic recordings of muscle spindle afferent activity in cats or monkeys awake during unrestrained movements, particularly locomotion, and (3) recordings of spindle activity in human subjects during the performance of voluntary movements. The conclusions drawn from these studies often are at variance with each other.

1) In expiratory and inspiratory cat muscle, α- and γ-activity may be rhythmically modulated in a coupled fashion, but not invariably so, because some γ-efferents also show sustained firing. From spindle afferent and motor efferent recordings in cat jaw muscles during cyclic movements, Appenteng et al. (1980), A. Taylor and Appenteng (1981) and Gottlieb and Taylor (1983) suggested that *static* γ-motoneurones be modulated rhythmically in conjunction with α-motoneurones, thus possibly yielding a temporal template of the

intended movement (as in a model-reference scheme: Sect. 4.6), whilst *dy-namic* γ-motoneurones be activated tonically to ensure sensitive monitoring of muscle length changes (see Fig. 55 C). About the same conclusion was drawn by Bessou et al. (1986) from recordings of efferent and afferent activity to and from the gastrocnemius muscle during spontaneous locomotion in thalamic cats. This suggestion complies with Matthews' servo-assistance scheme. The reverse pattern for the two types of fusimotor efferents was proposed by Murphy et al. (1984) from recordings in high decerebrate cats during locomotion, static γ-fibres essentially being tonic and dynamic ones modulated in phase with α-activity. Thus, whilst there may be α-γ-coactivation in reduced preparations, this holds for some γ-efferents only, and no unique scheme has emerged. These results are also at variance with those obtained in freely moving animals.

2) Since chronic recordings from muscle spindle afferents in freely moving animals have become available, the relation of fusi- to skeletomotor activity patterns could be studied indirectly during various kinds of natural movements. Besides Prochazka and co-workers, the group around Loeb performed extensive studies on this issue. Based on this work, Loeb and Marks (1985) proposed that fusimotor innervation serves to adjust the bias and sensitivity of spindles so as to optimise their transducing properties in response to perturbations. This idea is akin to model-reference control (Sect. 4.6) which is supported by Loeb et al. (1985 b) who suggest that "... voluntary changes in the extrafusal motor program were accompanied by compensatory changes in the intrafusal program, consistent with the servo-control notion of using sensory feedback to provide an error signal to the motor controller" (p. 562). However, the resemblance is only superficial. The spindle afferent signals which Loeb et al. (1985 b) recorded from cat thigh muscles during normal locomotion exhibited "well-modulated" activity between zero and more than 200 pulses/s. This is consistent with many other studies in which spindle discharge was dominated by its strong response to muscle stretch and by unloading during muscle shortening, dependent on shortening velocity (Hulliger 1984). If spindle afferents were to provide "conditional error signals", their discharge should be nearly constant during unperturbed movement (Sect. 4.6). But perhaps, muscle length is not the true variable measured by spindles (next section)? Be as it may, Loeb et al. (1985 b), Loeb and Hoffer (1985), and Loeb et al. (1985 a) suggested that α- and γ-motoneurones are not invariably co-activated, but that even during different phases of a muscle's cyclic movement the fusimotor program including the static/dynamic relation may be changed for adjustment of instantaneous spindle sensitivity.

α-γ-Independence, but not variability of the fusimotor program during a step cycle was inferred from simulation studies, upon which Hulliger et al. 1985 b based their new "fusimotor set hypothesis" (see also Prochazka et al.

1985). They tried to reproduce spindle afferent discharge patterns obtained from chronic recordings in awake cats by subjecting spindles in acute experiments to the chronically recorded muscle length inputs and various fusimotor program. They suggest that, during normal locomotion or similar routine movements, spindles are predominantly subjected to low-intensity tonic input from static fusimotor fibres. In other instances in which animals are faced with unexpected or unfamiliar (nonroutine) motor tasks, static fusimotor input is largely replaced with tonic dynamic fusimotor input. That is, whilst during each particular motor task the fusimotor input is nonmodulated (not linked to α-activity), the balance of static/dynamic input is set according to the demands of a motor task (see Fig. 55 D). These suggestions run counter to those drawn from results obtained in reduced preparations (see item 1). They emphasise α-γ-independence and task dependence of fusimotor activation patterns. The *general implication* of these results is that caution should be taken against any experimental findings obtained in reduced preparations.

3) Chronic recordings from muscle spindle afferents in man are fraught with a number of technical difficulties (Hulliger 1984) and have so far been limited to a small set of motor paradigms.

During isometric muscle contractions and slow high-precision (tracking) movements in finger positioning (in man), the mean firing frequencies (rates over an extended period of time; see App. G) of muscle spindle afferents are not correlated with finger position or muscle length (Hulliger et al. 1982), but are rather related to the effort needed to counterbalance the external load (Vallbo and Hulliger 1982). This is consistent with the concept that, in model-reference control systems, the error signal is zero in the absence of disturbance inputs (Sect. 4.6). α-γ-Linkage, or at least α-γ-coactivation is required in this case (see Fig. 55 A and D). However, spindle discharge may also indicate the *occurrence of movement*, its direction and – possibly – its velocity during the dynamic movement phases (Hulliger et al. 1985 a). This is reminiscent of the concept that primary spindle endings are mere motion detectors, but do not signal velocity precisely (Houk et al. 1981 b). Information on muscle length (limb position) might still be extracted from these afferent discharge patterns originating in the prime movers if use were made of some intricate mechanisms involving, for instance, corollary discharges (efference copy: Evarts 1972; Hulliger et al. 1982; Matthews 1981). For example, in this case, if the discharge of the Renshaw cells coupled to the prime movers were dominated by the excitatory input from motor axon collaterals and recurrent inhibition then reflected the amount and time course of extrafusal muscle contraction, the *difference between (constant) spindle feedback and recurrent inhibition* could signal muscle shortening. Note, however, that this possibility depends crucially on the strength and time course of the descending inputs to Renshaw cells bypassing the α-motoneurones, i.e.

on the inputs labelled $\underline{u}_1$ (t) in Fig. 53. Another possibility of obtaining length information is provided by spindle afferents from antagonist muscles (Hulliger et al. 1982). The advantage of this source would be the higher precision in spindle signals. The spindle afferents from the prime movers show a relatively high variability of discharge. In contrast, in motor tasks in which the antagonist muscle is not co-activated, the spindle signals from this muscle would be free of disturbances introduced by extrafusal and intrafusal activity (hence show little variability), so that mean rates of discharge are well defined. (Note that here, finally, the organisation of the motor system in *groups of muscles* including antagonists is emphasised.)

The general impression from such studies has long been that of α-γ-co-activation, although evidence to the contrary is slowly accumulating (see Hulliger 1984). More evidence of possible α-γ-independence under some behavioural conditions was recently obtained by Ribot et al. (1986), who managed to record the discharge of presumed fusimotor fibres in man.

The most general conclusion that may be drawn from the studies briefly reviewed above is that skeleto- and fusimotor activation patterns during natural movements are not stereotypically linked, but that they vary according to the demands made by a specific task. Judging from the general similarity of the two systems (as outlined above) it is a distinct possibility that – along the same lines – such task dependence also applies to the patterns of Renshaw cell activation in relation to those of α-motoneurones during various natural movements, although hardly anything is known from direct recordings during such movements. Hence, the output signals from muscle spindles and Renshaw cells should be viewed as containing task-dependent information on variables which – in turn – also vary with task and context.

4.12 γ-Motoneurones and Renshaw Cells as Integrative Centres

In the foregoing discussion, to γ-motoneurones and Renshaw cells an important role has been ascribed in the genesis and guidance of muscle contraction, a role quite apart and different from that of α-motoneurones.

The group around Appelberg has recently proposed a new "final common input hypothesis" concerning the role of γ-motoneurones in motor control. This hypothesis was briefly described by Johansson (1985). They view the muscle spindle not solely as a receptor with efferent control, but conceive of the γ-motoneurone-spindle complex as a pre-motoneuronal integrative system. Johansson (1985) argues that the differences found in reflex actions on α- and γ-motoneurones render a rigid parallelism between α- and γ-motoneurone firing patterns unlikely in motor acts which are adjusted by feedback from peripheral receptors. This and the individualised receptive profiles of the γ-motoneurones nearly exclude that reflex inputs to them merely provide general facilitation of the descending inputs. Instead, on this view, the γ-motoneurones integrate information from descending pathways and a wide range of skin, joint and muscle afferents which are functionally connected during active movements. These polymodal signals which depict the kinaesthetic state are first synthesised by the fusimotor neurones. This synthesis is then submitted to the spindles which perform a second integration whose result is adjusted according to the actual extrafusal muscle fibre status. The central neural networks therefore receive a polymodal feedback via the primary muscle spindle afferents (final common input).

One of the important features of γ-motoneurone input appears to be its "individual receptive profiles", whatever the precise reasons. This type of organisation is reminiscent of the input distribution to nonreciprocal Ia inhibitory interneurones (Sect. 2.9.1.1). That is, if the inputs projecting to γ-motoneurones (via interneurones) were distributed randomly according to their projection frequencies, each individual γ-motoneurone would be activated in some but not all contexts defined by certain patterns of input. In some sense, then, the γ-motoneurone system as well as the interneuronal system of nonreciprocal inhibition would act as a pattern discrimination and separation filter not unlike the mode of operation which was proposed for the cerebellar cortical network (see Marr 1969).

The individuality of input patterns to γ-motoneurones would ultimately bear physiological importance only if it were also transmitted to spindle afferents. Indeed, one of the basic assumptions of the final common input hypothesis is that each muscle spindle afferent, too, has a polymodal receptive field. This is not self-evident, because many fusimotor efferents usually converge on any one muscle spindle, whereby much of the individuality in γ-motoneurones could be blurred. Recently, Sojka et al. (1986) provided preliminary evidence that, in fact, the fusimotor input to different spindle afferents from different muscles, but also from the same muscle, can be different in response to various reflex inputs. But these studies have to be further extended systematically.

A fundamental question is to what use the individualised reflex inputs to γ-motoneurones are put. Interestingly, Johansson (1985) suggests a relation to muscle and reflex compartmentalisation. He argues that the γ-motoneurones are part of a complex network which deals with the co-ordination between different muscles as well as between *different parts of single muscles*. Thus, if the spindle output depends on both the reflex input to the γ-motoneurones and on the localisation of the spindle, a single spindle afferent cannot convey complete and unambiguous information about the mechanical state of a muscle or a limb. Rather, this information is composed of the firings of a large number of spindle afferents. But, in general, this afferent picture is supposed to show more differential structure than often presumed before.

It is now suggested that the Renshaw cells serve as a parallel integrative system in about the same way as proposed for the γ-motoneurone-spindle combination, thereby establishing a central model of the "peripheral" γ-loop. The segmental inputs to Renshaw cells have not been studied as extensively as those to γ-motoneurones (see Haase et al. 1975, for review), but it appears not unlikely that in Renshaw cells, too, polymodal (or multisensory) signals depicting the kinaesthetic state are synthesized and then transmitted to various spinal neuronal systems, although the effects of segmental inputs are perhaps weaker on Renshaw cells than on γ-motoneurones (see Bergmann-Erb 1985). Generally, Renshaw cells are known to show "individualised receptive profiles" in response to various inputs, in analogy to γ-motoneurones. From the differences that exist between the descending and segmental input systems to α- and γ-motoneurones (see above), which certainly have some physiological role, differences might also be expected in the input systems to γ-motoneurones and to Renshaw cells, in addition to some important parallels.

4.13 Suggestions for Further Research

The motor control scheme proposed here is based on a considerable amount of experimental data. However, this body of evidence is, in fact, split into two piles: one concerning recurrent inhibition, the other proprioceptive feedback, and there is hardly any connexion between them. Therefore the proposal put forward here could be viewed, at least in part, as a *research program* aimed at providing a common data base. Some of the items of comparison (Sect. 4.3) could be reinvestigated from the viewpoint of an integrated system. In detail, the following questions should be studied.

1) The fine structure (micro-anatomy) of the relation between spindle and recurrent feedback within single muscles should be studied to check the possibility of isomorphy (Sect. 4.3.1). This could be done by a combination of presently available experimental techniques (W. Koehler et al. 1984a; Hamm et al. 1987a). Such a project would be ambitious because the experiments are technically demanding. They could run as follows. (a) In an initially unparalysed preparation, record intracellularly from an α-motoneurone (e.g. of the cat medial gastrocnemius muscle), stimulate repetitively a homonymous motor unit axon (best by intramuscular stimulation), average motoneurone potential with respect to the stimuli. (b) Repeat (a) with the preparation paralysed. The first step would yield average potential changes effected via the afferent *and* the recurrent pathways, the second only those effected via the recurrent pathways. The difference between the two trajectories should reflect the contribution of the afferent pathway alone (assuming linear summation of afferent and recurrent effects). Of interest are common topographical relations: How do the membrane potential trajectories depend

upon the compartmental relations between the motor unit axon stimulated and the axon of the recorded motoneurone? Whenever possible, the location of these two axons within nerve branches innervating different compartments of the muscle under investigation should be determined. According to the present hypothesis, one would expect a common dependency of "recurrent" and "afferent" trajectories on topographical relations.

2) In the same paradigm as above, the influence of motor unit type should be studied by determining the types of the motoneurones recorded and those stimulated. There should be a positive correlation between the amplitudes of "recurrent" and "afferent" trajectories effected by the different contraction forces of the different motor unit types stimulated.

3) The same experimental paradigm can also be used to study the static and dynamic properties of signal transmission from motor axons to motoneurones via the recurrent and proprioceptive pathways.

4) How do input systems to γ-motoneurones and Renshaw cells compare, i.e. to which degree are they similar or dissimilar? This problem should best be investigated in the same preparations; that is, the inputs should be studied for γ-motoneurones (or indirectly for fusimotor effects on spindle afferents), and Renshaw cells (or indirectly for recurrent inhibition on α-motoneurones) identified with respect to the muscle or muscle group to which they belong. But again, this is technically difficult.

4.14 Summary

Based on the structural and functional analogies between recurrent inhibition and proprioceptive feedback, particularly via Ia afferents, a new interpretation of their relative functional roles in motor control is attempted. The idea to interpret recurrent inhibition as an efference copy in a scheme based on the reafference principle is attractive, but does not appear to account for all the features of the two feedback systems, except perhaps under very restricted conditions. Another approach is taken on the basis of model-reference and adaptive control schemes from systems theory, which have already been used in motor control physiology, but in too restricted a form. Recurrent inhibition and proprioceptive feedback are integrated in a common scheme which features both the capability of recognition of peripheral system (particularly muscle) properties and the adaptation of the parameters of central circuits to these peripheral properties.

Skeletal muscle properties can be monitored by measuring its neural input and mechanical output. The first is done by Renshaw cells, the second by muscle spindles. However, both systems get additional input from descending fibre and segmental afferent systems, the Renshaw cells via interneuronal pathways circumventing the α-motoneurones, the muscle spindles via γ-motoneurones. The integration of these signals in the two model systems determines which properties of skeletal muscle are monitored. This is task-dependent as already evident from the relation of skeleto- and fusimotor activation patterns during natural movement, which are normally not stereotypically linked, but vary with motor task. Similar variety can be presumed to occur in the relation of Renshaw cell discharge to skeletomotor (and probably fusimotor) activation patterns. This scheme is advanced to propose a series of experimental investigations.

5 Problems in Motor Control

> "Hence an important role for theory in neurobiology is not merely trying to create correct and detailed theories of neural processes (which may be an extremely difficult task) but pointing to which features it would be most useful to study and in particular to measure, in order to see what kind of theory is needed"
>
> F. H. C. Crick[8]

> "If the world were good for nothing else, it is a fine subject for speculation"
>
> W. Hazlitt[9]

Whilst the preceding chapter discussed some large-scale motor control schemes which tend to assemble data bases of varying size under a unifying concept, this chapter outlines open problems.

5.1 On the Modules of Motor Control

Many pages of this book have been concerned with fine features of neural function. The general issue has been the working of cell assemblies. In several concepts of such assemblies, single neurones are assumed to group and regroup for varying tasks by dynamically changing the effectiveness of their connectivities. Conceivably, this could be achieved using synaptic plasticity and various forms of correlated discharge. A concept that is loosely related to these ideas is that the single constituent cells receive various, possibly polymodal inputs, but in different proportions. How far do these features extend into the motor periphery? It was shown that correlated discharge, which could participate in generating functional cell assemblies, can be found at all levels of the nervous system, down to motor units and muscle receptors (Chaps. 1 and 2). Synaptic plasticity occurs in various forms in the spinal cord and at peripheral synapses (Sect. 3.5). "Fractionation" of spinal interneuronal pools (Sect. 2.9.1.1), of a muscle's γ-motoneurone pool and the associated muscle spindles (Sect. 4.12) and possibly of α-motoneurones by differential input patterns is increasingly being demonstrated. This fractionation raises the question of the entities which the nervous system uses to solve its motor problems. The answer is probably related to the more basic ques-

---

[8] Crick FCH (1979) Thinking about the brain. Sci Am 241: 181–188, p 184
[9] Quoted from Stanley A (ed) (1934) The bedside book. Victor Gollancz, London, p 328

tion of the degree of detail that the motor problems involve in various circumstances.

What could the function of fractionation at peripheral levels be? An attempt at outlining its importance in regard to stability problems has been made in Chap. 3. The example considered were the effects of subtle organisation of anatomy and neural discharge – in connexion with muscle and reflex partitioning – on physiological and pathological tremors. Except to this stability issue, partitioning could also be related to other functional issues, which derive from details of peripheral motor organisation (Windhorst 1979b). This will be discussed now. Although it has been argued in the preceding chapter that recurrent inhibition should be viewed as acting in conjunction with proprioceptive feedback, the following discussion will again focus on localised stretch reflexes, as a "leitmotiv". But in the back of our mind, we should always keep thinking in parallel of recurrent inhibition as something supplementing the proprioceptive system.

Localised stretch reflexes could be viewed as functional units or modules which operate at a level intermediate between that established, on the efferent side, by motor units and whole muscles and, on the afferent side, by single afferents and the entire sets of afferents emerging from a muscle. Their establishment and function may be related to a number of factors whose consideration may give clues to some basic problems, which the nervous system has to deal with in order to solve motor problems and which we have to deal with in order to understand those solutions.

5.1.1 Differential Activation of Muscle Compartments

There is very sparse information on the activation patterns of muscle compartments during natural movements; but some reports have presented evidence that different compartments within a muscle may be activated differentially depending upon as yet unrevealed initial and boundary conditions and probably on motor tasks.

Ter Haar Romeny et al. (1982, 1984) demonstrated that motor units in the multifunctional human biceps brachii muscle (long head) are activated differentially depending on the task to be performed. Some motor units were only activated by a flexion command, these units being located laterally in the muscle. Some units were activated only during supination, these units being located medially. A large proportion of units were activated by the combination of flexion and supination, these units being located centrally and medially. Herring et al. (1979) showed that, in the multipinnate pig masseter muscle, activity in different portions of the muscle varies systematically during different phases of mastication, i.e. exhibits a spatio-temporal pattern based upon differential innervation of different muscle regions. Kandou and

Kernell (1986) reported "localization of intramuscular activity" in the cat peroneus longus muscle upon stimulation of the contralateral motor cortex as compared to stimulation of the superficial peroneus nerve. English (1984) observed different activation patterns of cat lateral gastrocnemius compartments in normal unrestrained locomotion. Russell et al. (1982) reported that compartments of the cat lateral gastrocnemius muscle can be activated in various combinations with each other and with other muscles in response to postural perturbations. These authors wrote that "these data indicate that the subcompartments of cat LG can be activated independently by the CNS during natural movements" (p. 948). The possibility of independent action is also suggested by the structural complexity of some muscles. For example, with respect to the complicated structure of the cat flexor carpi radialis muscle, Gonyea and Ericson (1977) speculated: "The close association of muscle spindles and SO fibers in the cat FCR and the differential distribution of muscle fibers into slow-twitch and fast-twitch regions might allow these regions to function independently of one another when called upon to perform complex behavioral tasks" (p. 337). The problem in these suggestions is the meaning of "independent". Very probably, what the authors had in mind was a "soft" definition meaning that the relative degree of activation of different compartments can be varied in different tasks. This "incomplete" independence might well be useful functionally. A stricter definition of "independence" is provided by multivariable control theory. It is worth expanding a bit upon this notion of independence because it may yield new insights and ideas for further research.

In technical multivariable control systems, one input to the plant may influence more than one output by internal interactions. This is often undesirable, and theoretical and practical solutions to the problem of "decoupling" the output variables have been investigated (Mesarovic 1960). An application of some basic notions to the control of different compartments of a muscle was proposed by Windhorst (1978 b, 1979 b). If muscle compartments are assumed to possess some variable (such as internal compartmental length, force development, work, etc.) that is to be controlled independently by the respective compartmental inputs, the problem is how to achieve this independence in view of mechanical coupling between the compartments. This coupling implies that the control signal directed to one compartment would not only influence the corresponding compartmental variable, but also those in other compartments. (At a larger scale, the same problem arises for groups of synergistic muscles.) The problem and a solution will now be explained by recourse to a very specific example. This is the nonlinear mechanical interaction between parallel muscle compartments and its compensation, under particular conditions, by partitioned reflex connexions. The explanation given is not meant to be a unique one for all patterns of partitioning found experimentally, as other factors may also be important for establishing partitioned reflex connexions (Sects. 5.1.3–5.1.6). (The following approach is different from that taken in Windhorst 1979b, because it has been adapted to the new theoretical scheme outlined in the previous chapter.)

5.1.2 Nonlinear Mechanical Interaction Between Muscle Compartments

In many instances, different parts of a muscle interact mechanically. Contraction in one part alters the length of muscle fibres in adjacent parts and thereby changes their contractility, i.e. the initial and the boundary conditions of muscular action. This nonlinear "cross-effect" depends on the spatial relations of the muscle parts in question, e.g. on whether and how they share a common tendon, etc. For example, in a reasonably fusiform muscle with parallel fibres sharing common tendons of insertion and origin, contraction in one part will shorten the muscle fibres in the noncontracting (or less contracting) part. If these fibres contract too, they meet a mechanical boundary condition different from that without contraction in the parallel part. Similar general considerations apply, though with different mechanical consequences, if the muscle parts (compartments) are situated in series to each other (Sect. 5.1.2.6). As already outlined in Sect. 3.2.2.2, this interaction is a complicated spatio-temporal stochastic process. Nonetheless, Demiéville and Partridge (1980) argued that these effects could be accounted for by the length-tension and force-velocity relations of skeletal muscle. Of these two, during movement, the force-velocity relation is probably of greater importance (Loeb 1987). However, for simplicity of the argument, let us consider mean values, i.e. temporal averages of the pertinent variables, and steady-state conditions, i.e. length-tension characteristics, only.

5.1.2.1 A Formal Representation of the Force–Length Relationship

The force produced by a muscle under isometric conditions upon constant tetanic excitation depends on the amount of excitation and the length of the muscle. The form of these well-known length-tension characteristics in turn depends on muscle structure and ultrastructure (i.e. parallel-fibred vs pinnate; see Woittiez et al. 1983; McMahon 1984), besides on excitation level (e.g. tetanic vs subtetanic stimulus rate; see Rack and Westbury 1969; Partridge and Benton 1981). Very roughly, at a given level of neural excitation, muscle force declines with muscle length, with the possible exception of a local maximum around muscle resting length, l_0, in parallel-fibred muscles (McMahon 1984). Since most skeletal muscles are pinnated, though to different degrees, and neatly parallel-fibred muscles are less frequent (Sacks and Roy 1982), the following argument is based on the assumption of a monoton-

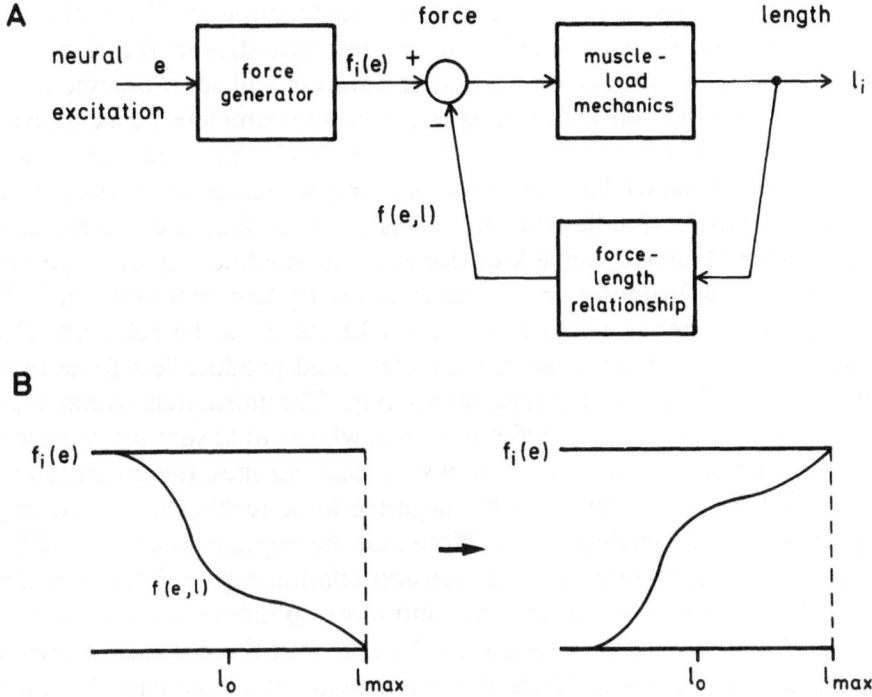

Fig. 56 A, B. A formal scheme of the dependence of muscle force production on length. **A** It is assumed that the force generator converts a neural excitation, e, into an intrinsic force, $f_i(e)$, whose magnitude solely depends on e. Any force leaving the summing point is transformed by the muscle-load mechanics into an internal muscle length, l_i. This length is in turn converted into a length-dependent force, $f(e, l)$, by a box labelled "force-length relationship". The characteristic of this ("inverted") force-length relationship is displayed in the *lower left graph* of **B**. The summing point builds the difference $f_i(e) - f(e, l)$, which is depicted in the *lower right graph* of **B** and represents the known force-length relationship of skeletal muscle

ic length-tension curve, which does not affect its generality (though see paragraph after the next).

In a formal, if somewhat unusual, way, the length-dependence of force production can be represented by the block diagram in Fig. 56A. The force generator in the muscle fibres is here assumed to produce, in response to a constant excitation, e, a constant internal force, $f_i(e)$, which depends on the amount of excitation, but not on muscle length. The block representing the muscle-load characteristics of skeletal muscle transforms muscle force (at the summing point) into internal (muscle fibre) length, l_i. (Note that this is the variable measured by muscle spindles.) This variable, in turn, is transformed by another block representing "inverse" length-tension characteristics into a length-dependent force, $f(e, l)$. The inverse length-tension relationship is shown in the left plot of Fig. 56B. At a summing point, the length-dependent force, $f(e, l)$, is subtracted from $f_i(e)$. The influence of length on muscle force output is thus represented by a negative feedback loop. The entire system then yields the familiar length-tension characteristics in the right plot of Fig. 56 B. The advantage of this respresentation will soon become obvious.

The following consideration is based on the completely speculative idea that at least part of the negative force feedback that yields an important nonlinearity of skeletal muscle could be compensated by proprioceptive feedback via muscle spindle afferents. This requires that the central nervous system should have a vague knowledge of the nonlinearity of its peripheral plants and, via fusimotor spindle innervation, try to compensate for it. Thus, the fusimotor signal would command a length, l_i, to be achieved. Due to negative force feedback, skeletal muscle would produce less force than required to achieve the designed shortening. The mismatch would increase spindle discharge above its reference level, which could support, via excitatory reflex connexions to homonymous α-motoneurones, the excitation to the plant and, hence, compensate for negative force feedback, at least in part, depending on the feedback gain. Note that the argument can be modified as follows for nonmonotonic length-tension relationships, i.e. those exhibiting a local maximum around resting length (corresponding to a local minimum in the inverse length-tension curve). Assume that the designed muscle-fibre shortening has to proceed from the local trough (at longer length) to the local maximum (at shorter length). This shortening would increase the force output of the muscle fibres (with constant neural excitation). The muscle fibres would then shorten to a shorter than the desired length, l_i, which is set by the fusimotor signal. This would reduce spindle firing from the reference level and, via exitatory synaptic connexions to homoymous α-motoneurones, reduce force output. This extension generalises the argument also to mixed cases.

As a result, this type of feedback might act to render the motor apparatus less dependent on muscle properties and to linearise system performance (see Nichols and Houk 1976).

5.1.2.2 Independent Activation of Two Muscle Compartments

Let us now consider two muscle compartments in parallel which mechanically interact with each other (see Fig. 57). The effect of contraction of one compartment on the contractility of the other is then represented by a negative cross-feedback. In Fig. 57, each compartment is represented by a block diagram like that in Fig. 56 A (enclosed by dashed lines and labelled MC_1 and MC_2). The cross-influence is represented by blocks labelled C_{12} and C_{21}. The static characteristics of these blocks are assumed to be given by inverse length-tension relationships similar to those in Fig. 56 B (left lower plot). That is, contraction of compartment MC_1 adds, via block C_{21}, negative force

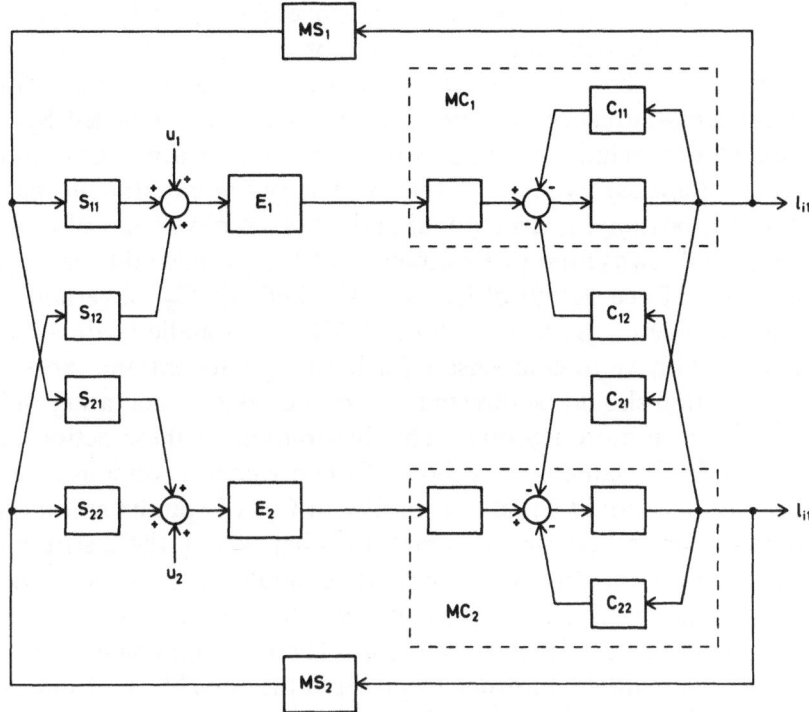

Fig. 57. Two-loop model of muscle reflex system with peripheral (*right*) and central (*left*) coupling. Muscle compartments, MC_1 and MC_2 (enclosed by *dashed lines*) are represented by the same formal systems as depicted in Fig. 56 A (compare structure of MC_2 with Fig. 56 A). The two compartments are assumed to interact such that contraction of one diminishes force production of the other. This is represented by the cross-connexions labelled C_{12} and C_{21}, which have principally the same effect as the feedback loops C_{11} and C_{22} in each single compartment, although qualitative and quantitative differences may exist. In this configuration, the two compartments are thus in parallel to each other. Internal lengths, l_{i1} and l_{i2}, are sensed by muscle spindles MS_1 and MS_2, respectively. The spindle afferent signals are relayed via synapses (and possibly interposed interneurones), represented by *blocks* labelled S_{ij}, to α-motoneurones which are represented by summing points and the *blocks* labelled E_1 and E_2 (representing encoders). Other (e.g. command) inputs to the motoneurones are labelled u_1 and u_2

to the force production of compartment MC_2 in roughly the same way it does, via block C_{11}, to its own force production, although the precise form of the curves, their slopes etc. may be different; and vice versa. In terms of multivariable systems, the structure of the plant as depicted on the right side of Fig. 57 is said to be of V-canonical form, which can easily be decoupled by means of feedback loops (Mesarovic 1960).

The decoupling can be easily understood intuitively, without much mathematics. The basic idea is that the peripheral mechanical cross-influence, e.g. via C_{21}, is compensated by a central cross-reflex connexion, i.e. via S_{21}. Muscle spindles (labelled MS_1 and MS_2) sense the respective internal lengths,

l_{i1} and l_{i2}. For compensation of the peripheral cross-influences (decoupling) to be achieved, their afferents must not only feed back to their "own" motoneurones, but also to those innervating the parallel compartment. This central cross-influence is represented by the blocks labelled S_{12} and S_{21}. These blocks include the systems between afferent fibres and motoneurone encoders (labelled E_1 and E_2), that is, synapses and motoneuronal dendritic trees and, possibly, interneuronal pathways. In physiological terms, decoupling would run as follows. Contraction of MC_1 would add, via C_{11}, negative force to the force output of its own MC_1 and, via C_{21}, a certain amount of negative force to the force output of MC_2. If spindle feedback from, say, MC_1 would have to compensate for both the inherent and cross-effects, its central action should be directed to motoneurones innervating MC_2 as well as to the own motoneurones. The distribution of these actions should be weighted in the same way as those of the peripheral couplings, i.e. the ratio of S_{11} to S_{21} should equal that of C_{11} to C_{21} (if spindle and motoneurone properties are otherwise the same). In other words, the distribution of the strengths of the central cross-connexions should be the same as the distribution of strengths of the peripheral cross-influences. The control inputs to the α-motoneurones are labelled u_1 and u_2. These are, and have to be, treated as independent entities, in order to preserve the possibility of independent or differential control. The latter also applies to the inputs to the related sets of γ-motoneurones and, according to the motor control scheme proposed in Chap. 4, to the related sets of Renshaw cells.

The advantage of this simple decoupling scheme, which assumes a V-canonical peripheral structure, is that it can easily be extended to more than two compartments without change in the above decoupling conditions. This is not the case if another peripheral structure is assumed as done by Windhorst (1979b). The present scheme would allow independent control (via u_1 and u_2) of the contractions of different parallel muscle compartments under the conditions outlined above. Although complete and perfect independence is unlikely to take place under physiological conditions (the requirements on quantitative relations simply are too sharp), even an imperfect implementation could already bring about partial linearisation of muscle behaviour. This was suggested before (Windhorst 1979b) on the basis of the proposal of Nichols and Houk (1976) that the stretch reflex functions to linearise muscle behaviour and thereby regulates stiffness (next section). The scheme also provides a rationale for the existence of central cross-connexions between spindle afferents and motoneurones, i.e. for the absence of complete reflex partitioning.

The main drawback of the scheme as outlined here is that it relies only on proprioceptive feedback and does not take into account recurrent feedback. What is recurrent inhibition needed for? A cursory answer is that the misalignment signals carried by muscle spindles are not uniquely defined by the relation of fusi- and skeletomotor inputs, but also depend on the type of load, which in turn is co-determined by the co-activation of agonists and antagonists. So is the nonlinearity itself (see Rack and Westbury 1969). Hence, the nonlinearity can only be captured by knowing muscle inputs and outputs (Sect. 4.9). Also, it is important to remember that the output of Renshaw cells is co-determined by descending and segmental inputs, whose patterns during natural movements in relation to α-motoneurone input are hardly known so far. Fusimotor action is more approachable experimentally, but its relation to skeletomotor activity patterns is not fixed. Instead, it probably varies with the motor task performed (Sect. 4.11). This task-dependence of fusimotor output might also imply task-dependence of the degree of linearisation, unless the gains of central connexions (within and across muscles) are adjusted accordingly. This problem remains to be solved.

At the time when the decoupling scheme was first proposed (Windhorst 1979 b), it may have appeared to be too rigidly matched to a technical template. With respect to that scheme, Agarwal and Gottlieb (1985) wrote that "anatomical, physiological, and functional evidence for such multiloop segmental systems are limited" (p. 71). Whereas this still holds true, anatomical and physiological data keep accumulating (see D. G. Stuart et al. 1988). However, the further evidence needed to support the functional significance of multiloop segmental systems will be obtained slowly due to the technical difficulty of the required studies. Yet, as a guideline, the present scheme may nevertheless be helpful in designing appropriate experiments, in interpreting the obtained results in terms of well-defined notions and in gathering and unifying scattered suggestions under a precise concept.

5.1.2.3 Linearisation of Muscle Properties

With regard to whole muscles (lumped systems), Nichols and Houk (1976) proposed that the stretch reflex linearises nonlinear muscle responses to stretch and release (external inputs!). They further suggested that no single variable, such as length or tension (force), was regulated, each on its own, but rather a "compound" variable, namely stiffness (see also Crago et al. 1976; Houk et al. 1981 a; Houk and Rymer 1981; Allum and Mauritz 1984). This hypothesis made a shy attempt at abandoning the arbitrary selection of one possible set of basic (Newtonian) variables, such as length, velocity and force, used most often to describe motor action (see Sect. 5.2; Partridge 1982). Moreover, it appeared to offer the chance of integrating muscle spindle and Golgi tendon organ feedback into a common scheme; but, unfortunately, this promise has not held true (see Houk and Rymer 1981), because the linearisation action of the stretch reflex has been shown to be solely explicable by the nonlinear behaviour of primary spindle endings. (Hasan 1986, has theoretically shown the importance of stiffness regulation for optimisation of simple movements.)

As demonstrated above, it is conceivable that sensory feedback could also, at least partially, compensate for nonlinear mechanical interactions within the muscle tissue and thereby homogenise the inhomogeneous distributions of muscle properties. If proprioceptive feedback has such a "microregulatory" role, it should probably be matched to fine features of muscle architecture, and, hence, organised in a spatially distributed manner, at least in muscles of complex internal structure (Sects. 5.1.2.5 and 5.1.2.6). If so, this function would also bear on optimisation of certain variables, e.g. energy expenditure (next section).

5.1.2.4 Optimisation of Energy Expenditure

In regard to his observations on compartmental activation patterns in the cat lateral gastrocnemius during locomotion, English (1984) mentioned energy expenditure as a possible factor of importance in organising these patterns. Motor units could be activated differently in different compartments during the performance of different tasks. Locomotion is mostly considered a single task, but it could indeed be composed of more than one behavioural component. Regarding fine details, locomotion is not such a stereotyped movement as might be presumed. In response to varying characteristics of the ground, a walking cat must continually adjust its posture to maintain balance. This could well happen so as to conserve energy while producing forward movement. "If motor units are activated differently in these tasks, then different activation patterns might be observed when even subtle changes in postural, energetic, or movement tasks are demanded" (English 1984; pp. 124–125). Hatze and Buys (1977) proposed that "the specific patterns of motor unit recruitment and stimulation frequency observed in mammalian skeletal muscle under static isometric contractions are determined by a minimum-energy principle" (p. 9). According to these authors, feedback from muscle stretch receptors plays an important role in optimally adjusting muscle state variables such as motor unit recruitment and firing rates. If it is recalled that state variables are nonhomogeneously distributed throughout the muscle, this nonhomogeneity being reflected in the discharge patterns of the relevant muscle receptors, it becomes probable that the spatial organisation of motor output is pertinent to the minimisation of energy expenditure, not only in isometric but in other types of muscle contractions as well.

5.1.2.5 Asymmetries of Muscle Structure

The mechanical interaction between muscle compartments may be asymmetrical. Consider a muscle with a largely parallel-fibred compartment in parallel to a pinnate compartment, such as the cat flexor carpi radialis (Gonyea and Ericson 1977, Richmond and Stuart 1985; see Sect. 5.1.5). Active length–force relationships are spread over a wider (normalised) length range for parallel-fibred than for pinnate muscles, as shown by Woittiez et al. (1983) for the rat semimembranosus and medial gastrocnemius muscles, respectively. Thus, if the parallel-fibred compartment (and, hence, the whole muscle) contracts by a certain amount, the length (and angle of pinnation) of the shorter muscle fibres in the pinnate compartment change, too. This change affects the force-producing capacity and mechanical efficiency of the pinnated fibres much more than those of the parallel fibres. Conversely, the same amount of whole-muscle shortening effected by contraction of the pinnate muscle compartment would not affect the contractility of the parallel fibres in the adjacent compartment as much. Whether such factors are taken into account by differential motor innervation is unknown, but it is likely with respect to energy conservation principles (previous section).

5.1.2.6 Spatial Arrangement of Muscle Compartments

A further factor that very probably is important for the specific organisation of partitioned reflex connexions is the spatial relation of muscle compartments. These relations may be largely "in parallel" (Sect. 3.3.1), "in series" (like in the human and cat M. rectus abdominis; and the cat M. semitendinosus; Sect. 3.3.2) or any combination of the former (like in the cat M. lateral gastrocnemius: English and Letbetter 1982; the dog M. tibialis anterior: Wilder et al. 1953; or the cat M. splenius muscle: Richmond and Abrahams 1975a; Brink et al. 1981; Sect. 3.3.2).

The discussion of decoupling focussed on parallel muscle compartments (Sect. 5.1.2.2). The same arguments are harder to apply to strictly in-series compartments (like those in the semitendinosus muscle). For, if the same V-canonical model of peripheral interactions including the same inverse force–length characteristics (left plot in Fig. 56 B) is applied unaltered to the in-series case, the peripheral cross-influences via C_{21} and C_{12} would change signs (from negative to positive at the right summing points in Fig. 57), whilst the central cross-connexions would retain their positive signs. However, it is a matter of conjecture whether the same kind of mechanical interaction prevails in the in-series arrangement. Probably, the neat in-series arrangement of muscle compartments poses very special motor control problems

(Bodine et al. 1982a, b; Botterman et al. 1983b; see also Sect. 3.3.2; for in-series muscle fibres see Loeb et al. 1987). Botterman et al. (1983b) argue that this arrangement has consequences for the effective development of force. A great degree of synergy of activation of both compartments would be required for efficient contraction to ensue because, in the case of separate activation, the passive visco-elastic properties of the inactive compartment would result in marked shortening of the active compartment and, hence, in a diminished force production (Botterman et al. 1983b). This argument would in principle also apply to the less extreme and more natural case of unbalanced co-activation of both compartments.

The same argument cannot be applied to cat neck muscles like the splenius which show a strong monosynaptic reflex partitioning (Sect. 3.3.2). Firstly, muscle fibres are arranged in parallel as well as in series. Secondly, there are more than just one tendinous inscription between the in-series muscle fibres and, finally, the tendons of origin and insertion are distributed spatially. The situation in the splenius is therefore much more complex, and the partitioning may have to do with this complexity, which probably requires other factors to be taken into account by the nervous system.

5.1.3 Different Mechanical Output Actions of Muscle Compartments

The relationship between a muscle's contraction and its motor effect is defined rigidly only for a muscle acting on a hinge joint, and only when the sites of origin and insertion are sharply confined. As pointed out by Partridge and Benton (1981), the situation becomes more complicated when a muscle's action involves more than one muscle head, a distributed muscle attachment, or compound internal organisation. The above relationships are then described either as multiple functions or as functions continuously varying with different lever ratios and muscle lengths. Loeb (1987) emphasises that in the case of muscles with broad origins or insertions, and hence heterogeneous skeletal lever arms, individual muscle units tend to be spatially restricted to narrow longitudinal strips within which skeletal action and muscle fibre architecture tend to be homogeneous (Loeb et al. 1987).

The term "mechanical *output* actions" in the section heading is to distinguish immediately observable actions on joint constellations from muscle-inherent actions (without joint movement) on other muscle parts (Sect. 5.1.2). Different parts of a muscle may have different actions on joint constellations and movements depending on their gross anatomical structure. This is particularly evident in biarticular muscles which, for instance, have flexor effects on one joint and extensor actions on another joint. As with groups of muscles, this would require different neural control of the two parts, possibly including reflex partitioning. Consider some examples.

The cat biceps femoris has a dual function: hip extension and knee flexion, exerted by its anterior and posterior parts, respectively. The reflex organisation of this muscle is partitioned (see Sect. 3.3.1). Similar patterns were found in the cat semimembranosus muscle.

Whereas these patterns are probably explained simply by the bifunctionality of the two muscles, localised homonymous monosynaptic Ia connexions in cat medial gastrocnemius (Lucas and Binder 1984) and lateral gastrocnemius (Vanden-Noven et al. 1986) have probably another major cause. Although both are biarticular muscles, too, their main action undoubtedly is ankle extension. Yet they are compartmentalised (Letbetter 1974; English and Letbetter 1982), and the compartments in the lateral gastrocnemius muscle are differently and variably activated during locomotion in patterns that do not always comply with the usual recruitment order of skeleto-motoneurones (English 1984).

The partitioned monosynaptic Ia connexions in the first group of muscles (semimembranosus and biceps femoris) should therefore not be treated as equivalent to those in the second group (medial and lateral gastrocnemius).

5.1.4 Spatial Density Gradients for Different Types of Muscle Fibre

As with groups of muscles (e.g. synergists; see Burke 1981; Maier 1981), different muscle fibre types are often not distributed homogeneously throughout a muscle's volume (reviews: Botterman et al. 1978; Armstrong 1980; more recent single examples include: cat flexor carpi radialis, Gonyea and Ericson 1977; rat masticatory muscles, Maier 1979; cat lateral gastrocnemius muscle, English and Letbetter 1982; cat tibialis anterior, Iliya and Dum 1984; turtle ambiens muscle, Hermanson et al. 1986; rat sternomastoid muscle, Zenker et al. 1986). Loeb (1987) points out that many muscles contain high concentrations of FF-type motor units in superficial layers, perhaps because these muscle parts have longer skeletal lever arms or require less blood supply. Density gradients may be continuous or rather discontinuous. In the latter case, internal muscle structure resembles that of the triceps surae group or the quadriceps femoris. Such a segregation often entails local differences in densities of muscle spindles (and probably Golgi tendon organs) as with groups of muscles, such that muscle spindles tend to be concentrated in muscle regions rich of oxidative fibres (Botterman et al. 1978; also Iliya and Dum 1984; Richmond and Stuart 1985; Hermanson et al. 1986; Zenker et al. 1986). However, there are exceptions (Botterman et al. 1978; Maier 1981). For example, spindle density may be high in muscle regions undergoing relatively large length changes during natural movements (Maier 1979). These regions may or may not coincide with regions of high oxidative index (Hermanson et al. 1986). Anyway, in case of such a coincidence, reflex partitioning would imply that, because of the strong peripheral association of oxidative muscle fibres and spindles, Ia afferents would have a stronger monosynaptic connexion to motoneurones of nonfatiguable (S- and FR-type) motor units than to FF-motoneurones, and this weighted connexion should not simply result from type-dependent differences in motoneurone input resistance, but be genuine (Fleshman et al. 1981). The quantitative strength of this type of reflex partitioning might be assumed to depend on the strength of peripheral coupling between motor units and muscle spindles. Now this coupling in turn depends on at least two factors, namely, firstly, the relation of spindle density to oxidative muscle fibre density, and, secondly, the strength of influence that any single motor unit exerts on the discharge of each single afferent (Sect. 3.2.3). For a localised reflex, the low gain of signal transmission from S-type motor units to spindle afferents should be compensated for by a high projection frequency and strength of connexion from Ia afferents to the related (S-type) motoneurones. This appears to be the case in the medial gastrocnemius muscle (Fleshman et al. 1981), although there is only a slight trend towards segregration of slow from fast muscle fibres (Burke et al. 1977). More investigations on other muscles are desirable.

5.1.5 Muscle Fibre Architecture

An important factor relating to muscle structure is *pinnation*. Its functional implications for muscle function have been discussed before (e.g. Beritoff 1925; Benninghoff and Rollhäuser 1952; Gans and Bock 1965; Partridge and Benton 1981; see also Herring et al. 1979; Muhl 1982; Woittiez et al. 1983; Loeb 1987). As compared to parallel-fibred muscles, pinnated muscles exhibit a number of mechanical differences. For example, the active length–force relations have different forms, and the maximal contraction amplitudes and speeds as well as the work done are less in pinnated muscles than in parallel-fibred muscles depending on pinnation angle. Different angles of pinnation may also be associated with different organisations of tendons or aponeuroses (see Proske and Morgan 1984). The important point here is that the angle of pinnation and, hence, muscle fibre length may vary within a single muscle. It is reasonable to hypothesise that muscles with such mechanical heterogeneity require differential neural control as well. Let us discuss some specific examples, namely the rabbit digastric muscle, the cat sartorius and the cat flexor carpi radialis muscle.

The rabbit digastric is used here as a characteristic example because some interesting mechanical data are available (Muhl 1982), although it may not contain significant numbers of muscle spindles as suggested by data on the cat (Maier 1979). Muscles of similar structure containing spindles are, for example, the human M. flexor digitorum superficialis and the M. flexor pollicis longus. The digastric is a unipinnate muscle whose muscle fibres have widely varying length (Muhl 1982). When the muscle tendon is stretched, anterior (shorter) muscle fibres increase on average by 0.72 mm and posterior (longer) fibres by 0.82 mm for each mm of total muscle length increase. When these values are normalised through division by the initial length, the respective values are about 10.4% for the anterior and 4.6% for the posterior fibres per mm length change of the entire muscle. That is, the anterior muscle fibres are stretched relatively more than twice as much as the posterior fibres. (The same would apply to muscle spindles arranged alongside these different extrafusal muscle fibres.) The converse should hold for muscle contraction. That is, for a given overall muscle shortening to be achieved, anterior muscle fibres must shorten much more than posterior ones.

In the case of a muscle with such a structure but also having a significant number of muscle spindles, changes in spindle length would differ in different regions of the muscle as well. This mechanical arrangement would necessitate a differential neural command input to the muscle fibres in different regions, with each neural command being adjusted to the mechanical conditions in each region. A prerequisite would be the existence of motor unit territories restricted to small transverse regions (see below). Now consider the monosynaptic stretch reflex mediated by Ia fibres and α-motoneurones. If a disturbance stretch input of 0.5 mm amplitude were applied to the common muscle tendon starting at the initial length, anterior muscle fibres would be stretched by 0.36 mm (5.2%) and posterior fibres by 0.41 mm (2.3%) (see above). If monosynaptic Ia feedback to motoneurones were to compensate for only 50% of the length increase, it would have to cause contraction of anterior muscle fibres by 2.6% and of posterior fibres by 1.15%. This would most economically be achieved by channelling back the stronger Ia signal from anterior muscle regions to anterior motoneurones, and the smaller Ia signal from posterior regions to posterior motoneurones, i.e. by partitioned reflex connectivity. The above prerequisite of the confinement of motor unit territories is fulfilled in the sheet-like cat hindlimb sartorius muscle, whose characteristic design has motor units restricted to narrow longitudinal strips. Loeb et al. (1987) point out that this organisation ensures that the forces, generated by each motor unit, are evenly distributed over a small transverse domain. The motor units would then be segregated according to the gradually changing lever arms from medial to anterior muscle portions across the widespread insertion of the muscle. Whereas all parts of the sartorius contribute to hip flexion, its action on the knee changes from flexion to neutral to extension along this direction (Loeb et al. 1987). A certain reflex localisation, at least regarding the division of sartorius into an anterior and a medial part, is likely (Eccles and Lundberg 1958), but more refined features have still to be investigated.

Similar considerations apply to the cat flexor carpi radialis muscle. This forearm muscle has a complicated internal structure (Gonyea and Ericson 1977; Richmond and Stuart 1985), its cross-section being divided incompletely by two tendon sheets (tendon of insertion and deep tendon of origin). A muscle "core" (Richmond and Stuart 1985) situated between these sheets is made up of short oblique muscle fibres, the angle of pinnation being large. In contrast, muscle fibres in the other regions are longer and less pinnated (Richmond and Stuart 1985). Hence, upon stretching the whole muscle, the muscle fibres in the core will be stretched relatively much more than those in the other regions. Conversely, to achieve a certain amount of muscle contraction (shortening), the former fibres will have to shorten more extensively than the latter.

Benninghoff and Rollhäuser (1952) point out that, during ongoing whole muscle contraction, pinnated muscle fibres contribute decreasing portions to overall muscle force (and work), because, due to increasing pinnation angle, their force vector in the direction of main muscle pull becomes smaller. This effect of relative "loss of force" is the more accentuated, the greater the original pinnation angle. Its magnitude also depends on initial and final muscle length. The authors conclude that strongly pinnated muscles would be suited best to exert strong forces over short distances (also because of their higher fibre package density), whereas parallel-fibred muscles are optimal for small-load movements over larger distances. Benninghoff and Rollhäuser also suggest that in muscles with fibres of different pinnation (such as the human M. flexor hallucis longus), differently pinnated muscle fibres could divide labor and tasks among them. But needless to say that would also require differential innervation. However, it appears unlikely that this division be totally strict (but see Sect. 5.1.1). Yet proper coordination of activities is desirable if energy expenditure and other variables are to be optimised. It may be assumed that, in order to optimise muscle performance with respect to (no doubt variable) criteria (such as minimisation of energy expenditure), motor input should be distributed nonhomogeneously to the various areas, depending upon prevailing conditions in the muscle. In this task, partitioned feedback from Ia fibres to α-motoneurones might be helpful. But it is still unknown whether a differential monosynaptic distribution from homonymous Ia fibres to cat flexor carpi radialis and human flexor hallucis longus motoneurones indeed exists. In cat flexor carpi radialis, most spindles (and Golgi tendon organs) are situated in the muscle core (Gonyea and Ericson 1977; Richmond and Stuart 1985), but differential central connectivity still remains to be demonstrated.

5.1.6 Ontogeny

In part, muscle partitioning may evolve during ontogeny (possibly based on phylogenetic dispositions) without having any functional "meaning". At the level of whole motor nuclei, there is a certain somatotopic relation between their rostro-caudal location in the spinal cord and the location of the related muscles in the limb (for the cat hindlimb see Romanes 1951; for the dog forelimb see Mutai et al. 1986). Intramuscular somatotopic relations may then represent intrapolations of this coarser whole-muscle pattern. For example, Donselaar et al. (1985) try to explain their findings of somatotopic relations between rostro-caudal motoneurone location and muscle fibre location in the cat's peroneus longus muscle as follows. The neural tube generally tends to differentiate in a rostro-caudal direction from the brainstem downwards. Muscle fibres of the mouse hindlimb tend to differentiate in a proximo-distal and in an antero-posterior direction. Probably, this is also the case within the cat's hindlimb. Hence, the rostro-caudal distribution of motoneurones may be reflected in an antero-posterior as well as a proximo-distal distribution of the muscle fibres they innervate. Likewise, reflex partitioning might, at least to a certain extent, have emerged from such a trend (see Lüscher et al. 1980). Whether this is likely or whether reflex partitioning rather yields a benefit to the organism putting a selection pressure on its evolution remains to be established. In certain cases, at least, the evolution by chance has been rendered unlikely by the – as yet sparse – experimental demonstration that different muscle compartments may be activated differentially during certain movement tasks (Sect. 5.1.1). This also applies to the cat peroneus longus muscle (Kandou and Kernell 1986).

5.1.7 The Task Group Notion

A concept that might have some relation to that of localised stretch reflexes is Loeb's (1984) *task group* notion. Loeb conceives of a task group as consisting of an assembly of α- and γ-motoneurones, as well as spindle and other proprioceptive afferents, which performs a kinematically homogeneous motor task. He was led to develop this notion following the observation that α-motoneurones of the multifunctional cat anterior sartorius muscle are recruited during only one or the other of the two periods of activation of the muscle in locomotion (Hoffer et al. 1987a, b). These two phases of muscle activation during a step cycle occur under different kinematic conditions, shortening or lengthening, which might constitute different control problems or tasks within one behavioural pattern (Loeb 1987). According to Loeb, the tasks should therefore be defined kinematically, and the task group would then denote a *functional association* of neuromuscular elements geared to the demands of specialised control circuits in the spinal cord. Task groups may be restricted to parts, or they may transcend the limits, of anatomically defined muscles. It is important in this concept that constituent elements of a task group may participate in other task groups, which are established to perform either kinematically different tasks or the same tasks with different goals, such as different speed or different synergies for joint stabilisation. This is reminiscent of the cell assembly concept, in which one cell may contribute its activity to various assemblies under different conditions. Within each task group, certain organisation principles may prevail, such as rank-ordered recruitment of motoneurones. Loeb (1984) then emphasises that the task group is a functional notion independent of details of micro-anatomical organisation, such as mechanical compartmentalisation, but it is probably correlated with detailed synaptic projections and different patterns of muscle fiber type distribution. The neural connectivity is presumably subject to ongoing plastic changes, the purpose being optimisation of system performance by learning from experience (Loeb 1984). Loeb's concept must still be put on a firm footing and supported by more experimental evidence. But it makes the important point that motor control strategies are selected from a (time-variable) store, possibly by composing neuromuscular assemblies from smaller units ("modules").

Loeb's task group notion is conceptually also loosely related to that of a set of interconnected *central pattern generators*, which may be combined in various ways to perform different types of locomotion and possibly other tasks, such as scratching and what have you. Grillner and Wallén (1985) suggested that the locomotor system in each animal could be composed of a number of "unit pattern generators" grouped together in larger entities controlling individual limbs or the trunk. By changing their interconnexions, they would yield a very versatile system which could produce backward and forward walking, trotting and galloping. These unit pattern generators can be envisaged as central modules (neural networks) whose number probably increases with the versatility and precision of movements executable by an organism. Although they are able to generate rhythms on their own, their precise action depends on peripheral sensory feedback (Grillner and Wallén 1985). It appears likely that the peripheral feedback is also somewhat fractionated according to the modular organisation of the central pattern generators. The larger units, each comprised of a central pattern generator, its related motor output (motoneurones are considered not to belong to the locomotor generator) and sensory feedback, then bear some resemblance to Loeb's task groups. The differences probably lie in scale and range.

How localised stretch reflexes are specifically integrated with the above two concepts (for examples as modules) has still to be clarified, but it is desirable to eventually arrive at a unified concept.

5.2 The Problem of the Variable(s) Controlled in Motor Acts

A general difficulty in studying motor control arises as soon as one tries to define the variables which are controlled by the nervous system and should hence be used for an adequate description of motor control processes (see review by Stein 1982, and commentaries). In his article: *How was movement controlled before Newton?*, Partridge (1982) has outlined concisely the problems with any simple suggestions as to the controlled variables, such as length, velocity, force or stiffness. His objections merit some consideration.

He starts by stating that physiological motor control has usually been treated in terms most commonly used in texts on Newtonian mechanics, such as force, position, velocity and mechanical impedance. Other equally valid coordinates with dimensions, such as momentum, power, work, kinetic energy and potential energy, are almost never employed, but would be as limited for describing motor control as the elementary Newtonian variables. This usage is encouraged by the – possibly unconscious – assumption that evolution has chosen to operate in one of the Newtonian coordinate systems while ignoring the variety of other equivalent systems. (A similar argument is advanced by Pellionisz and Llinás, 1982, who point out that movements are externally describable by classical Newtonian mechanics, but that this does not imply that the inner workings of the CNS use the same Newtonian mechanics.)

In motor control physiology, it is frequently assumed that the neural control of either muscle force or muscle length or velocity would – by way of the relations laid down in the laws of mechanics – result in the control of motor output actions. Partridge argues that this could happen in highly restrictive experimental situations, but usually not in normal activity, because the motor action of a muscle is related to the above variables not in a fixed, but in a variable way, to the extent that some motor actions might be entirely independent of them. It would not help to replace the above measures with system properties such as stiffness or reflex gain, because this would shift the controlled variable even further away from the motor output.

For example, if the motor response were to be determined by varying muscle force, the following complications would arise. Tension developed by a muscle not only is nonlinearly related to system geometry and neural input, but it is co-determined by external forces. Muscle force results in acceleration depending on changing gravity, synergist and antagonist tensions, centrifugal and coriolis forces (at the moving joint), and finally depending on load, whose properties in turn change with the state of the limb. Since (unavoidable) force errors accumulate through the integrative process from acceleration to displacement, Partridge (1982) considers control by way of force as close to impossible (see also Loeb 1987).

On the other hand, control of motor output by way of varying muscle length or velocity is compromised by the complicated relationship between the three variables. At many limb joints, the degrees of freedom of motion are overdetermined by numbers of muscles in excess of those needed to move the joint (see also Gielen and van Zuylen 1986). This provides additional options of coordinated muscle activation for particular limb positionings, but disposes of a fixed relationship between motor action and individual muscle lengths. [Indeed, the length of an individual muscle is not uniquely related to the position of a particular joint. For example, as noted by Burgess et al. 1982, the cat ankle has three degrees of freedom: flexion-extension, abduction-adduction and twist. But there are twelve muscles crossing the joint, most of which cross yet other joints, i.e. either the knee (two) or the toe joints (two) or both (two), and only two muscles span only the ankle joint (soleus and tibialis anterior). Equivalently, the discharge of muscle spindles in an individual muscle (even in soleus and tibialis anterior) are usually influenced by movement in more than one degrees of freedom; see Burgess et al. 1982.]

Ultimately, motor control is accomplished by providing muscles with appropriate spatio-temporal distributions of neural inputs. Partridge (1982) argues that no known evolutionary or computational advantage can be derived "from an intermediate conversion of the neural representation of desired motor response into textbook dimensions before final conversion into multimuscle drive signals" (p. 561). It is difficult enough for the nervous system to perform the conversion of the former representation (desired motor output) into the latter (multimuscle signal), that it should not be complicated by introducing yet another stage of intermediate conversion.

Very similar considerations are propounded by Loeb (1987). He emphasises that there has been a common presumption in motor control physiology that individual muscles are sensible entities of analysis and that they also form the building blocks in sensorimotor control. This view has favoured the traditional emphasis on variables, such as length, force, stiffness, etc., which are more easily defined with respect to single muscles. In a multi-articulated limb, however, the relationship between a muscle's tension and the movement trajectory is very complicated (see Partridge's comments; also J. L. Smith and Zernicke 1987). As an example, Loeb (1987) presents the case of the dorsiflexion of the cat's (hind) foot at the end of the locomotor swing phase. This flexion movement may predominantly result mechanically from contraction of hamstring muscles: "The late swing-phase activation of the hamstrings to arrest the forward momentum of the shank results in a whip-like torque at the ankle" (Loeb 1987, pp. 111). Similarly, J. L. Smith and Zernicke (1987) state that ankle muscles produce torques controlling paw segments directly, whereas knee muscles produce torques controlling inter-segmental dynamics between the paw and the shank. According to Loeb

(1987), therefore, it is not clear which use the nervous system could make from control at the low level of single muscles or even muscle compartments or task groups (see above).

It is thus obvious that motor control is not achieved by any simple control configuration with one (or at most a few) regulated variables. Even Houk (1976), although very much inclined in favour of stiffness regulation, expressed this insight by writing: "Both the decerebrate stretch reflex and the automatic reflex process in man act to compensate for differences in muscle properties which depend on the direction of length change. One would also expect compensation for other variations in muscle properties, such as those related to muscular fatigue, the frequency of motor unit discharge, the level of recruitment and the length of the muscle" (p. 311). More generally, there are a large number of variables that span a multidimensional space from which, depending on the circumstances and the task to be performed, certain subspaces may be selected. At present, little that would be more specific can be said. It remains for future research to identify these subspaces, but it will be a hard job.

5.3 Which Are the Measured Variables?

Conventionally, afferents from certain receptor systems are assumed to carry information about specific, defined variables. For instance, muscle "stretch" receptor afferents from muscle spindles and Golgi tendon organs are considered as providing information about muscle length and tension, respectively. With the "softening" of concepts of the variables controlled in motor systems (previous section), these notions should perhaps also be reconsidered. Some examples for the different kinds of afferent information in different motor tasks are therefore briefly discussed here, selecting spindles and Golgi tendon organs as specimens.

In slow high-precision tracking movements in man (see Sect. 4.11, item 3), the discharge patterns of spindle afferents from the prime movers are relatively irregular. If length information is indeed obtained from spindles in antagonist muscles, spindle afferents from agonists (prime movers) could be endowed with other functions. Ia fibres in particular, with their high sensitivity to small-amplitude inputs, can signal variables related to motor unit activity, such as its spatial and temporal distribution, discharge frequencies, amount of synchronisation etc. (see quotation by Houk 1976; previous section). Remember that the spatio-temporal pattern of motor unit activity leads to nonlinear phenomena in force production (Sect. 3.2.2), which should be of interest to the CNS. (These phenomena are not fully represented in stretch receptor discharges; for comparison, a central model representing motor output patterns, i.e. the Renshaw cell system, would therefore be of help to estimate more precisely the degree of nonlinearity; Sects. 4.8, 4.9.) This information would not be coded in any (artificially defined) long-term "mean firing rate", but rather in the moment-to-moment fluctuations (about means) of the discharge of sets of afferents (probably including Golgi tendon organs) where correlations play an important role. During the performance of many "high-precision" tasks (including quiet stance), the maintenance of system stability, i.e. the avoidance of large-amplitude tremor, is of the utmost importance. What has to be controlled in this situation is the *population statistics* of the firings of large populations of neurones and they have to be prevented from following their propensity for oscillating in gross rhythms. For it is ultimately the same network that produces rather smooth low-amplitude or large-amplitude movements. However, the feedback from the periphery is different in these two conditions, and the variables fed back are different. Thus, Ia afferents, and possibly also Golgi tendon organ afferents, have – at least – a dual function using different codes. Of course, these two kinds of information could be, and probably are, differentially evaluated by neuronal networks at different levels of the CNS.

One simplification in the foregoing considerations is the emphasis laid on Ia afferents which are assumed to measure muscle length and its changes. However, muscle length, velocity and force, apart from not being a unique set of base vectors sufficient for the description of motor performance, are also not independent parameters that – by means of proprioceptive and other afferent feedback – could be regulated or controlled separately in each single muscle (in the strict sense). This is already evident in the signals carried by the best studied muscle stretch receptors, i.e. muscle spindle group Ia and II and Golgi tendon organ group Ib afferents (disregarding here mechanosensitive group III and IV afferents).

For instance, under certain conditions, the *mean discharge frequency of spindle afferents* may rather accurately reflect muscle length (or even joint angle: Fellows and Rack 1987; but cf. Rack 1985) and its changes. But very often, to different degrees in different muscles and in different motor tasks, it probably also reflects muscle tension (disregarding the complicating effect of fusimotor inputs for the moment). This results from the fact that many muscles, particularly in larger animals (such as humans), possess fairly long tendons or aponeuroses, by which the force produced by the muscle fibres is transmitted to the bones (Rack and Westbury 1984; Rack and Ross 1984; Rack 1985; Proske and Morgan 1984). Thus, total muscle length is composed of the in-series contributions of internal muscle and tendon fibre lengths. Since tendons behave like (nonlinear) springs (see Rack 1985), their length varies monotonically with muscle force, and so does – in a complementary fashion – muscle fibre and muscle spindle length.

The series elasticity can vary enormously from muscle to muscle depending on their anatomical structure and the use made of them in natural movements (Proske and Morgan 1984). Indeed, in some cases, series compliance can be so significant quantitatively that "... one cannot assume that the muscle fibres and muscle spindles see all the movement that is imposed on the joint" (Rack and Ross 1984; p. 99). Therefore Rack (1985) proposed that "the location of the spindles enables them to 'see' changes in the length of the muscle fibres, so that monosynaptic stretch reflexes arising from the spindles would tend to prevent or reduce sudden changes in the lengths of these fibres ..." (Rack 1985; p. 227), but not in total muscle length, one may add. Rack (1985) also suggested that "the more precise control of joint position presumably depends on signals from other receptors acting through longer neural pathways" (p. 227). A candidate (not mentioned by Rack) might be the Golgi tendon organs. Instead of viewing them as muscle tension monitors, they might also be considered as devices measuring the length of the in-series elastic component, thus complementing the measurement of muscle fibre length by spindles. But the reconstruction of total muscle length from these two afferent signals would require an appropriate central connectivity, i.e. a convergence of both types of afferent on the same central neurones with the same sign (see paragraph after the next two).

The problem of series compliance also arises in muscles composed of in-series compartments, where each compartment represents an in-series compliance of varying strength (depending on its motor activation) to the other. But it also arises in parallel-structured muscles or synergistic muscle groups as a function of their relative activation. Fellows and Rack (1987) compared changes in muscle fibre length of human biceps brachii with those in elbow joint angle at different movement (flexion-extension) frequencies when the muscle was activated either electrically or voluntarily. With electrical activation, muscle fibre length changed in antiphase with elbow movement at frequencies between about 2.2 and 5 Hz, i.e. *muscle fibres shortened when the elbow extended*. This did not occur with voluntary elbow movements, the reason being that in this situation synergists (brachialis and brachioradialis) and probably also the antagonist (triceps) were partially co-activated which increased the in-series stiffness and hence the natural frequency of the mass-spring system composed of the tendons and the lower arm. Fellows and Rack (1987) concluded that in this instance where the flexing effort was shared among several muscles, muscle spindles in the biceps brachii presumably "saw" a fairly straightforward representation of the joint movement (although fusimotor innervation might complicate the matter). But what if the distribution of skeleto-motor activity to different synergists or parts of a muscle is uneven, i.e. "... the motoneurone activity is concentrated into some part of a muscle group which acts through a relatively compliant tendon?" (Fellows and Rack 1987; p. 411). In this case, the muscle spindles might move in no direct relation to overall muscle length. Such instances of differential muscle activation are well known (Sect. 5.1.1). This example shows that the mechanics of the muscle-tendon-limb system and their dependence on particular skeleto-motor activation patterns play an important part for the kind of information conveyed by muscle spindle discharges.

Golgi tendon organ afferents, commonly assumed to be monitors of overall muscle tension (Crago et al. 1982), respond with stepwise changes in firing rate during very slow, smooth changes in overall muscle tension (Appenteng and Prochazka 1984; see also Vallbo 1974). This may be attributed to the recruitment of motor units contributing fibres to the tendon organ and is compatible with the fact that Ib fibre discharge may be "driven" by unfused contractions, especially those of fast motor units coupled to them, so that they respond dynamically to the force oscillations rather than statically to the mean level of force produced (Jami et al. 1985a), like spindle primary endings do by way of their early discharges (Jami et al. 1985b; Sects. 2.5.2, 3.2.3.2 and 3.2.3.3). In addition, Golgi tendon organs may also be unloaded by motor unit contractions (like spindles; for discussion and an anatomical explanation see Zelena and Soukup 1983; also Hulliger 1984; Jami et al. 1985a, with some more references), so that the pattern of tendon organ discharge orginating from a muscle during natural muscle contraction is complicated. Even the fact that each tendon organ senses the contractions of several (roughly ten) motor units and contractions of a motor unit are sensed by several tendon organs (convergence–divergence structure) does not provide for an accurate mean tension measurement (L. Jami, pers. comm.; see also Appenteng and Prochazka 1984). Moreover, for this information (which has a local spatial aspect; Binder and Osborn 1985) to be regained – if only in a mutilated form – these Ib afferents would have to converge onto common postsynaptic neurones. But these interneurones are fractionated into subgroups by differentiated input from Ia afferents and other systems (Sect. 2.9.1.1 and below), so that the summing function they are supposed to perform is by no means guaranteed. Thus, sensory partitioning of tendon organ input and fractionation of the Ib interneurone pool may result in pictures of muscle state other than that described in terms of Newtonian force. In this context, it deserves mention that it has also been suggested that Golgi tendon organs can be controlled indirectly by fusimotor inputs (see Discussion in Jankowska and McCrea 1983).

Finally, the variables effectively measured by the afferents depend on the central connectivity. For instance, consider again the partial convergence of Ia and Ib fibres on nonreciprocal Ia inhibitory interneurones, which in turn project to homonymous (and many other) α-motoneurones and to cells of origin of the dorsal spinocerebellar tract (Sect. 2.9.1.1). From such convergence quite non-Newtonian representations of the muscle mechanical state may well result. Assume for the moment and for simplicity that muscle spindles measure muscle length and Golgi tendon organs monitor muscle tension (but see below). The excitation of the above interneurones by convergent input from both receptor types might then, under some conditions (e.g. in actively contracting but lengthening muscle), represent the *work done on the muscle during movement* (see Partridge 1982). To achieve this, the interneurones would have to perform multiplication of the input signals – no impossible requirement (see Sect. 1.4.1; for an implementation using a succession of logarithmic and exponential input-output relationships see Pellionisz and Llinás 1982). By their "slow velocity responses", Ia afferents would signal the distance moved (Lennerstrand 1968; Windhorst et al. 1976; Windhorst and Schmidt 1976). However, under these conditions, Golgi tendon organs "signal a dynamic, nonlinear function of whole muscle force over a range encompassing movements involving very low to very high force levels" (Appenteng and Prochazka 1984; p. 91). Another interpretation would not even require multiplication to be performed by the nonreciprocal Ia inhibitory interneurones. If the signals carried on the two types of afferent are both viewed as providing information on the length of muscle and tendon fibres, respectively, the nonreciprocal Ia inhibitory interneurones could reconstruct the information on total muscle length (see above). But these ideas are completely speculative and merely discussed here to demonstrate that the variables measured by spindles and Golgi tendon organs are not easily identified.

A further task-dependent modulating factor is presynaptic inhibition. In regard to the widespread convergence onto the above interneurones from Ia and Ib afferents, Brink et al. (1984) discuss the possible importance that the asymmetry of presynatic inhibition of Ia and Ib fibres may have in different motor tasks. Ib fibres generally evoke a strong presynaptic depolarisation (PAD) in Ia fibres, but do not receive PAD from Ia afferents. This may alter the balance of Ia and Ib input to the above interneurones. For instance, in motor tasks in which Golgi tendon organs are strongly activated and muscle spindle afferents only weakly so (fast isotonic movements), Ia afferents may be strongly depolarised presynaptically, whereby Ib input to those interneurones may dominate and select certain fractions of them (see Sect. 2.9.1.1). In motor tasks, in which muscle spindles are activated more strongly than Golgi tendon organs (slow isometric contractions), Ia afferent terminals are subject to weaker presynaptic inhibition. Ia input to the above interneurones is then more important and may select other fractions of them to co-determine motor output patterns. Although it was and could not be stated clearly by Brink et al. (1984) what all this might be good for, their considerations emphasise two points. Firstly, varying cell assemblies may be involved in different motor tasks (see Sect. 5.1.7). Secondly, the muscle state variables fed back to neural circuits of the CNS are not fixed, but time-variable and modifiable according to motor tasks; indeed, these state variables might be represented in different cell assemblies.

Mutatis mutandis, similar considerations may also hold for the processing of the information reaching supraspinal sensory centres from muscle spindles and Golgi tendon organs. There is accumulating evidence that the latter's signals contribute to muscle sense (McCloskey 1978; Burgess et al. 1982; Matthews 1982). Matthews (1982) suggests that, predominantly, spindle primaries provide information on movement (due to their velocity sensitivity), spindle secondaries supply positional information and Golgi tendon organs convey information on tension. However, referring to data about a new illusion reported by Roland and Ladegaard-Pedersen (1977), and Rymer and D'Almeida (1980), he also spots a new "enigma" to be solved: ". . . there seems to be a confounding of movement information, presumably derived from spindles, and information on changing contractile force, whether derived from tendon organs or corollary discharges" (p. 214). Thus, even at higher CNS levels, which are probably concerned with more global sensorimotor aspects, the separation of length, velocity and tension may obtain only under particular conditions and not otherwise.

In general, then, the separation of length, velocity, force or other components in the compound afferent signals from a muscle is at best a heuristic (and simplistic) measure. The de-composition of these signals into the above components is just one possibility of spanning a multidimensional space which could also be spanned by another set of base vectors (see Partridge, previous section). Note that this does not preclude the possibility of experimentally testing the system behaviour in subspaces of lower dimension, for instance by applying length or force inputs (see Rosenthal et al. 1970; Berthoz et al. 1971). It may then respond in a way expected of simple systems (as if a dynamic system were excited at one of its modes or in the direction of one of its "eigenvectors" since tested appropriately; see Pellionisz and Llinás 1979). The multidimensionality does however provide the complex system with a large versatility of operation including adaptability. Finally, it must be stressed that limb position and its dynamic changes usually result from a complex multidimensional interplay of the actions of several to many muscles involved. This adds to the complexity and requires that the control problem be described and treated, if in any technical terms, by means of multivariable control theory (a tiny application of which was outlined in Sect. 5.1.2.2), perhaps in combination with tensor analysis which has recently become fashionable (Pellionisz and Llinás 1979, 1980; 1982; Gielen and van Zuylen 1986).

5.4 Spatial Transformations

The sensory signals flowing from the periphery to the CNS, and the efferent signals dispatched by the latter towards the motor periphery bear "place tokens". The CNS has to perform a spatial transformation between these signals. For the problems arising, consider a "simple" example. Burgess et al. (1982) wondered how muscle spindle afferent discharge could contribute to kinaesthesia (joint and cutaneous afferents probably do not contribute significantly; see also Matthews 1982). Their specific question was whether (cat) soleus and tibialis anterior spindles could specify ankle position in the flexion-extension dimension without participation of receptors in other muscles. The spindles in these two muscles should be best suited for this task because soleus and tibialis anterior are the only two muscles spanning only the ankle joint. However, soleus spindle discharge is additionally influenced by ankle abduction, adduction and twist in both directions, and tibialis anterior spindles are influenced by ankle abduction and twist in one direction. Hence, the spindle signals are not uniquely determined by movement in one dimension (disregarding all the complications arising from extra- and intrafusal activity for the moment). The spindle signals arising from soleus and tibialis anterior could be "disambiguated", i.e. be made more closely related to ankle flexion-extension, if tibialis anterior spindle activity were centrally inhibited by tibialis posterior spindle activity, and if soleus spindle signals were inhibited by tibialis posterior and peroneal (?) nerve activity. (Another example referring to the signals originating from multi-articulate muscles is also discussed by Burgess et al. 1982.) Quite apart from its speculative character, this example illustrates which role the central connectivity probably plays in the transformation of signals from one spatial reference frame to another.

It is perhaps a common feeling that complicated spatial transformations are performed preferentially in phylogenetically young and highly organised structures such as the cortices. Particularly the cerebellum, having evolved largely in parallel to the neocortex (Eccles 1977), is known to be concerned with the fine and precise control of motor coordination, i.e. with the temporal and spatial features of posture and movement. Several models have been proposed to account for its remarkably uniform "lattice-type" structure (e.g. Marr 1969; Eccles 1977; Albus 1971; Kornhuber 1974; Braitenberg 1967, 1977; Pellionisz and Llinás 1979, 1982). Whereas Marr and Albus emphasised the learning capabilities in pattern recognition, Pellionisz and Llinás suggest that the cerebellum performs, firstly, a transformation of spatial reference systems in sensory and motor systems (1979, 1980, 1982), and, secondly, a prediction from the incoming signals to foresee the course of movement (1979) and to compensate for delays in neuromuscular signal transmission (1982). The transformation of spatial coordinates needs to take place, because sensory and motor systems very probably use different frames of reference with different numbers of dimensions. However, details are still largely unknown or, if known, of extreme complexity (Pellionisz and Llinás 1980, 1982).

Whereas the cerebellum appears to be an organ particularly specialised for spatial transformations, other structures may also be involved. Among these, the spinal cord can and must also perform such conversions. Berkinblit et al. (1987) have recently reviewed some aspects of spatial motor coordination. They point out that, despite a point in space being determined by three independent coordinates, the configuration of a limb in space is characterised by a larger number of coordinates (see also Pellionisz and Llinás 1980). If the limb's tip is to reach a certain point in space, this can be achieved via different paths. In every actual trial, however, a particular individual path among those available is selected by the nervous system. The underlying problem (usually referred to as Bernstein's problem) is to understand how the nervous system reduces the redundant degrees of freedom in order to execute a particular movement under special conditions. Berkinblit et al. (1987) refer to the approach used by Pellionisz and Llinás (1979, 1980), who consider the neural transformations of coordinates of a target point into a limb's multi-joint coordinates as a tensorial operation (transformation of co-variant coordinates into contravariant coordinates by means of a metric tensor), in which the cerebellum plays a key role. However, on the basis of their experiments with spinal frogs (without a cerebellum), the Russian authors emphasise that the spinal cord *per se* is capable of performing such metric transformations. The emphasis on the autochthonous capabilities of the spinal cord of performing complex spatial transformations is very important. Although with the progressing "cephalisation" in higher vertebrates some of these capabilities may have been shifted to supraspinal structures, the spinal cord will probably still be left with some residual capacity to the extent of refining spatial coordination by setting initial and boundary conditions for descending commands using afferent feedback.

Kornhuber (1974), envisaging the cerebellum as a function generator for fast movements (saccadic eye and ballistic limb movements), has ascribed a conceptually similar role to feedback from mechanoreceptors. He proposes that a function of mechanoreceptive afferent input to the cerebellar cortex is to contribute to selecting the correct granular cells. This information from the periphery is needed to adjust the energies to be raised for movements of different starting positions. For example, different amounts of energy are needed in saccadic eye movements from the mid-position to a lateral position and for saccades from the lateral to the mid-position. Thus, although granular cells are selected predominantly on account of afferent input from the cerebral cortex, mechanoreceptive afferents should be important in defining the initial condition (starting position) of the limb to be moved. If the suggestion of Berkinblit et al. (1987) holds true, that both the spinal cord and the cerebellum can perform spatial transformations, Kornhuber's argument may also apply to the spinal cord (see also Gottlieb and Agarwal 1980; Sect. 4.7.2).

The precise neural implementation of spatial transformations at the spinal level are not yet known (neither are they at the cerebellar level). On the way from the sensory periphery to spinal neural circuits, the three-dimensional coordinates (in external Euclidian space) of the positions of body and limbs are transformed at least twice into other coordinate systems (disregarding the vestibular system). Firstly, they are transformed into discharge patterns of various kinds of afferents which, in a complicated pattern (see Burgess et al. 1982), reflect joint angles, the lengths and situations of muscles, relations of cutaneous receptive fields and so forth. One might think of this mapping as a (divergent) splitting of the three independent space coordinates into multidimensional (i.e. overcomplete) sets of other base vectors which also differ in their sensory modality. This mapping is then followed by a second one consisting of the projection patterns of the afferents onto spinal (or other) neuronal circuits. There might be some reduction of dimensionality through the "multisensory" convergence of afferents of different modality onto common interneurones, but this is not certain. As the simplest example, consider the monosynaptic Ia projection from different muscles onto motoneurones innervating different muscles. Classifying muscles *functionally* as flexors, extensors, adductors etc. is almost, though not completely, equivalent to specifying their *spatial* situation with respect to joints, the sites of their origin(s) and insertion(s) and, hence, the spatial effect on the body parts moved. The Ia-motoneurone mapping follows functional ("species-specific") and topographical principles. There is a certain relation between the positions of muscles, e.g. of the limbs, and the positions of the corresponding motoneurone pools within the spinal cord, and the afferent Ia connexions establish a complicated mapping between these positions of muscles and motoneurones. It may well be that the responsiveness of spindle afferents to local muscular events and their partitioned reflex connectivity contribute details to the body image which is needed even at the spinal level.

Less well known are spatial projection principles in oligo- and polysynaptic reflex (and other) pathways. It appears as if projections in these pathways were fairly widespread and hence diffuse, but this impression may result from our current ignorance and/or inability to understand the complexity of the networks and of the signal transmission occurring through them. Whilst the connexions may be diffuse *anatomically*, they may not be so *functionally* in a behaving organism, because pathways may be opened and closed differentially depending upon motor task and a variety of poorly understood principles, such as signal transmission through neuronal chains by correlated discharge. This is an area of further research.

5.5 Temporal Aspects

Intimately related to the aforementioned spatial aspect is the temporal dimension (see Pellionisz and Llinás 1982). As an example, consider again muscle responses evoked by muscle stretch. (This paradigm should be recognised as isolating a very special aspect, thus artificially reducing the dimensionality of the system in question.) It was argued before (Sect. 4.7.2) that the function of the early stretch reflex component, set going by the initial discharge of Ia fibres, should be reconsidered. Its role may be to cause a state of "alertness" in the cord, not only in the α-motoneurones, but also in the interneurones, in order to prepare the system both for further changes to be expected in the periphery and for the first long-latency response. This "predictive" capacity was assigned to primary muscle spindle endings by Houk et al. (1981a). Dufresne et al. (1978, 1979), who investigated the EMG responses of arm biceps and triceps muscles in man to random position perturbations of the forearm, developed a model in which the EMG was a weighted sum of response components to velocity, acceleration and position, these components having latencies of 25 ms, 45 ms and 86 ms, respectively. Thus the predictive components occur at the shortest latencies. If the first (segmental) predictive component were to last until the onset of the longer-latency responses, some sort of information storage (short-term memory) would be required. There are relevant circuits in the spinal cord which could fulfil this function, and some information would also be stored by the long contraction times of those motor units discharged by the initial reflex. Note that the phasic reflex response (which may have even longer-lasting after-effects) is superimposed upon the ongoing afferent and motor activity which reflects the current state of affairs and co-determines the reflex responses of all kinds.

This possibility of prediction is again reminiscent of similar properties of the cerebellum. As mentioned above, Pellionisz and Llinás (1979) assume that one of the main functions of the cerebellum is prediction, and they also propose a mechanism. Prediction is presumed to be executed by means of a Taylor expansion such that a stack of Purkinje cells, which are contacted by a common "beam" of parallel fibres and project onto common nuclear cells, produce different time derivatives (e.g. from zero-th to second order) of the input. The nuclear cells, upon which these different Purkinje cell signals converge, then produce a prediction of the probable future course of the input signal. (Experimental evidence for this predictive capacity of the cerebellum was adduced by Vilis and Hore 1980; Hore and Vilis 1984; see Sect. 3.4.2.2.) Pellionisz and Llinás (1979) themselves mention that such prediction probably occurs in other neural networks as well, e.g. the retina or the auditory system.

From these cursory comparisons, it may be suggested that the spinal cord performs functions similar to those of the cerebellum, though probably in more rudimentary form. Whereas the cerebellum (in conjunction with other supraspinal structures) executes its functions at a more global level, the spinal cord is more deeply involved in details. For instance, it is unlikely that the cerebellum can cope with such "fine grain" information as that required for localised stretch reflexes (see Sects. 3.3.1 and 3.3.2). The similarities and differences in structure and function of spinal cord and cerebellum are worth being elaborated on with a view towards an integrated theory.

5.6 Multimodal Organisation of Spinal Neural Assemblies

As stressed in the preceding sections, the features of the peripheral organisation of the motor system are immensely complicated. This may be a reason for the complexity of the spinal (and other central) neuronal networks which have to deal with the periphery (see Loeb 1987). A particular aspect of this complexity is the polymodal or "multisensory" convergence of many input systems on interneurones intercalated in reflex pathways to α-motoneurones (Lundberg 1979), but also to γ-motoneurones and probably to Renshaw cells. The same applies to the many tracts ascending from the spinal cord, which may provide higher centres with information on the activity in spinal interneuronal systems (see Oscarsson's note in Evarts 1972).

It is sometimes convenient to think of a main unimodal pathway as being "modulated" by other afferent modalities or descending inputs. For example, it is still customary to envisage the classical monosynaptic reflex loop constituted by motor units and spindle Ia afferents as a single-loop negative feedback system whose gain may be altered by other inputs (e.g. by fusimotor input; Hulliger et al. 1985). Essentially the same applies to the recurrent inhibitory system (Hultborn et al. 1979; Pompeiano et al. 1985b). This view ultimately receives its strength from the engineering control theory which supplies powerful theoretical and practical tools for understanding and analysing simple physiological systems. The application of this theory is relatively simple in the case of single-loop feedback systems. This fact may explain the common practice of conceptually isolating unimodal pathways. This, however, is justified only when the pathway in question has the highest gain among those of importance, but it is usually wishful thinking.

It is worth considering the possibility of applying concepts of "population neural coding" (Erickson 1974) to the above problems. These concepts were originally developed for the case of unimodal sensory information processing in order to account for the fact that many sensory neurones are broadly tuned over the qualitative dimension of concern. With respect to colour coding,

Erickson (1974) explained this concept as the idea "that hue is represented by certain *relative amounts of activity within a small population* of broadly tuned color-coded neurons" (here of three receptor types; Erickson 1974; p. 159). This scheme of coding is usually briefly referred to as an "across-fibre or across-neurone pattern" code (see also D. V. Smith et al. 1983; Yamamoto et al. 1985). It is often contrasted with the coding scheme of "labelled or dedicated lines", in which a sensory quality is represented in the activity of small groups of neurones responding "best" to the particular stimulus quality (for taste see D. V. Smith et al. 1983; and Yamamoto et al. 1985). The across-neurone pattern scheme avoids the probable confusion of stimulus intensity and quality in a single neurone's discharge. Erickson applied this idea to various modalities: temperature, kinaesthesis, somesthesis, hue, taste, and feature detection. Interestingly, he argued that topographical mappings could also be regarded in the same way in that location is a qualitative dimension, too. For feature detectors in the visual cortex he claimed that groups of them could simultaneously code for various qualitative and quantitative dimensions of visual stimuli, e.g. line orientation, colour, intensity and further parameters. Thus, each single neurone would have an individual response profile without being able to tell the difference between different stimulus parameters. The emphasis is on the operation of *populations of neurones*. Processes of synaptic plasticity or modulation were not discussed in this context by Erickson (1974). D. V. Smith et al. (1983) showed that, in the taste system, the labelled-line concept is not completely disjunct to the across-neurone pattern concept, because the same cells which could be termed a "labelled line" by one point of view also determine the similarities and differences in the "across-neurone patterns". Although the special properties of certain sets of neurones are essential for coding the tastes of particular stimulus classes, activity in more than one set of neurones is necessary for the unambiguous coding of taste quality. Yamamoto et al. (1985) argued that distributed spatial (chemotopic) representations of different tastes in the cortex could contribute to the discrimination of tastes. Thus spatial coordinates would expand the dimensionality of the neural space in which sensory stimuli are represented. Note that the notions used in the foregoing are nearly the same as those used by Abeles (Sect. 1.3.1).

Although the preceding discussion is concerned with the "intramodal" detection of various stimulus features, an extension of the above approach to the "intermodal" situation found for the convergence patterns of many neuronal systems in the spinal cord could be stimulating. Due to differentiated (fractionated) input distributions, many neurones exhibit a broad but individual response profile. This provides for the possibility that certain (multidimensional) input configurations can be adequately detected, represented and processed by across-neurone patterns in (relatively) small populations of cells. Such an extension to the "primitive" spinal reflex level does not appear to be far-fetched. Erickson (1974) proposed a parallelism of neural coding in sensory and motor systems. For neural classification of movements, many parallel neurones would be required. Different movements are distinguished by different sets of active efferent motor fibres to flexors and extensors, so that the "across-fibre pattern of activity" shifts with change of movement.

Erickson's justification for suggesting the parallelism is interesting. First, it would be surprising intuitively if the nervous system had developed different encoding schemes for sensory and motor functions; that would be uneconomical because the same machinery (neurones and their interconnexions) is used. Second, in performing various functions (sensation, movement, memory) the nervous system is faced with similar problems, the basic one being the need for economy. For example, in the colour system, a few types of colour-coded neurones represent the total wavelength range. In the motor system, many different movements can be executed with a limited number of motoneurones. Third, in both cases, the function is executed in a distributed manner by the cooperative action of many neurones, so that "each function has the neural safety and weight of being expressed by a large portion of the brain" (Erickson 1974; p. 164).

The notion of "safety" refers to the issue of redundancy, which is a basic problem in any coding of information. Information transfer through a system of multiple parallel channels depends on interchannel correlations (e.g. synchronisations) in such a way that the absence of correlation (independence) is associated with the highest information transfer, and increasing correlation with decreasing information transfer (see Rudomín 1980; Sect. 2.9.2.2). Thus, the broad rhythmical correlations involving vast neural assemblies (as during locomotion) correspond to states of low information transfer and content. Such states are rather regular and, hence, easy to recognise and describe. On the other hand, the "attentive" states of posture and slow movement possibly involve short-term synchronisations which might temporarily activate synfire chains and set up neural assemblies, possibly with the help of synaptic modulation or similar processes. These cooperative structures would be associated with medium (not the highest possible) levels of information transfer and, hence, they would exhibit some redundancy. Lundberg (1979) speculates that the multimodality of reflex pathways has ultimately developed because of its economy (see Erickson, above). If so, the possible functional restitution of specificity in reflex path ways by means of synchronised discharges, i.e. redundancy, concurs with economy as well as with the safety which derives from using a redundant code.

We are left, then, with dynamically changing neural assemblies of multifunctional cells which are grouped and regrouped according to momentary needs and conditions (see Loeb's task groups; Sect. 5.1.7). At this level of generality, the preceding description is similar to that given by Abeles (1982; Sect. 1.3.1) for cortical networks.

5.7 Summary

The operation of the spinal cord cannot be understood in terms of single-loop feedback or single-line reflex arc models. There are several reasons for this impossibility.

First, it is not clear how far the parcellation of functional units goes: For example, which roles do localised stretch reflexes play in natural movement? What is their organisational rationale? To which anatomical or physiological properties of the neuromuscular system are they linked? The task of understanding their organisation and determine their function is aggravated by the numerous possible factors that can play a role in establishing such modules. Among these factors are the differential activation of muscle compartments, nonlinear mechanical interactions between them, their various spatial arrangements, their different mechanical output actions, spatial gradients of different types of muscle fibre, muscle fibre architecture, and simply ontogenetic propensities. The combination of such factors renders interpretation of existent partitioning difficult. This opens a whole new field of research. The notion of localised stretch reflexes has some, although not yet fully specified, similarity with Loeb's (1984) concept of task group. This, too, requires qualification and further experimental and theoretical work.

Second, even in the motor control of single muscles, there is no single variable which is regulated in any simple manner by servos, let alone in the cooperative operation of many muscles (or muscle compartments) during natural movements. The signals from each category of muscle afferents are not uniquely and permanently defined as related to a single variable each. They change with task, and their significance for motor control is co-determined by spinal neural networks with time-varying characteristics. Thus, formally, the motor control strategies as well as the multimodal afferent feedback span a virtual multidimensional variable space, from which subspaces are selected depending upon the task to be performed.

In a rudimentary form, spinal circuits should be capable of pattern recognition because the polymodal input to many neuronal systems in the spinal cord is fractionated, thus providing for individualised receptive properties of the neurones. These may then participate in various combinations in cell assemblies varying with input pattern and task. These spinal networks, including reflex circuits, perform some functions such as spatial transformations and predictions, which are also executed by supraspinal structures, particularly the cerebellum.

These general ideas certainly require specification which, however, needs some time for evolution. Ultimately, a basic demand that any theory will have to fulfil is to specify the role of different neurone types within its framework, if only to render theoretical predictions amenable to experimental verification. Marr's in many ways fascinating model of the cerebellum (Marr 1969) may well have defined the standard in this respect, although, as is the fate of the best models, it may well be wrong in more than one of its predictions, and Marr himself (1982) found it deficient in yielding a sketch for a computer simulation.

5.8 Concluding Remarks

In considering spinal cord functions, it is worth recalling some of the warnings of Pellionisz and Llinás (1979) against widely applied simplistic concepts. One such concept is that of "loops", in which the complexity of parallel neuronal networks is reduced to a chain of some serially connected individual neurones. This concept may mislead us to believe that such a "reflex" is conceptually equivalent to the entire neuronal assembly. An opposite concept is that of "mass action", in which diffuse "random" connexions between the neuronal elements are envisaged as the principle of organisation. Marr (1982) derides "Lashley's mass-action ideas (according to which the brain was a kind of thinking porridge whose only critical factor was how much was working at the time)" (pp. 339–340).

These warnings are equally valid when applied to the spinal cord. The neuronal circuitry of the spinal cord may be rather more complex than that of the cerebellum, and so too must be its signal processing. It is almost certainly too simplistic to think in terms of simple lumped reflex arcs in which information about one modality is conveyed in the form of mean neuronal discharge rates. Rather, the multidimensionality or parallel organisation of neural networks together with their intricate signal processing capabilities (including more elaborate codes) must be taken into account. On this view, spatio-temporal patterns (variations of discharge around mean frequency) of cell assemblies are important, possibly or probably in conjunction with plastic synaptic processes, which would in turn depend upon the neural activity pattern, as was hypothetically outlined, for instance, by von der Malsburg (1981, 1985).

In this conceptual framework, redundancy introduced by correlated discharge of parallel neural elements is of functional importance. This would run counter to the third attack that Pellionisz and Llinás (1979) launch against the concept of "redundancy". They admit that this concept "... reflects more carefully upon the inevitability of reckoning functionally with the parallel features of networks" (p. 324), but criticise that it "... combines the loop-view with the assertion that the large network is basically many such loops together" (p. 324). However, this criticism is based on a notion of redundancy that is perhaps a bit too narrow. In Pellionisz and Llinás' (1979) usage, the meaning of redundancy appears to be related to static anatomical rather than functional signal-processing aspects. And this relates to the fact that their model of cerebellar function is formulated in terms of (time-variable) mean discharge rates in fixed arrays of neurones without synaptic plasticity. However, in a later paper (1982) they loosened their resistance to redundancy because they envisaged the external physical invariants to be represented in terms of co-variant projections into overcomplete (unknown) coordinate systems (see Sect. 5.4). This notion of redundancy is still different from that used in this book, which argued that correlation patterns are of paramount importance in the functioning of nerve cell assemblies.

In general, however, it is worth pondering whether any attempt at a theory of spinal motor function that starts from neuronal networks is not seriously and fundamentally flawed. To explain this, I am referring to Marr's approach to understanding human visual perception (1982). Marr's contribution of interest here is his methodology. Starting from the assertion that vision is primarily a complex information-processing task, he distinguishes three levels of understanding and investigating this task. The uppermost and most abstract level is called "computational theory". This theory is concerned with two questions: *What* is computed in the information-processing task, and *why*? The answer to the latter question selects that solution to the information-processing task at hand that is most appropriate with respect to constraints originating from the physical structure of the external world. The next lower level is concerned with the problem of the *algorithm* by which the computation might be performed. This algorithm, i.e. the particular procedure by which any solution is computed, depends very much on the "representation" in which the pertinent information is cast. And only at the lowest level appears the question of *how* the representation and algorithm could be implemented by a machine such as a neuronal network. Often, even in research concerned with visual perception, the upper level in particular is neglected, leading to confusion at the lower levels. To illustrate his point by an example, Marr refers to the fast Fourier transform. He argues that it is no use trying to understand the fast Fourier transform in terms of transistors as it runs on an IBM 370. It is simply too difficult. What has to be done is to study the transform at the theoretical level, in terms of the equations involved etc. To be sure, the lower levels are of importance and interest, too. But there is no hope of understanding them without the upper level. As a theoretician, Marr (1982) is rather pessimistic about the prospects of elucidating – in terms of neuronal networks – the basic mechanisms carrying out complex information-processing tasks. But still this type of work at the lower level has to continue.

It may be worth while to pursue a similar top-down approach to the motor control system. There is no doubt that motor control is a highly complex information-processing task, too. The computational theory to be developed should again deal with the two basic questions: What is to be achieved (computed), and why does it need to be done? The next lower level would have to deal with possible representations and algorithms by which the theoretical goals may be attained. It is at this level that the question of which types of information (e.g. motor control variables; see Sects. 5.2 and 5.3) are represented in which form, should be considered. And the lowest level should finally consider possible neuronal networks and peripheral devices executing the tasks at hand (i.e. the implementation). There is perhaps a special feature in this approach as applied to motor control. Whereas visual perception in Marr's treatment is essentially a unidirectional cascade of processes, motor control implies at least one other direction (above and beyond that from the sensory periphery to central neural networks), namely the efferent side, which needs its own computational theory, representations, algorithms and implementations. More concretely, the instruments available for performing a motor act, such as the mechanical devices (limbs, joints, muscles, tendons, etc.) should themselves be subject to a theoretical analysis since they provide peripheral constraints for any theoretical and practical solution to the problems to be solved. The theoretical treatments of both motor and sensory sides therefore have to undergo a feedback, too. Pieces of such interwoven considerations have been presented in Sect. 5.1, where the peripheral mechanical structure of the executing devices was suggested to partly determine the structure of nervous control networks (here of reflex loops). It is certainly true, as argued by Sacks and Roy (1982), that the architecture and biochemistry of skeletal muscle are adapted to the mechanical circumstances and tasks they encounter, which may to some extent reduce the complexity of the programming to be performed by the CNS in movement control. But the peripheral devices themselves are subject to mechanical and metabolic constraints limiting their adaptability. The nervous system therefore has to take account of the physical constraints inherent in the organism's own peripheral devices. In summary, therefore, theorising is of the essence in understanding motor control. What finally has to be done, in vision as in motor control, is to bridge the gap between our knowledge of the behaviour of single neurones and that of the intact animal: "One great challenge is to bridge this gap to understand how individual CNS neurons interact to generate different patterns of behavior, or make us perceive the environment, or memorize what we experience" (Grillner and Wallén 1985; p. 233).

Now, then, how brain-like is the spinal cord? An answer to this question depends on the definition of a brain. If we resorted to simply and elusively defining the brain anatomically as comprising everything from the brainstem upwards, our question would be meaningless. Only if anatomy is put aside for a moment and functional considerations are put in the foreground, can we make sense of the question. The problem is then shifted to defining the typical attributes of brains. Which criteria have to be met by a brain to pass as one, what must it be able to do, and which tools are typically used by a brain? *Consciousness* is unsuitable as a *general criterion*. *Perception* is probably a valid and useful criterion although its underlying mechanisms are not yet elucidated. Some of these basic mechanisms might also, though perhaps at a more rudimentary level, occur in the spinal cord without leading to perception. With *recognition* the situation is less certain. Pattern recognition processes could very well be at work in many spinal networks. Progress in this field depends not only on the experimental elucidation of neural networks and their signal processing capacities, but also on the development of appropriate theoretical frameworks to guide the search. For this purpose, seemingly crazy concepts are as valuable as proposals which seem much more serious, as long as they enable us to make progress, if only by exclusion. *Pattern formation* processes (e.g. rhythmogenesis) occur widely throughout the nervous system and are probably based upon the same or similar principles of neural activity.

These remarks may suffice to re-emphasise that many basic mechanisms which are employed by supraspinal structures to solve complicated problems of sensation and motor control may also be used by the spinal cord to solve similar problems, though at a different level. This has hopefully become apparent, though sometimes implicitly, throughout the book. Some suggestions, which may have appeared hazardous, have been presented to help trigger new ideas. If some of these proposals turn out to be inconsistent, it would not be too bad, for, as Miguel de Unamuno says: "If a person never contradicts himself, it must be that he says nothing".[10]

---

[10] Quoted from Hofstadter DR (1980) Gödel, Escher, Bach: An eternal golden braid. Vintage Books, New York, p 698

Appendix A: Auto- and Cross-Correlations

It is very often of interest to determine the degree to which a time-varying signal x (t) is correlated with itself, i.e. whether, and if so, how strongly the value at time t determines that at time t + τ. This question can be answered by computing so-called "auto-correlations". For this purpose, two signal values separated in time by an interval τ are multiplied, and all these products are then averaged over time t according to the formula (see Bendat and Piersol 1971):

$$R_{xx}(\tau) = \lim_{T \to \infty} (1/T) \int_0^T x(t) \cdot x(t + \tau) \, dt \tag{A1}$$

Analogously, "cross-correlations" quantitatively express the degree of coupling between two different signals x (t) and y (t). Here one of the factors in the integrand of Eq. (A1) is replaced by the corresponding expression for the second signal:

$$R_{xy}(\tau) = \lim_{T \to \infty} (1/T) \int_0^T x(t) \cdot y(t + \tau) \, dt \tag{A2}$$

These formulae are valid for continuous analogue signals x (t) and y (t). In neurophysiology, auto- and cross-correlations are most commonly computed by constructing histograms which estimate the probability of occurrence of action potentials in temporal relation to each other. The relation between the two representations of correlations becomes clear when the amplitude and duration of action potentials are disregarded and the latter are represented by idealised, so-called "Dirac (delta) functions" (or impulses). The theory and practice of "point processes" can then be applied (see Perkel et al. 1967a, b). The Dirac function, δ (t), is identically zero except where its argument is zero (t = 0); at t = 0, δ (t) is infinite, but the area underneath the delta impulse equals 1. The delta function can be imagined as one to be obtained in the limit when a rectangular pulse of width Δt (from $t = 0$ to $t = \Delta t$) and height $1/\Delta t$ becomes ever narrower, i.e. $\Delta t \rightarrow 0$.

A sequence of action potentials is then represented by a sum of Dirac delta functions:

$$x(t) = \sum_{i=0}^{\infty} \delta(t - t_i),$$

where the t_i are the instants at which the action potentials occur. With this definition of a signal the above formulae A1 and A2 are applicable (see ten Hoopen 1974).

As usually as applied to spike trains, auto- and cross-correlations are defined directly in terms of probabilities. Accordingly, the auto-correlation $R_A(\tau)$ signifies the probability W of a spike occurring at $t_0 + \tau$ when a spike has occurred at t_0 (see Perkel et al. 1967a):

$$R_A(\tau) = \lim_{\Delta T \rightarrow \infty} (1/\Delta \tau) \cdot W \quad \{\text{spike in the interval}$$
$$(t_0 + \tau, t_0 + \tau + \Delta \tau)$$
$$\text{if spike at } t_0\} \qquad\qquad (A3)$$

The action potential at $t_0 + \tau$ can be the first, the second, the third, and so forth. The probability for the occurrence of the first spike (after a preceding one) is given by the first-order interval density $f^{(1)}(\tau)$, the probability for the occurrence of the second spike by the second-order interval density $f^{(2)}(\tau)$, which represents the probability density of the sum of two subsequent intervals, and so forth. Hence:

$$R_A(\tau) = \sum_{i=0}^{\infty} f^{(i)}(\tau) \qquad\qquad (A4)$$

If successive intervals between action potentials are mutually independent, that is, if they constitute a so-called "renewal process", interval densities of higher order can be defined by so-called "convolution integrals":

$$f^{(n+1)}(\tau) = \int_{0}^{\tau} f^{(n)}(t) \cdot f^{(1)}(\tau - t)\, dt, \qquad\qquad (A5)$$

that is, the probability for $n + 1$ successive intervals to sum up to τ equals the joint probability that n intervals together have length t and one interval is equal tc $\tau - t$ integrated over all t from 0 to τ.

It is worthwhile to mention that

$$\lim_{\tau \rightarrow \infty} R_A(\tau) = 1/\mu, \qquad\qquad (A6)$$

where μ symbolises the mean interval length (Perkel et al. 1967a).

In the same way as the auto-correlation describes the conditional probability for the occurrence of a spike at $t_0 + \tau$ after a spike at t_0 in the same spike sequence, the cross-correlation does so for action potentials from different spike trains A and B (Perkel et al. 1967b):

$$R_{AB}(\tau) = \lim_{\Delta T \to \infty} (1/\Delta\tau) \cdot W \quad \{\text{spike in B in the interval} \\ (t_0 + \tau, t_0 + \tau + \Delta\tau) \\ \text{if spike in A at } t_0\} \qquad (A7)$$

The time shift τ can assume positive and negative values. In Eq. (A7) the spike train B is correlated with respect to spike train A as reference train. If these trains are independent of each other, spikes in B occur randomly with respect to those in A. Then

$$R_{AB}(\tau) = 1/\mu_B, \qquad (A8)$$

where μ_B is the mean interval between spikes in B. That is, the probability of encountering a spike in B at an interval τ after a spike in A is given by the mean discharge rate in B. Usually this also holds for correlated spike trains A and B at large τ, since correlation commonly fades away over time. Hence:

$$\lim_{\tau \to \infty} R_{AB}(\tau) = 1/\mu_B \qquad (A9)$$

Conversely, when A is correlated with respect to B as reference:

$$\lim_{\tau \to \infty} R_{BA}(\tau) = 1/\mu_A \qquad (A10)$$

We thus obtain the following symmetry relation for the cross-correlation (Perkel et al. 1967b):

$$R_{AB}(\tau)/\mu_A = R_{BA}(-\tau)/\mu_B \qquad (A11)$$

Because of this symmetry relation, only one cross-correlation need be computed (e.g. B with respect to A as reference), although this must then be done for both positive and negative τ.

It should be mentioned that the above auto- and cross-correlation histograms are sometimes also called "renewal densities" (as by Abeles 1982), or, "auto"- or "cross-covariance" after subtraction of mean bin contents (as by Schultz et al. 1985). Nomenclature is thus not uniform, unfortunately.

The proper interpretation of cross-correlation histograms is often no simple matter because the firing pattern characteristics of the reference as well as of the correlated spike trains enter into the cross-correlogram and co-determine its time course (see Moore et al. 1970). Auto-correlograms should therefore usually be calculated as well for comparison.

Appendix B:
Cross-Correlations Between Three Spike Trains

The experimental techniques for recording spike trains from many neurones (at least more than two) have been developed and used (see review by Krüger 1983). The main problem in dealing with the data remains the processing which has been handicapped by the lack of quantitative methods for the estimation of higher-order cell interactions. In this and the following two appendices, some methods to this end are described.

A technique for displaying the interactions of three neurones was first presented by Perkel et al. (1975). Abeles (1982) used it to investigate such interactions between cortical neurones (Sect. 1.3.1). He also presented a quantification of this scheme (1983), which will be briefly described here. Three questions are of interest when investigating the interactions between three neurones: (1) What is the probability of occurrence of a particular sequence of spikes discharged by three neurones, A, B and C? (2) Which effect has the firing of one neurone on the combined firings of the other two? (3) Which effect has the combined firing of two neurones on the discharge probability of the third? As pointed out by Abeles (1983), these questions are related but not identical, which is already evident from the units in which the answer is given. If the firing rate of a neurone is given in units of s^{-1}, the frequency of a sequence of three spikes (question 1) is given in s^{-3}, the unit for the second question is s^{-2}, and that for the third question is s^{-1}. The answer to the first question is a "covariance function", that to the second a "compound renewal density function", and that to the third a "simple renewal density function".

To estimate the probability of occurrence of triplets of spikes, Perkel et al. (1975) proposed the following scheme (see Fig. 58). Suppose the discharge times of the three neurones, A, B and C, have been measured. For a triplet of spikes occurring at instants T_A, T_B and T_C, three intervals are defined: t_{BA}, t_{CA} and t_{CB}, as indicated in Fig. 58 A. It should be noted that $t_{BA} = -t_{AB}$, and so forth. As obvious, these intervals are not independent of each other

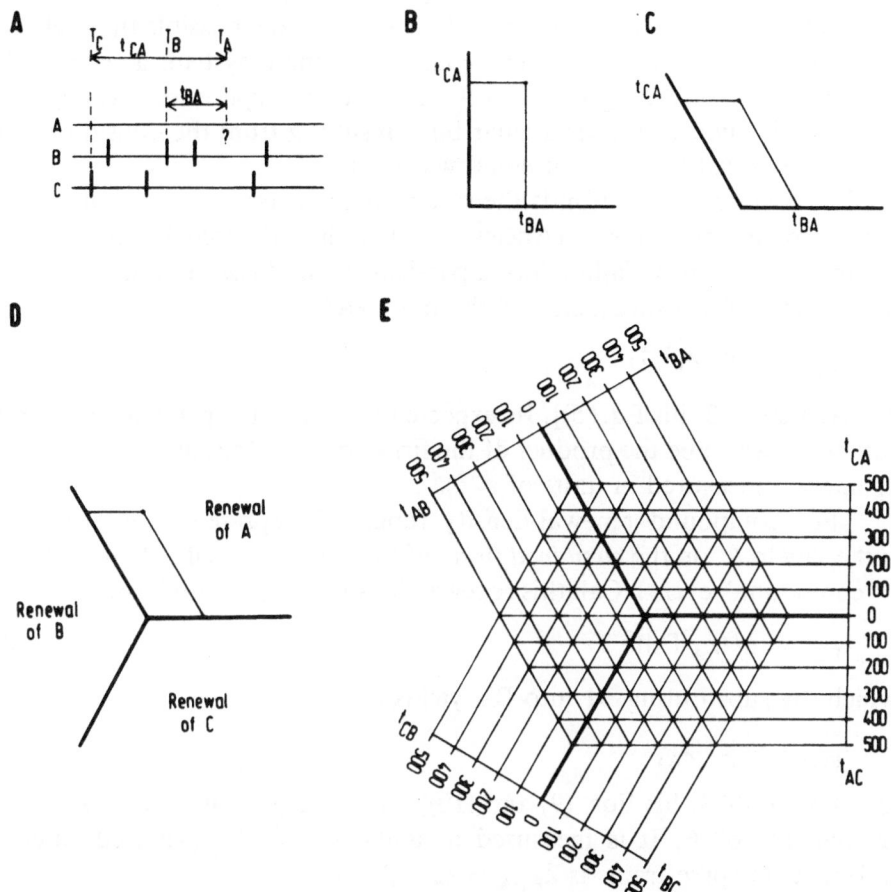

Fig. 58 A–E. Coordinate system for the three-cell density. **A** The firing times of neurones B and C are known and the probability of firing of neurone A in relation to these trains is investigated; **B** Regular coordinate system used to plot the firing rate of neurone A as a function of times elapsed from firing of neurone B and neurone C; **C** Same coordinate system as in **B**, but the angle between the axes increased to 120°; **D** Three coordinate systems such as in **C** are attached together to give all the possible time relations between the firings of the three neurones; **E** The scales of the three axes are projected out and extended to occupy the entire range of the three graphs. The three intervals (t_{CA}, t_{BA}, t_{BC}) associated with any point on the graph can be read by projecting the point onto the three external axes. (With permission from Abeles 1982; his Fig. 19)

because $t_{CA} = t_{CB} + t_{BA}$. This suggests the plotting of all three spike intervals as one point in a common coordinate system. If this coordinate system is rectangular as in Fig. 58 B, the time of occurrence of a spike in A can be represented by the two intervals, t_{BA} and t_{CA}, elapsed from spikes in B and C, respectively. The probability of discharge ("renewal") of A can then be estimated by plotting many such interval pairs. However, this plot would so far only represent the effect of spikes in B and C on *following* spikes in A. For a more complete description, other time relations would also be of interest, including intervals $-t_{CA}$ and $-t_{BA}$. To display all the possible time relations between spike occurrences, a triangular coordinate system, as sketched in Fig. 58 C–E, is advantageous. For the estimation of spike density, the resulting plane is divided into triangular bins, resulting from the division of time for each interval into bins of equal width Δt.

The covariance function is the rate of occurrence of spike triplets, λ_{ABC}, measured in units of s^{-3} (Abeles 1983). It is estimated by counting the number, n, of triplets falling into a particular bin of size Δt and normalising by the time of measurement and the bin area:

$$\lambda_{ABC} = n/(T \, \Delta t^2/2) \tag{B1}$$

[see Abeles 1983; his Eq. (5)]. Its expected value for independent spike trains can be shown to be the product of the firing rates of the single neurones, λ_A, λ_B and λ_C (measured in units of s^{-1}).

The "compound renewal-density function" represents the effect of a spike, say in A, on the *combined firing* of the other two cells, B and C (i.e. on a "compound" event). Call this renewal density $\lambda_{BC/A}$. It can be estimated by:

$$\lambda_{BC/A} = n/(\lambda_A \, T \, \Delta t^2/2) \tag{B2}$$

which, by substituting B1 into B2, yields:

$$\lambda_{BC/A} = \lambda_{ABC}/\lambda_A, \tag{B3}$$

[see Abeles 1983; his Eqs. (9) and (10)], i.e. the covariance function divided by the rate of A. It is measured in units of s^{-2}. Its expected value for independent spike trains is $\lambda_{BC/A} = \lambda_B \cdot \lambda_C$.

The "simple renewal-density function" represents the way by which the firing times of two cells, say A and B, affect the firing rate of the third, C. Call this renewal density $\lambda_{C/AB}$. It can be estimated by:

$$\lambda_{C/AB} = n/(m \cdot \Delta t/2) \tag{B4}$$

where n is the number of counts in the bin, and m is the number of occurrences of the triggering events, i.e. of the doublets in A and B, over the time of measurement, T. With $m = \lambda_{AB} \cdot T \cdot \Delta t$, we get:

$$\lambda_{C/AB} = \lambda_{ABC}/\lambda_{AB}, \tag{B5}$$

[see Abeles 1983; his Eqs. (12) and (14)], i.e. the covariance function divided by the rate of occurrence of doublets in A and B. It is measured in units of s^{-1}. Its expected value for independent spike trains is $\lambda_{C/AB} = \lambda_C$.

The display of these measures is limited to three neuronal spike trains. Alternative methods to study the interaction of more than two spike trains are explained in the next two appendices.

Appendix C: Event-Related Cross-Correlations

(With W. KOEHLER and C. SCHWARZ)

Conventional correlation methods are usually limited to two point event trains (App. A). The application of the "snow-flake" technique for correlating three point event trains (App. B) is still an exception (Sect. 1.3.1). The design of evaluation techniques depends upon the question asked under the specific conditions. For example, the "snow-flake" technique for three event trains is suited for triples of neurones whose interrelations are not yet known. If, on the other hand, we are faced with a system in which multiple inputs and outputs can be clearly distinguished (and are accessible experimentally), we could study the interaction of two (or more) input channels with respect to their effects on one output channel. The underlying theory of performing this kind of analysis is briefly reviewed in App. D. Or else, since common inputs may correlate two (or more) outputs of the system under study, it would be of interest to know how the output correlation is temporally related to input events. A method to tackle this problem is described here.

The method used here to correlate discharge patterns of two neurones in temporal relation to events (stimuli in the present case) is illustrated in Fig. 59. Within a digital computer, the reference stimulus train and the spike trains to be correlated with each other are represented as points in time and are symbolised in Fig. 59 as sequences of brief vertical pulses (labelled S, and A and B, respectively). Each spike in A and B is converted into (convolved with) a rectangular pulse of selectable width Δt and arbitrary but constant height. The spike trains A and B are thus transformed into the respective pulse trains A' and B', i.e. essentially into analogue signals. Note that subsequent pulses overlapping in time superimpose linearly. For each chosen interval τ from the reference events in S, pairs of values designated x_i and y_i are taken from the so constructed pulse trains A' and B', respectively, as indicated in the figure. Linear correlation coefficients are then computed for the array of values x_i and y_i collected for all the reference events. The

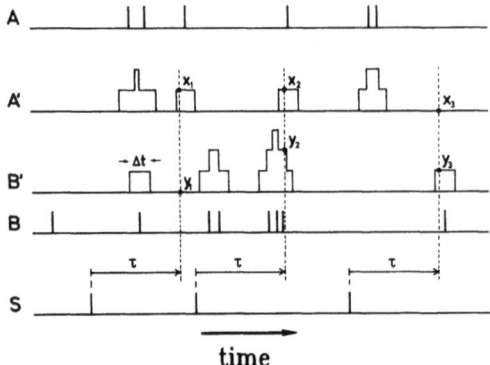

time

Fig. 59. Scheme of the technique used to calculate stimulus-related correlation coefficients. The spike sequences *A* and *B* are convolved by rectangular pulses of selectable width Δ*t* and arbitrary but fixed amplitude, whereby analogue signals *A′* and *B′*, respectively, result. For intervals τ before or after each event in the stimulus train *S*, values x_i and y_i are read to calculate conventional linear correlation coefficients. (With permission from Windhorst, Koehler and Schwarz 1987; their Fig. 2)

correlation coefficient is defined as

$$r_{A,B}(\tau) = \frac{\sum\limits_{i=1}^{n}(x_i - \bar{x})(y_i - \bar{y})}{\sqrt{\sum\limits_{i=1}^{n}(x_i - \bar{x})^2 \cdot \sum\limits_{i=1}^{n}(y_i - \bar{y})^2}} \tag{C1}$$

These coefficients are determined for a series of different τ's sampled mostly at steps of 2 ms, and are then plotted as a function of the τ.

The correlation coefficients in Eq. (C1) are determined for "synchronous" x_i and y_i which are located at equal intervals τ from the reference pulses. For a more complete description, they should also be computed for unequal $τ_A$ and $τ_B$. This can be done by shifting in time one spike train with respect to the other before computation of (C1). For example, if channel B in Fig. 59 is initially shifted to the right by ϑ, the subsequent cross-correlation calculation is for A′ at τ and B′ at τ − ϑ, i.e. it determines the correlation between activity fluctuations of the two afferents ϑ ms apart.

A similar method was proposed by Schneider et al. (1983). But they used half cosine functions to convolve the spikes with. While this procedure would appear more complicated than the use of rectangular pulses, its consequences with respect to the quality of signal representation are also less clear. Note that the convolution of a signal in the time domain with rectangular pulses of width Δt corresponds to filtering the spectrum of the signal with a sinc-function of the form W (f) = Δt · (sin π f Δt/π f Δt), where f is the frequency in

Hz (e.g. Jenkins and Watts 1968; pp. 243 ff.). [The term "sinc-function" denotes a sine function divided by its argument, i.e. (sin x)/x.)] The first zero crossing of W (f) occurs at a frequency of $1/\Delta t$, e.g. for $\Delta t = 30$ ms at about 33 Hz. Frequencies below half of this value are well preserved, whereas those above it are more or less strongly attenuated.

Also, Schneider et al. (1983) applied their technique only to the case of regularly recurring reference events (stimuli) far enough apart to exclude subsequent reference events from each analysis period. Our method is generalised insofar as this restriction no longer holds. Event-related cross-correlation addresses the same problem as Abeles' "compound renewal density function" (App. B).

The event-related cross-correlation presented here can be extended to more output and input channels. On the output side, the calculation of correlation coefficients is not restricted to two variables, but can be extended to more variables (multivariate analysis; see, e.g. Yamane 1973). For example, if there are three random (output) variables X, Y, and Z, whose interrelationship is to be studied, a so-called "partial" correlation coefficient can be computed. This examines the correlation between two, e.g. Z and Y, of the three variables after removal of the effect of the third variable, e.g. X. The underlying assumption is that X might have a correlating effect on the other two variables, Z and Y. By keeping X constant, the change in correlation between Z and Y against the "simple" correlation coefficient $r_{Z,Y}$ would give an indication of the influence of X. This is checked by the partial correlation coefficient which is denoted by $r_{zy \cdot x}$ and can be expressed in terms of the three possible simple correlation coefficients for pairs of variables, i.e.

$$r_{zy \cdot x} = \frac{r_{zy} - r_{zx} r_{xy}}{\sqrt{(1 - r_{zx}^2)(1 - r_{xy}^2)}} \tag{C2}$$

(see Yamane 1973). There are three such coefficients which are obtained by permutation of the variables in Eq. (C2). For these three variables, still another type of correlation coefficient can be defined, the so-called "multiple" correlation coefficient. It examines the correlation between one of the variables and the other two (for its definition and computation see Yamane 1973). These notions can be extended to more than three variables. The computation becomes cumbersome, however, and we do not go into more detail.

It might appear as a disadvantage of these extensions that the (output) variables are somewhat arbitrarily classified into two groups, one containing the independent and the other the dependent variables. However, in many cases, this distinction may make physiological sense. Suppose, for example, that we are faced with a system like that depicted in Fig. 4 C (without a priori knowing its structure), and we manage to stimulate (or record) one or both of the inputs A and B and to record the output of cell 3 (at level b) as well as the outputs of cells 2 and 6 (at level c). Since cell 3 at level b provides input to cells 2 and 6 at level c, it will have a correlating effect on them, which can be removed by computation of the partial correlation coefficient $r_{26.3}(\tau)$. The remaining correlation between cells 2 and 6 at level c is then mediated through pathways circumventing cell 3 at level b. Such situations are quite common in multi-unit recordings from CNS structures, so that the computation of event-related partial or multiple correlation coefficients is of importance and may disclose to which extent the correlation between two neurones is mediated by a third.

Another extension of the method concerns the input channels. Instead of averaging the system output with respect to all the events in one input (reference) channel, they can be averaged with respect to pairs of events that occur in selected time constellations in two input channels, i.e. with respect to randomly occurring compound events. The theoretical background for this procedure is the Wiener-Volterra theory applied to spike trains (Krausz 1975; Marmarelis and Marmarelis 1978; Windhorst et al. 1983) and is explained in the following appendix. The method allows calculation of "second order" effects and thereby the determination of nonlinear behaviour of the system under study (see Niemann et al. 1986; App. D).

The extension to two input channels can theoretically be carried even further. Time constellations of input events (triplets, etc.) can also be extracted from more than two input channels; but the amount of data needed to compute reliable estimates increases rapidly, and so does the time of computation. Moreover, the above time constellations can be extracted from a single input channel (see Windhorst et al. 1987a).

These extensions on the output and input side exceed the number of three spike trains underlying Abeles' techniques (App. B).

It is noted in passing that Gerstein and Aertsen (1985) have described a method of detecting and representing the cooperative firing of groups of neurones and the temporal change of their near-simultaneous discharge. Also, Ebner and Bloedel (1981a) calculated event-related auto-correlograms.

Appendix D:
Statistical Analysis of Nonlinear Physiological Systems

(With U. NIEMANN and W. KOEHLER)

Most physiological systems are more or less nonlinear except perhaps over limited input amplitude ranges. A method to analyse and describe this nonlinearity is the white noise approach where (band-limited) Gaussian random signals are applied to a system's input(s) and its output is then described in terms of a series expansion of Wiener kernels (for review see Marmarelis and Marmarelis 1978). A simplification of the involved computations results when the random input signals consist of Poisson impulse trains (Krausz 1975). This is of particular interest in neurophysiology where many systems have spike trains as input(s) and analogue signals (e.g. muscle tension and length) or again spike trains as output(s), and for many statistical purposes nerve action potentials are indeed idealised as point processes, i.e. as instants occurring sequentially in time (see App. A).

The simplification proposed by Krausz (1975) is readily extended to multi-input systems as he noted. Here we illustrate the extension to two inputs. Moreover, in addition to analogue outputs, spike train outputs of neuronal systems are considered (thereby addressing the same problem, with a different technique, as approached by Abeles' "simple renewal density function"; App. B). One input–one output systems are considered first. The input $z(t)$ is assumed to be a Poisson impulse train, i.e. a sequence of Dirac delta functions:

$$z(t) = \sum_i \delta(t - t_i),$$

where the t_i denote the random instants of impulses occurring at a mean rate of λ. From these definitions results the "true" input signal $x(t) = z(t) - \lambda$, which, together with kernels $h_i(\tau_1, \ldots, \tau_i)$ of various orders i, is used to describe the system output $y(t)$ as the expansion:

$$y(t) = G_0 + G_1 + G_2 \ldots, \tag{D1}$$

where

$$G_0 = h_0$$
$$G_1 = \int h_1(\tau) x(t - \tau) d\tau$$
$$G_2 = \iint_{\tau_1 \neq \tau_2} h_2(\tau_1, \tau_2) x(t - \tau_1) x(t - \tau_2) d\tau_1 d\tau_2 \qquad (D2)$$

.

.

.

$$G_n = \iint_{\tau_1 \neq \tau_2 \dots \neq \tau_n} \dots \int h_n(\tau_1, \dots, \tau_n) x(t - \tau_1) \dots x(t - \tau_n) d\tau_1 \dots d\tau_n .$$

Restricting the τ_i to be unequal (i.e. $\tau_i \neq \tau_j$; $i \neq j$) enables the important simplification of Krausz' approach. (It should be noted that this restriction is intuitively reasonable since it prevents two successive impulses from occurring simultaneously, which in neurones is impossible anyway.) The kernels can now be determined by cross-correlations:

$$h_0 = \overline{y(t)}$$
$$h_1(\tau) = (1/\lambda) \overline{y(t) x(t - \tau)} \qquad (D3)$$
$$h_n(\tau_1, \dots, \tau_n) = (1/(n! \, \lambda^n)) \overline{y(t) x(t - \tau_1) \dots x(t - \tau_n)},$$
$$\tau_1 \neq \tau_2 \neq \dots \neq \tau_n,$$

where the horizontal bars symbolise averaging over time. Resubstituting $x(t) = z(t) - \lambda$ yields the first three kernels:

$$h_0 = \overline{y(t)}$$
$$h_1(\tau) = (1/\lambda) \overline{y(t) z(t - \tau)} - h_0 \qquad (D4)$$
$$h_2(\tau_1, \tau_2) = (1/(2\lambda^2)) \overline{y(t) z(t - \tau_1) z(t - \tau_2)} - (1/2) h_1(\tau_1)$$
$$- (1/2) h_1(\tau_2) - (1/2) h_0, \ \tau_1 = \tau_2 .$$

h_0 is the average level around which $y(t)$ fluctuates. Since h_0 is often of no interest, further analysis is simplified by subtracting it from $y(t)$ prior to the computation of the first- and second-order kernels. $h_1(\tau)$ represents the average system response at time t to an input impulse occurring a time interval τ before. In a linear system, $h_1(\tau)$ would be the impulse response; $h_2(\tau_1, \tau_2)$, the second-order kernel, is the first nonlinear response component (at time t) obtained by averaging $y(t)$ with respect to two preceding impulses occurring τ_1 and τ_2 time units before, and by subtracting twice the first kernel after appropriate time shifts (and finally by dividing by 2).

The preceding approach analyses nonlinear features in a single input–single output system. As pointed out in App. C, one question regarding multiple input–multiple output systems concerns the interaction between several inputs with respect to their effects on one output. This problem can be studied by an extension of the above approach to systems with two inputs $z_1(t)$ and $z_2(t)$ in that a second-order "cross-kernel" is defined by taking a second input train as "conditioning" input. The kernels run as follows:

$$h_0 = \overline{y(t)}$$

$$h_1^{(z_1)}(\tau) = (1/\lambda_1)\overline{y(t)z_1(t-\tau)} - h_0$$

$$h_1^{(z_2)}(\tau) = (1/\lambda_2)\overline{y(t)z_2(t-\tau)} - h_0 \qquad\qquad (D5)$$

$$h_2^{(z_1,z_2)}(\tau_1,\tau_2) = (1/(2\lambda_1\lambda_2))\overline{y(t)z_1(t-\tau_1)z_2(t-\tau_2)} - (1/2)h_1^{(z_1)}(\tau_1)$$
$$- (1/2)h_1^{(z_2)}(\tau_2) - (1/2)h_0 \,.$$

The first-order "auto"-kernels $h_1^{(z_1)}(\tau)$ and $h_1^{(z_2)}(\tau)$ are now obtained by averaging $y(t)$ with respect to impulses in $z_1(t)$ and $z_2(t)$, respectively. $h_2^{(z_1,z_2)}(\tau_1,\tau_2)$ results from averaging $y(t)$ with respect to impulses in $z_1(t)$ occurring τ_1 time units before *and* to impulses in $z_2(t)$ occurring τ_2 time units before, and by subtracting the respective first-order kernels after appropriate time shifts (and finally dividing by 2). It is obvious that there is now no need to exclude the possibility of simultaneous pairs of stimuli in the two input channels, i.e. τ_1 can be equal to τ_2 (see Windhorst et al. 1983).

The preceding analysis requires averaging with respect to selected time constellations of stimuli. In the data presented in Sects. 3.2.2 and 3.2.3, these selected time constellations were sorted out (by a computer program) as illustrated in Fig. 60. Here two stimulus patterns are represented by the upper two traces (S_1 and S_2). From train S_2 are selected out and written on a reference channel ("Ref.") those stimuli which are preceded by stimuli in channel S_1 within time windows extending from -35 to -25 ms, i.e. stimuli in S_1 precede those in S_2 by on average $\delta = -30$ ms. If the two stimulus channels 1 and 2 are identical, the effects of conditioning stimuli on stimuli in the same channel are studied. If the channels are different, the average delay variable δ can vary from negative to positive values (see Windhorst et al. 1983).

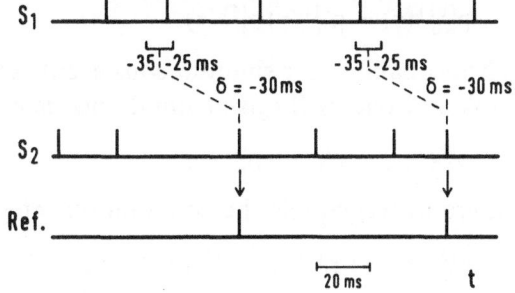

Fig. 60. The process by which selected time constellations of stimuli are sorted out from two stimulus trains. Those stimuli of train S_2 are written on a reference channel (*Ref.*) which, in this case, are preceded by ("conditioning") stimuli in channel S_1 occurring within a time window from -35 to -25 ms (mean value $\delta = -30$ ms). The parameter δ can vary from negative to positive values. If the two stimulus channels are identical, the procedure selects time constellations of stimuli in one stimulus train. (With permission from Niemann et al. 1986; their Fig. 1)

Appendix E: On Spurious Synchronisation of Two Semiregularly Firing Neurones

Consider two motor units, each firing stationarily, fairly regularly, and one independently of the other.

1) Renewal processes. Several investigators maintain that the discharge of motor units corresponds to a renewal process in which each interspike interval is produced independently of all preceding and following ones (e.g. Clamann 1969; Shiavi and Negin 1975). Let us first consider this case. Assume two point processes A and B, whose first-order interspike intervals are distributed according to probability density functions (pdf's) denoted respectively by $p_A^{(1)}(\tau)$ and $p_B^{(1)}(\tau)$, whose second-order intervals are distributed according to pdf's denoted by $p_A^{(2)}(\tau)$ and $p_B^{(2)}(\tau)$, and so forth. These pdf's can be interpreted as describing the conditional probability for the occurrence of the first (second, third, and so forth) event at time $t_i + \tau$ (after a delay τ) given an event at t_i. If two events from A and B happen to occur simultaneously at t_i, the probability for two simultaneous events at $t_i + \tau$ after only one interval in each process is given by the joint pdf:

$$p_{AB}^{(11)}(\tau) = p_A^{(1)}(\tau) p_B^{(1)}(\tau). \tag{E1}$$

The probability for simultaneous events to occur at $t_i + \tau$ after two intervals in A and one in B (given simultaneous events at t_i) is

$$p_{AB}^{(21)}(\tau) = p_A^{(2)}(\tau) p_B^{(1)}(\tau), \tag{E2}$$

where $p_A^{(2)}(\tau)$ equals the convolution integral

$$p_A^{(2)}(\tau) = \int_0^\tau p_A^{(1)}(\sigma) \cdot p_A^{(1)}(\tau - \sigma) \, d\sigma. \tag{E3}$$

The latter can be generalised to higher-order interval pdf's:

$$p_A^{(k)}(\tau) = \int_0^\tau p_A^{(k-1)}(\sigma) \cdot p_A^{(1)}(\tau - \sigma) \, d\sigma, \tag{E4}$$

which are thus defined recursively starting with (E3).

Thus, assuming that the kth interval in A is independent of the lth in B, the generalised joint pdf $p_{AB}^{(kl)}(\tau)$ is given by:

$$p_{AB}^{(kl)}(\tau) = \left(\int_0^\tau p_A^{(k-1)}(\sigma) \cdot p_A^{(1)}(\tau - \sigma)\, d\sigma \right) \cdot \left(\int_0^\tau p_B^{(l-1)}(\vartheta) \cdot p_B^{(1)}(\tau - \vartheta)\, d\vartheta \right). \tag{E5}$$

Finally, the probability for the occurrence of simultaneous events at $t_i + \tau$ (given simultaneous events at t_i) for *any* combination of intervals in A and B between the simultaneous occurrences is given by the sum of all possible joint probabilities $p_{AB}^{(kl)}(\tau)$, which determine the auto-correlation of the process of simultaneous events, $R_{AB}(\tau)$:

$$R_{AB}(\tau) = \sum_{k,l} p_{AB}^{(kl)}(\tau)$$

$$= \sum_{k,l} \left(\int_0^\tau p_A^{(k-1)}(\sigma) \cdot p_A^{(1)}(\tau - \sigma)\, d\sigma \right) \cdot \left(\int_0^\tau p_B^{(l-1)}(\vartheta) \cdot p_B^{(1)}(\tau - \vartheta)\, d\vartheta \right)$$

$$= \left(\sum_k \int_0^\tau p_A^{(k-1)}(\sigma) \cdot p_A^{(1)}(\tau - \sigma)\, d\sigma \right)$$

$$\cdot \left(\sum_l \int_0^\tau p_B^{(l-1)}(\vartheta) \cdot p_B^{(1)}(\tau - \vartheta)\, d\vartheta \right) = R_A(\tau) \cdot R_B(\tau), \tag{E6}$$

i.e., the terms in brackets are the auto-correlation of the processes A and B, respectively. Now assume that the two auto-correlations have similar shapes exhibiting damped oscillations with peaks at similar delays τ, because the two motor units fire at approximately the same mean rates and with low variability (coefficient of variation of the order of $0.1-0.2$). The product, $R_{AB}(\tau)$, of these auto-correlations then replicates their oscillations to an extent depending on the similarity of mean rates and variabilities.

2) Nonrenewal processes. Some authors found negative correlations between adjacent interspike intervals in the discharge of motor units (e.g. Kranz and Baumgartner 1974; Andreassen and Rosenfalck 1980). This situation cannot be formalised as before. But intuitively it is clear that if only two adjacent intervals are negatively correlated, the rhythmic recurrence of each second spike (after a second-order interval) is strengthened (see Perkel et al. 1967a, for examples). Thus the argument at the end of the last paragraph holds even more strongly in this case. It is interesting to note in passing that Andreassen and Rosenfalck (1980) suggested three possible mechanisms for the negative serial correlation: summation of afterhyperpolarisations in α-motoneurones, spinal recurrent inhibition via Renshaw cells, and proprioceptive feedback via muscle stretch receptors. If the latter mechanisms were to work in case of asynchronous independent motor unit activity (as assumed here), this would imply "private" (localised) feedback of the activity of each single motor unit onto itself, through spinal recurrent and/or proprioceptive pathways (see Sect. 4.3.1) (see Windhorst 1984).

Appendix F: Spectral Analysis

(With C. N. CHRISTAKOS)

In this appendix, functions of frequency characterising random signals and systems are presented.

A) Auto- and Cross-Spectra

The spectral density function (auto-spectrum), $S_x(f)$ (with f denoting frequency in Hz), of a random signal, $x(t)$, displays the distribution of the power of the signal over a certain frequency range. It thus enables the recognition of frequency components that may be prominent in the signal, reflecting periodicities and other rhythms, and can give clues to the function of the system generating them. The auto-spectrum is the Fourier transform of the auto-correlation function [Eq. (A1)]. Analogously, the cross-spectrum, $S_{xy}(f)$, of two random signals, $x(t)$ and $y(t)$, is the Fourier transform of the cross-correlation function [Eq. (A2)]. Examples of auto-spectra computed for Ia afferent discharges are shown in Figs. 29.

B) Coherence Function

For two random signals $x(t)$ and $y(t)$, i.e. input and output to a system under examination, respectively, the coherence function is defined as:

$$\gamma_{xy}^2(f) = |S_{xy}(f)|^2/[(S_x(f) \cdot S_y(f)] \tag{F1}$$

where $S_x(f)$, $S_y(f)$ are the auto-spectra of $x(t)$ and $y(t)$, and $S_{xy}(f)$ is their cross-spectrum (see above). For a linear, single-input system, $\gamma_{xy}^2(f)$ equals 1 in the absence of noise. With noise present either at the input or at the output or both, its value is less than 1, thereby providing an index of the extent to which the output is due to the input. The coherence function can be used to check for the frequency range of linear operation of a physiological system. Confidence intervals for the estimates of the coherence function are given by Bendat and Piersol (1971).

C) Frequency Response

The frequency response of a time-invariant system describes the steady-state transfer characteristics of the system in the range of linear operation. It can be estimated from the auto- and cross-spectra of the input and the output to the system, x (t) and y (t), as:

$$H(f) = S_{xy}(f)/S_x(f), \tag{F2}$$

and is a complex function of frequency. Using the (real) gain factor $|H(f)|$ and the phase factor $\phi(f)$ (angle), $H(f)$ may be written:

$$H(f) = |H(f)| \exp[-j\phi(f)] \tag{F3}$$

where j is the square root of -1.

It is easy to show that the frequency response can be directly estimated from linearly filtered versions of the input and the output to the system, provided that the two signals have been processed in the same way.

The way in which spike trains can be treated for computation of spectra and related measures was described in detail by Christakos et al. (1984). Since one is usually interested in the low-frequency content of spike trains, up to some hundred Hz, it is convenient to low-pass filter the impulse sequence and to re-sample the ensuing signal at an appropriate new rate so as to avoid aliasing. The filtering can be done digitally (on a digital computer) and involves, in this case, only summation instead of convolution that is required for the filtering of the original signal. Further, the re-sampling is now more "economical" because of the reduced frequency content of the signal and the new lower sampling rate.

In practice, the times of spike occurrences, recorded for example using a threshold device, are stored as multiples of a certain, small enough sampling period, T_s. A sinc function of the form $\sin(\pi t/kT_s)/\pi t$, sampled at $1/T_s$, is centred about every such point on the time axis. The sinc functions, which may overlap, are superposed, resulting in a signal sampled also at $1/T_s$. This operation is equivalent to ideal low-pass filtering of the original sequence. Further, if kT_s represents the interval between two successive zero-crossings on one side of the sinc function, then the cut-off frequency of the filter will be given in terms of T_s as

$$F_c = 1/2kT_s. \tag{F4}$$

The desired value of k can therefore be determined using Eq. (F4). Equation (F4) also indicates the appropriate new sampling period according to the Nyquist criterion, namely $T'_s = kT_s$. The sinc function is in practice truncated on each side to include a certain selected number of oscillations.

Caution is warranted and experience is required in interpreting spectra and further measures based on them. It is relatively easy to misinterpret spectra if basic notions are not rigorously used in their strict sense (see Boyd and Rosenberg 1985).

Appendix G: A Remark on Correlation Patterns

The following is just a brief note. A more systematic and detailed discussion of neural codes and related problems is presented by Perkel and Bullock (1969).

For conceptual clarity it is worthwhile to distinguish between the low-frequency rhythmic correlation and the short-term synchronisation patterns as the extremes of a continuum.

The former rhythmic correlation could come about through the common modulation of the discharge rates of two cells. For example, if both cells discharge at relatively high average rates determined over long time periods, say 300 pulses/10 s, a low-frequency (say 1 Hz) modulation of moderate depth (say 50%) could be regarded as a correlated (e.g. synchronised) change of mean firing rates of the cells. These mean rates would then be defined over time intervals (say 200 ms) *always comprising several spikes*. The exact positions of single spikes within these time intervals would be of minor importance. What is correlated in this type of rhythmic correlation are the mean firing rates of the two cells. This view envisages the code of information transfer to be contained in the mean discharge frequency (hence "frequency code").

In short-term synchronisation, at the other extreme, average firing rates of the two cells (even if determined over the above short time intervals of 200 ms) are constant. What matters is the *exact relative positioning in time of single spikes* in the two cells, or, in other words, the fine structure of correlated fluctuations in cell firing rates around the respective means. The code used here may be different (see Abeles 1982).

Admittedly, this distinction is somewhat arbitrary since it depends on how "mean firing rate" is defined, i.e. with respect to which time ("integration") interval it is calculated. This definition should reasonably be adapted to the time constants of the postsynaptic neurone commonly contacted by the two cells in question. If the dominant time constant is short (e.g. 5 ms in α-motoneurones), the effects of the two correlation patterns distinguished above are very different and, hence, bear different significance. The two correlation patterns correspond to different codes. For longer time constants, the sharp features of the short-term synchronisation are low-pass filtered, and the distinction made above becomes blurred. Imagine that the short-term synchronised discharges of the two presynaptic cells also recur rhythmically at the same frequency as the rhythmic modulation of the first pattern. Given strong short-term synchronisation and sufficiently long time constants, the membrane potential changes of the postsynaptic neurone will be modulated smoothly at the recurrence frequency.

There are combinations of the correlation patterns and intermediate forms. Thus different codes may be superimposed.

References

Abeles M (1982) Local cortical circuits. An electrophysiological study. Springer, Berlin Heidelberg New York

Abeles M (1983) The quantification and graphic display of correlations among three spike trains. IEEE Trans Biomed Eng BME-30:235–239

Adal MN, Barker D (1962) Intramuscular diameters of afferent nerve fibres in the rectus femoris muscle of the cat. In: Barker D (ed) Symposium on muscle receptors. Hong Kong University Press, Hong Kong, pp 249–256

Adam D, Windhorst U, Inbar GF (1978) The effects of recurrent inhibition on the cross-correlated firing patterns of motoneurones (and their relation to signal transmission in the spinal cord-muscle channel). Biol Cybern 29:229–235

Agarwal GC, Gottlieb GL (1977) Oscillation of the human ankle joint in response to applied sinusoidal torque on the foot. J Physiol (Lond) 268:151–176

Agarwal GC, Gottlieb GL (1985) Mathematical modeling and simulation of the postural control loop. Part III. CRC Crit Rev Biomed Eng 12:49–93

Alberts WW (1972) A simple view of Parkinsonian tremor. Electrical stimulation of cortex adjacent to the Rolandic fissure in awake man. Brain Res 44:357–369

Albus JS (1971) Theory of cerebellar function. Math Biosci 10:25–61

Allum JHJ, Mauritz KH (1984) Compensation for intrinsic muscle stiffness by short-latency reflexes in human triceps surae muscles. J Neurophysiol 52:797–818

Allum JHJ, Dietz V, Freund HJ (1978) Neuronal mechanisms underlying physiological tremor. J Neurophysiol 41:557–571

Andersen P (1982) Cerebellar synaptic plasticity – putting theories to the test. Trends Neurosci 5:324–325

Andersen P, Andersson SA (1968) Physiological basis of the alpha rhythm. Appleton-Century-Crofts, New York

Andreassen S, Rosenfalck A (1980) Regulation of the firing pattern of single motor units. J Neurophysiol Neurosurg Psychiat 43:897–906

Appelberg B, Johansson H, Kalistratov G (1977) The influence of group II muscle afferents and low threshold skin afferents on dynamic fusimoneurones to the triceps surae of the cat. Brain Res 132:153–158

Appelberg B, Hulliger M, Johansson H, Sojka P (1982) Fusimotor reflexes in triceps surae elicited by natural stimulation of muscle afferents from the cat ipsilateral hind limb. J Physiol (Lond) 329:211–229

Appelberg B, Hulliger M, Johansson H, Sojka P (1983a) Actions on γ-motoneurones elicited by electrical stimulation of group I muscle afferent fibres in the hind limb of the cat. J Physiol (Lond) 335:237–253

Appelberg B, Hulliger M, Johansson H, Sojka P (1983b) Actions on γ-motoneurones elicited by electrical stimulation of group II muscle afferent fibres in the hind limb of the cat. J Physiol (Lond) 335:255–273

Appelberg B, Hulliger M, Johansson H, Sojka P (1983c) Actions on γ-motoneurones elicited by electrical stimulation of group III muscle afferent fibres in the hind limb of the cat. J Physiol (Lond) 335:275–292

Appelberg B, Hulliger M, Johansson H, Sojka P (1983d) Recurrent actions on γ-motoneurones mediated via large and small ventral root fibres in the cat. J Physiol (Lond) 335:293–305

Appenteng K, Prochazka A (1984) Tendon organ firing during active muscle lengthening in awake, normally behaving cats. J Physiol (Lond) 353:81–92

Appenteng K, O'Donovan MJ, Somjen G, Stephens JA, Taylor A (1978) The projection of jaw elevator muscle spindle afferents to fifth nerve motoneurones in the cat. J Physiol (Lond) 279:409–423

Appenteng K, Morimoto T, Taylor A (1980) Fusimotor activity in masseter nerve of the cat during reflex jaw movements. J Physiol (Lond) 305:415–431

Armstrong RB (1980) Properties and distribution of fiber types in the locomotory muscles of mammals. In:Schmidt-Nielson K, Bolis L, Taylor CR(eds) Comparative Physiology: Primitive mammals. Cambridge University Press, Cambridge, pp 243–254

Arnett DW, Spraker T (1981) Cross-correlation analysis of the maintained discharge of rabbit retinal ganglion cells. J Physiol (Lond) 317:29–47

Arshavsky YI, Gelfand IM, Orlovsky GN (1983) The cerebellum and control of rhythmical movements. Trends Neurosci 6:417–422

Baldissera F, Hultborn H, Illert M (1981) Integration in spinal neuronal systems. In: Brooks VB (ed) The nervous system. Am Physiol Soc, pp 509–595 (Handbook of Physiology, vol. II, part 1)

Baldissera F, Campadelli P, Piccinelli L (1982) Neural encoding of input transients investigated by intracellular injection of ramp currents in cat α-motoneurones. J Physiol (Lond) 328:73–86

Baranyi A, Fehér O (1981) Synaptic facilitation requires paired activation of convergent pathways in the neocortex. Nature 290:413–415

Barnes CD, Pompeiano O (1970) Presynaptic and postsynaptic effects in the monosynaptic reflex pathway to extensor motoneurones following vibration of synergic muscles. Arch Ital Biol 108:259–294

Barrett JN, Crill WE (1974) Influence of dendritic location and membrane properties on the effectiveness of synapses in cat motoneurones. J Physiol (Lond) 239:325–345

Bawa P, Calancie B (1983) Repetitive doublets in human flexor carpi radialis muscle. J Physiol (Lond) 339:123–132

Bawa P, Tatton WG (1979) Motor unit responses in muscles stretched by imposed displacements of the monkey wrist. Exp Brain Res 37:417–437

Bear MF, Singer W (1986) Modulation of visual cortical plasticity by acetylcholine and noradrenaline. Nature 320:172–176

Bell CC, Grimm RJ (1969) Discharge properties of Purkinje cells recorded on single and double microelectrodes. J Neurophysiol 32:1044–1055

Bell CC, Kawasaki T (1972) Relations among climbing fiber responses of nearby Purkinje cells. J Neurophysiol 35:155–169

Bendat JS, Piersol AG (1971) Random data: analysis and measurement procedures. Wiley, New York

Benecke R, Hagenah R, Henatsch H-D, Schmidt J (1975) Effects of stimulation of the substantia nigra on spinal interneurones. Pflügers Arch 355:R90

Benecke R, Meyer-Lohmann J, Guntau J (1976) Inverse changes in the excitability of Renshaw cells and α-motoneurones induced by interpositus stimulation. Pflügers Arch 365:R40

Bennett MR, Lavidis NS (1982) Development of the topographical projection of motor neurons to amphibian muscle accompanies motor neuron death. Dev Brain Res 2:448–452

Benninghoff A, Rollhäuser H (1952) Zur inneren Mechanik des gefiederten Muskels. Pflügers Arch 254:527–548

Bergmann-Erb D (1985) Physiologische und neuropharmakologische Untersuchungen zur Interaktion kleiner α-Motoneuronenpopulationen mit einzelnen Renshaw-Zellen. Math-Nat Dissertation. Universität Düsseldorf

Beritoff J (1925) Über die Kontraktionsfähigkeit der Skelettmuskeln. IV. Über die physiologische Bedeutung des gefiederten Baues der Muskeln. Pflügers Arch 209:763–778

Berkinblit MB, Feldman AG, Fukson OI (1986) Adaptability of innate motor patterns and motor control mechanisms. Behav Brain Sci 9: 585–638

Berthoz WJ, Roberts WJ, Rosenthal NP (1971) Dynamic characteristics of the stretch reflex using force inputs. J Neurophysiol 34:612–619

Bessou P, Cabelguen J-M, Joffroy M, Montoya R, Pagès B (1986) Efferent and afferent activity in a gastrocnemius nerve branch during locomotion in the thalamic cat. Exp Brain Res 64:553–568

Bilotto G, Schor RH, Uchino Y, Wilson VJ (1982) Localization of proprioceptive reflexes in the splenius muscle of the cat. Brain Res 238:217–221

Binder MD, Osborn CE (1985) Interactions between motor units and Golgi tendon organs in the tibialis posterior muscle of the cat. J Physiol (Lond) 364:199–215

Binder MD, Stuart DG (1980a) Response of Ia and spindle group II afferents to single motor-unit contractions. J Neurophysiol 43:621–629

Binder MD, Stuart DG (1980b) Motor-unit muscle receptor interactions: Design features of the neuromuscular control system. In: Desmedt JE (ed) Suprasegmental and segmental mechanisms. Karger, Basel, pp 72–98 (Prog Clin Neurophysiol, vol. 8)

Biscoe TJ, Krnjevic K (1963) Chloralose and the activity of Renshaw cells. Exp Neurol 8:395–405

Bliss TVP (1979) Synaptic plasticity in the hippocampus. Trends Neurosci 2:42–45

Bloedel JR, Ebner TJ, YU Q-X (1983) Increased responsiveness of Purkinje cells associated with climbing fiber inputs to neighboring neurons. J Neurophysiol 50:220–239

Bodine SC, Roy RR, Meadows DA, Zernicke RF, Sacks RD, Fournier M, Edgerton VR (1982a) Architectural, histochemical, and contractile characteristics of a unique biarticular muscle: the cat semitendinosus. J Neurophysiol 48:192–201

Bodine SC, Roy RR, Zernicke RF, Edgerton VR (1982b) Intracontractile length changes in the proximal and distal compartments of the semitendinosus. Soc Neurosci Abstr 8:948

Booth CM, Brown MC (1983) A demonstration that rostro-caudal location in the spinal cord dictates a motoneurone's antero-posterior distribution within the rodent gluteus muscle. J Physiol (Lond) 345:8P

Botterman BR, Binder MD, Stuart DG (1978) Functional anatomy of the association between motor units and muscle receptors. Am Zool 18:135–152

Botterman BR, Hamm TM, Reinking RM, Stuart DG (1983a) Localization of monosynaptic Ia excitatory post-synaptic potentials in the motor nucleus of the cat biceps femoris muscle. J Physiol (Lond) 338:355–377

Botterman BR, Hamm TM, Reinking RM, Stuart DG (1983b) Distribution of monosynaptic Ia excitatory post-synaptic potentials in the motor nucleus of the cat semitendinosus muscle. J Physiol (Lond) 338:379–393

Boyd IA (1981) The action of the three types of intrafusal fibre of isolated cat muscle spindles on the dynamic and length sensitivities of primary and secondary sensory endings. In: Taylor A, Prochazka A (eds) Muscle receptors and movement. Macmillan, London, pp 17–32

Boyd I, Rosenberg J (1985) Review. In: Boyd IA, Gladden MH (eds) The muscle spindle. Stockton, New York, pp 327–352

Boyd IA, Smith RS (1984) The muscle spindle. In: Dyck PJ, Thomas PK, Lombert EH, Bunge R (eds) Peripheral neuropathy, 2nd edn. Saunders, Philadelphia, pp 171–202

Boyle R, Pompeiano O (1984) Discharge activity of spindle afferents from the gastrocnemius-soleus muscle during head rotation in the decerebrate cat. Pflügers Arch 400:140–150

Braitenberg V (1967) Is the cerebellar cortex a biological clock in the millisecond range? In: Fox CA, Snider RS (eds) The cerebellum. Prog Brain Res, vol 25. Elsevier, Amsterdam, pp 334–346 (Progress in brain research, vol. 25)

Braitenberg V (1977) On the texture of brains. Springer, Berlin Heidelberg New York

Brink EE, Suzuki I (1987) Recurrent inhibitory connexions among neck motoneurones in the cat. J Physiol (Lond) 383:301–326

Brink EE, Jinnai K, Wilson VJ (1981) Pattern of segmental monosynaptic input to cat dorsal neck motoneurons. J Neurophysiol 46:496–505

Brink E, Jankowska E, McCrea DA, Skoog B (1983) Inhibitory interactions between interneurones in reflex pathways from group Ia and group Ib afferents in the cat. J Physiol (Lond) 343:361–373

Brink E, Jankowska E, Skoog B (1984) Convergence on to interneurons subserving primary afferent depolarisation of group I afferents. J Neurophysiol 51:432–449

Broman H, DeLuca CJ, Mambrito B (1985) Motor unit recruitment and firing rates interaction in the control of human muscles. Brain Res 337:311–319

Brooks VB, Wilson VJ (1958) Localization of stretch reflexes by recurrent inhibition. Science 127:472–473

Brooks VB, Wilson VJ (1959) Recurrent inhibition in the cat's spinal cord. J Physiol (Lond) 146:380–391

Brown AG (1981) Organization in the spinal cord. The anatomy and physiology of identified neurones. Springer, Berlin Heidelberg New York

Brown AG, Fyffe REW (1978) The morphology of group Ia afferent fibre collaterals in the spinal cord of the cat. J Physiol (Lond) 274:111–127

Brown MC, Booth CM (1983) Segregation of motor nerves on a segmental basis during synapse elimination in neonatal muscles. Brain Res 273:188–190

Brown TIH, Rack PMH, Ross HG (1982) Different types of tremor in the human thumb. J Physiol (Lond) 332:113–123

Buahin KG, Rymer WZ (1984) Renshaw cell desynchronization of motor output. Soc Neurosci Abstr 10, Part 1:329

Burgess PR, Wei JY, Clark FJ, Simon J (1982) Signaling of kinesthetic information by peripheral sensory receptors. Annu Rev Neurosci 5:171–187

Burke RE (1967) Composite nature of the monosynaptic excitatory postsynaptic potential. J Neurophysiol 30:1114–1137

Burke RE (1981) Motor units: anatomy, physiology, and functional organization. In: Brooks VB (ed) The nervous system. Am Physiol Soc, Bethesda, pp 354–422 (Handbook of Physiology, vol. II, part 1)

Burke RE, Tsairis P (1973) Anatomy and innervation ratios in motor units of cat gastrocnemius. J Physiol (Lond) 234:749–765

Burke RE, Fedina L, Lundberg A (1971) Spatial synaptic distribution of recurrent and group Ia inhibitory systems in the cat spinal motoneurones. J Physiol (Lond) 214:305–326

Burke RE, Strick PL, Kanda K, Kim CC, Walmsley B (1977) Anatomy of medial gastrocnemius and soleus motor nuclei in cat spinal cord. J Neurophysiol 40:667–680

Burne JA, Lippold OCJ, Pryor M (1984) Proprioceptors and normal tremor. J Physiol (Lond) 348:559–572

Burns D, Webb AC (1979) The corrrelation between discharge times of neighbouring neurons in isolated cerebral cortex. Proc R Soc Ser B 203:347–360

Calancie B, Bawa P (1985) Firing patterns of human flexor carpi radialis motor units during the stretch reflex. J Neurophysiol 53:1179–1193

Calvin H, Schwindt PC (1972) Steps in production of motoneuron spikes during rhythmic firing. J Neurophysiol 35:297–310

Calvin H, Stevens CF (1968) Synaptic noise and other sources of randomness in motoneuron interspike intervals. J Neurophysiol 31:574–587

Cameron WE, Binder MD, Botterman BR, Reinking RM, Stuart DG (1981) "Sensory partitioning" of cat medial gastrocnemius muscle by its muscle spindles and tendon organs. J Neurophysiol 46:32–47

Carli G, Diete-Spiff K, Pompeiano O (1967) Responses of the muscle spindles and of the extrafusal muscle fibres in an extensor muscle to stimulation of the lateral vestibular nucleus in the cat. Arch Ital Biol 105:209–242

Cavallari P, Edgley SA, Jankowska E, Skoog B (1986) Combined morphological and functional studies of neuronal pathways from group II muscle afferents. Neurosci Lett Suppl 26:S13

Chen WJ, Poppele RE (1973) Static fusimotor effect on the sensitivity of mammalian muscle spindles. Brain Res 57:244–247

Christakos CN (1982) A study of the muscle force waveform using a population stochastic model of skeletal muscle. Biol Cybern 44:91–106

Christakos CN (1986) The mathematical basis of population rhythms in nervous and neuromuscular systems. Int J Neurosci 29:103–107

Christakos CN, Windhorst U (1986a) Spindle gain increase during muscle unit fatigue. Brain Res 365:388–392

Christakos CN, Windhorst U (1986b) The information carried by spindle afferents on motor unit activity as revealed by spectral analysis. Brain Res 367:52–62

Christakos CN, Rost I, Windhorst U (1984) The use of frequency domain techniques in the study of signal transmission in skeletal muscle. Pflügers Arch 400:100–105

Christakos CN, Rost I, Windhorst U (1985) Quality of signal transmission from α-motoneurones to spindle afferents in the cat. In: Boyd IA, Gladden MH (eds) The muscle spindle. Stockton, New York, pp 403–407

Churchland PS (1986) Neurophilosophy. Toward a unified science of the mind. MIT Press, Cambridge (Mass)

Clamann HP (1969) Statistical analysis of motor unit firing patterns in human skeletal muscle. Biophys J 9:1233–1251

Clamann HP, Henneman E, Lüscher HR, Mathis J (1985) Structural and topographical influences on functional connectivity in spinal monosynaptic reflex arcs in the cat. J Physiol (Lond) 358:483–507

Clark FJ, Matthews PBC, Muir RB (1981) Motor unit firing and its relation to tremor in the tonic vibration reflex of the decerebrate cat. J Physiol (Lond) 313:317–334

Clement JD, Forsythe ID, Redman SJ (1987) Presynaptic inhibition of synaptic potentials evoked in cat spinal motoneurones by impulses in single group Ia axons. J Physiol (Lond) 383:153–169

Cleveland S (1977) Dynamische Eigenschaften der Renshaw-Zellen. Math-Nat Dissertation. Universität Düsseldorf

Cleveland S (1980) Verarbeitung spinal-motorischer Ausgangssignale durch die Renshaw-Zellen. Med Habilitationsschrift, Universität Düsseldorf

Cleveland S, Ross HG (1977) Dynamic properties of Renshaw cells: Frequency response characteristics. Biol Cybern 27:175–184

Cleveland S, Ross HG (1985) Recurrent inhibition and the input-output relations of motoneurones. Neurosci Lett Suppl 22:S131

Cleveland S, Kuschmierz A, Ross HG (1981) Static input-output relations in the spinal recurrent inhibitory pathway. Biol Cybern 40:223–231

Coggshall JC, Bekey GA (1970) A stochastic model of skeletal muscle based on motor unit properties. Math Biosci 7:405–419

Cohen LA (1953) Localization of stretch reflex. J Neurophysiol 16:272–285

Cohen LA (1954) Organization of stretch reflex into two types of direct spinal arcs. J Neurophysiol 17:443–453

Cohen MI (1979) Neurogenesis of respiratory rhythm in the mammal. Physiol Rev 59:1105–1173

Collins III WF (1983) Organization of electrical coupling between frog lumbar motoneurons. J Neurophysiol 49:730–744

Connell LA, Davey NJ, Ellaway PH (1986) The degree of short-term synchrony between α- and γ-motoneurones coactivated during the flexion reflex in the cat. J Physiol (Lond) 376:47–61

Cope TC, Nelson SG, Mendell LM (1980) Factors outside neuraxis mediate "acute" increase in EPSP amplitude caudal to spinal cord transection. J Neurophysiol 44:174–183

Crago PE, Houk JC, Hasan Z (1976) Regulatory actions of human stretch reflex. J Neurophysiol 39:925–935

Crago PE, Houk JC, Rymer WZ (1982) Sampling of total muscle force by tendon organs. J Neurophysiol 47:1069–1083

Creutzfeldt OD (1983) Cortex cerebri. Leistung, strukturelle und funktionelle Organisation der Hirnrinde. Springer, Berlin Heidelberg New York

Crick FHC (1984) Function of the thalamic reticular complex: The searchlight hypothesis. Proc Natl Acad Sci (USA), Proc Biol Sci 81:4586–4590

Crowe A, Matthews PBC (1964) Further studies of static and dynamic fusimotor fibres. J Physiol (Lond) 174:132–151

Cullheim S, Kellerth J-O (1978a) A morphological study of the axons and recurrent axon collaterals of cat α-motoneurones supplying different hind-limb muscles. J Physiol (Lond) 281:285–299

Cullheim S, Kellerth J-O (1978b) A morphological study of the axons and recurrent axon collaterals of cat α-motoneurones supplying different types of muscle unit. J Physiol (Lond) 281:301–313

Cullheim S, Kellerth J-O (1981) Two kinds of recurrent inhibition of cat spinal α-motoneurones as differentiated pharmacologically. J Physiol (Lond) 312:209–224

Cullheim S, Ulfhake B (1979) Observations on the morphology of intracellularly stained γ-motoneurons in relation to their axon conduction velocity. Neurosci Lett 13:47–50

Cullheim S, Ulfhake B (1985) Postnatal changes in the termination pattern of recurrent axon collaterals of triceps surae α-motoneurons in the cat. Dev Brain Res 17:63–73

Cullheim S, Kellerth J-O, Conradi S (1977) Evidence for direct synaptic interconnections between cat spinal alpha motoneurons via the recurrent axon collaterals: a morphological study using intracellular injection of horseradish peroxidase. Brain Res 132:1–10

Cullheim S, Lipsenthal L, Burke RE (1984) Direct monosynaptic contacts between type-identified alpha-motoneurons in the cat. Brain Res 308:196–199

Cussons PD, Hulliger M, Matthews PBC (1977) Effects of fusimotor stimulation on the response of the secondary ending of the muscle spindle to sinusoidal stretching. J Physiol (Lond) 270:835–850

Cussons PD, Matthews PBC, Muir RB (1979) Tremor in the tension developed isometrically by soleus during the tonic vibration reflex in the decerebrate cat. J Physiol (Lond) 292:35–57

Darton K, Lippold OCJ, Shahani M, Shahani U (1985) Long-latency spinal reflexes in humans. J Neurophysiol 53:1604–1618

Datta AK, Stephens JA (1980) Short-term synchronization of motor unit firing in human first dorsal interosseus muscle. J Physiol (Lond) 30:19P

Datta AK, Fleming JR, Hortobagyi T, Stephens JA (1985) Short-term synchronization of high-threshold motor units in human first dorsal interosseus muscle recorded during steady voluntary isometric contractions. J Physiol (Lond) 366:22P

Davey N, Ellaway P (1984) Patterns of discharge of γ-motoneurones and their tendency to synchronized firing. Neurosci Lett Suppl 18:S267

Davey NJ, Ellaway PH (1985) The nature of the reflex coupling between skin afferents and gamma motoneurones in the cat. J Physiol (Lond) 366:127P

Davey NJ, Ellaway PH, Friedland CL (1986) Correlated discharges of motor units in Parkinson's disease. Neurosci Lett Suppl 26:S194

Davis BM, Collins III WF, Mendell LM (1985) Potentiation of transmission at Ia-motoneuron connections induced by repeated short bursts of afferent activity. J Neurophysiol 54:1541–1552

Davis BM, Druzinski RE, Mendell LM (1987) Distribution of potentiation following short high frequency bursts to motoneurons of different rheobase. Exp Brain Res 65:639–648

Demiéville HN, Partridge LD (1980) Probability of peripheral interaction between motor units and implications for motor control. Am J Physiol 238 (Regulatory Integrative Comp Physiol 7):R119–R137

DeRibaupierre F, Abeles M, deRibaupierre Y (1985) Influence of tangential cortical distance on single units interactions in the cat auditory cortex. Neurosci Lett Suppl 22:S433

DeRusso PM, Roy RJ, Close CM (1965) State variables for engineers. Wiley, New York

Desmedt JE (1978) Cerebral motor control in man long loop mechanisms. Karger, Basel

Dietz V, Bischofsberger E, Wita C, Freund H-J (1976) Correlation between the discharges of two simultaneously recorded motor units and physiological tremor. Electroencephalogr Clin Neurophysiol 40:97–105

Donselaar Y, Kernell D, Eerbeek O, Verhey BA (1985) Somatotopic relations between spinal motoneurones and muscle fibres of the cat's musculus peroneus longus. Brain Res 335:81–88

Dufresne JR, Soechting JF, Terzuolo CA (1978) Electromyographic response to pseudo-random torque disturbances of human forearm position. Neuroscience 3:1213–1226

Dufresne JR, Soechting JF, Terzuolo CA (1979) Identification of short- and long-latency loops with individual feedback parameters. Neuroscience 4:1493–1500

Duncan R, Weston-Smith M (1977) Encyclopaedia of ignorance. Pergamon, Oxford

Durkovic RG (1983) Features of the reflex circuitry. Neurosci Lett 39:155–160

Durkovic RG (1985) Retention of a classically conditioned reflex response in spinal cat. Behav Neural Biol 43:12–20

Ebner TJ, Bloedel RJ (1981 a) Temporal patterning in simple spike discharge of Purkinje cells and its relationship to climbing fiber activity. J Neurophysiol 45:933–947

Ebner TJ, Bloedel RJ (1981 b) Correlation between activity of Purkinje cells and its modification by natural peripheral stimuli. J Neurophysiol 45:948–961

Ebner TJ, Yu Q-X, Bloedel JR (1983) Increase in Purkinje cell gain associated with naturally activated climbing fiber input. J Neurophysiol 50:205–219

Eccles JC (1973) The understanding of the brain. McGraw-Hill, New York

Eccles JC (1977) An instruction-selection theory of learning in the cerebellar cortex. Brain Res 127:327–352

Eccles JC (1979) Synaptic plasticity. Naturwissenschaften 66:147–153

Eccles JC (1983) Calcium in long-term potentiation as a model for memory. Neuroscience 10:1071–1081

Eccles RM, Lundberg A (1958) Integrative pattern of Ia synaptic actions on motoneurones of hip and knee muscles. J Physiol (Lond) 144:271–298

Eccles JC, Eccles RM, Iggo A, Ito M (1961) Distribution of recurrent inhibition among moto-neurones. J Physiol (Lond) 159:479–499

Edelman GM (1979) Group selection and phasic reentrant: A theory of higher brain function. In: Schmitt FO, Worden FG (eds) The neurosciences, fourth study program. MIT Press, Cambridge (Mass), pp 1115–1139

Egger MD, Freeman NCG, Proshansky E (1980) Morphology of spinal motoneurones mediating a cutaneous spinal reflex in the cat. J Physiol (Lond) 306:349–363

Ekerot C-F, Kano M (1985) Long-term depression of parallel fibre synapses following stimulation of climbing fibres. Brain Res 342:357–360

Eklund G, Hagbarth K-E, Hägglund JV, Wallin EU (1982) The resonance hypothesis versus the "long-loop" hypothesis. J Physiol (Lond) 326:79–90

Elble RJ, Randall JE (1976) Motor-unit activity responsible for 8- to 12-Hz component of human physiological finger tremor. J Neurophysiol 39:370–383

Eldred E, Maier A, Bridgman CF (1974) Differences in intrafusal fiber content of spindles in several muscles of the cat. Exp Neurol 45:8–18

Ellaway P(1971) Recurrent inhibition of fusimotor neurones exhibiting background discharges in the decerebrate and the spinal cat. J Physiol (Lond) 216:419–439

Ellaway PH, Furness P (1977) Increased probability of muscle spindle firing time-locked to the electrocardiogram in rabbits. J Physiol (Lond) 273:92P

Ellaway PH, Murphy PR (1980a) Autogenetic effects of muscle contraction on extensor gamma motoneurones in the cat. Exp Brain Res 38:305–312

Ellaway PH, Murphy PR (1980b) A comparison of the recurrent inhibition of α- and γ-motoneurones in the cat. J Physiol (Lond) 315:43–58

Ellaway PH, Murthy KSK (1982) The degree of synchrony between gamma motoneurone discharges in the cat. Neurosci Lett Suppl 10:S163

Ellaway PH, Murthy KSK (1983) The degree of synchrony of discharge between γ-motoneurones and its dependency upon firing rates in the cat. J Physiol (Lond) 338:33P

Ellaway PH, Murthy KSK (1985a) The origins and characteristics of cross- correlated activity between γ-motoneurones in the cat. Q J Exp Physiol 70:219–232

Ellaway PH, Murthy KSK (1985b) The source and distribution of short-term synchrony between γ-motoneurones in the cat. Q J Exp Physiol 70:233–247

Ellaway PH, Trott JR (1978) Autogenetic reflex action on to gamma motoneurones by stretch of triceps surae in the decerebrated cat. J Physiol (Lond) 276:49–66

Ellaway PH, Murphy PR, Trott JR (1979) Inhibition of gamma motoneurone discharge by contraction of the homonymous muscle in the decerebrate cat. J Physiol (Lond) 291:425–442

Ellaway PH, Murphy PR, Trott JR (1981) Autogenetic effects from spindle primary endings and tendon organs on discharge of gamma motoneurons in the cat. In: Taylor A, Prochazka A (eds) Muscle receptors and movement. Macmillan, London, pp 17–32

Ellaway PH, Murphy PR, Tripathi A (1982a) Closely coupled excitation of γ-motoneurones by group III muscle afferents with low mechanical threshold in the cat. J Physiol (Lond) 331:481–498

Ellaway PH, Murthy KSK, Pascoe JE (1982b) Correlations between discharges of gamma motoneurones in the cat. J Physiol (Lond) 328:2–3P

Emonet-Dénand F, Jami L, Laporte Y (1975) Skeleto-fusimotor axons in hind-limb muscles of the cat. J Physiol (Lond) 249:153–166

Emonet-Dénand F, Laporte Y, Tristant A (1980) Effects of slow muscle stretch on the response of primary and secondary endings to small amplitude periodic stretches in de-efferented soleus muscle spindles. Brain Res 191:551–554

English A (1984) An electromyographic analysis of compartments in cat lateral gastrocnemius during unrestrained locomotion. J Neurophysiol 52:114–125

English AW, Letbetter WD (1982) A histochemical analysis of identified compartments of cat lateral gastrocnemius muscle. Anat Rec 204:123–130

English AW, Weeks OI (1984) Compartmentalization of single muscle units in cat lateral gastrocnemius. Exp Brain Res 56:361–368

Erickson RP (1974) Parallel "population" neural coding in feature extraction. In: Schmitt FO, Worden FG (eds) The neurosciences, third study program. MIT Press, Cambridge (Mass), pp 155–169

Evarts EV (1972) Feedback and corollary discharge: a merging of concepts. In: Schmitt FO, Adelmann G, Melnechuk T, Worden FG (eds) Neurosci Res Symp Summ; MIT Press, Cambridge (Mass) London, pp 86–170

Evarts EV (1981) Sherrington's concept of proprioception. Trends Neurosci 4:44–46

Ezure K, Fukushima K, Schor RH, Wilson VJ (1983) Compartmentalization of the cervicocollic reflex in cat splenius muscle. Exp Brain Res 5:397–404

Fawcett JW, O'Leary DDM (1985) The role of electrical activity in the formation of topographic maps in the nervous systems. Trends Neurosci 8:201–206

Feenstra BWA, Tanke RHJ, Crowe A (1985) Multichannel transmission of proprioceptive input to motoneurons. Biol Cybern 52:53–58

Fellows SJ, Rack PMH (1987) Changes in the length of the human biceps brachii muscle during elbow movements. J Physiol (Lond) 383:405–412

Fetz EE, Gustafsson BG (1983) Relation between shapes of post-synaptic potentials and changes in firing probability of cat motoneurons. J Physiol (Lond) 341:387–410

Fetz EE, Jankowska E, Johannisson T, Lipski J (1979) Autogenetic inhibition of motoneurones by impulses in group Ia muscle spindle afferents. J Physiol (Lond) 293:173–195

Finkel LH, Edelman GM (1985) Interaction of synaptic modification rules within populations of neurons. Proc Natl Acad Sci 82:1291–1295

Fleshman JW, Munson J, Sypert GW (1981) Homonymous projection of individual group Ia-fibers to physiologically characterized medial gastrocnemius motoneurons in the cat. J Neurophysiol 46:1339–1348

Freeman WJ (1981) Dynamics of image formation by nerve cell assemblies. In: Başar E, Flohr H, Haken H, Mandell AJ (eds) Synergetics of the brain. Springer, Berlin Heidelberg New York, pp 102–121

Freeman WJ (1988) Nonlinear neural dynamics in olfaction as a model for cognition. In: Başar E (ed) Dynamics of sensory and cognitive processing in the brain. Springer, Berlin Heidelberg New York, pp 19–29

Freund H-J (1983) Motor unit and muscle activity in voluntary motor control. Physiol Rev 63:387–436

Freund H-J, Dietz V (1978) The relationship between physiological and pathological tremor. In: Desmedt JE (ed) Physiological tremor, pathological tremors and clonus. Karger, Basel, pp 66–89 (Prog Clin Neurophysiol, vol 5).

Friedman WA, Sypert GW, Munson J, Fleshman JW (1981) Recurrent inhibition in type-identified motoneurons. J Neurophysiol 46:1349–1359

Frohn H, Geiger H, Singer W (1987) A self-organizing neural network sharing features of the mammalian visual system. Biol Cybern 55:333–343

Fromm C, Haase J, Wolf E (1977) Depression of the recurrent inhibition of extensor motoneurons by the action of group II afferents. Brain Res 120:459–468

Fung SJ, Barnes CD (1987) Membrane excitability changes in hindlimb motoneurons induced by stimulation of the locus coeruleus in cats. Brain Res 402:230–242

Fung SJ, Pompeiano O, Barnes CD (1987) Suppression of the recurrent inhibitory pathway in lumbar cord segments during locus coeruleus stimulation in cats. Brain Res 402:351–354

Gans C, Bock WJ (1965) The functional significance of muscle architecture – a theoretical analysis. Ergeb Anat 38:115–142

Gauthier P, Barillot JC, Dussardier M (1980) Mise en évidence d'interactions d'origine centrale entre motoneurones laryngés. J Physiol (Paris) 76:647–661

Gelfand IM, Gurfinkel VS, Kots YM, Tsetlin ML, Shik ML (1963) Synchronization of motor units and associated model concepts. Biofizika 8:475–486

Gerstein GL, Aertsen AMHJ (1985) Representation of cooperative firing activity among simultaneously recorded neurons. J Neurophysiol 54:1513–1528

Gerstein GL, Michalski A (1981) Firing synchrony in a neural group: putative sensory code. In: Székely G, Lábos E, Damjavovitch S (eds) Neural communication and control. Pergamon, Akademiai Kiado, Budapest, pp 93–102 (Adv. Physiol. Sci. vol. 30)

Getting PA, Dekin MS (1985) Mechanisms of pattern generation underlying swimming in Tritonia. IV. Gating of central pattern generator. J Neurophysiol 53:466–480

Gielen CCAM, van Zuylen EJ (1986) Coordination of arm muscles during flexion and supination: application of the tensor analysis approach. Neuroscience 17:527–539

Glaser EM, Ruchkin DS (1976) Principles of neurobiological signal analysis. Academic, New York

Gogan P, Gueritaud JP, Horcholle-Bossavit G, Tyc-Dumont S (1977) Direct excitatory interactions between spinal motoneurones of the cat. J Physiol (Lond) 272:755–767

Gogan P, Gustafsson B, Jankowska E, Tyc-Dumont S (1984) On re-excitation of feline motoneurones: its mechanism and consequences. J Physiol (Lond) 350:81–91

Gonyea WJ, Ericson GC (1977) Morphological and histochemical organization of the flexor carpi radialis muscle in the cat. Am J Anat 148:329–344

Goodwin GM, Hulliger M, Matthews PBC (1975) The effects of fusimotor stimulation during small amplitude stretching on the frequency-response of the primary ending of the mammalian muscle spindle. J Physiol (Lond) 253:175–206

Goodwin GM, Hoffman D, Luschei ES (1978) The strength of the reflex response to sinusoidal stretch of monkey jaw closing muscles during voluntary contraction. J Physiol (Lond) 279:81–111

Gottlieb GL, Agarwal GL (1980) Response to sudden torques about ankle in man. II. Postmyotatic reactions. J Neurophysiol 43:86–101

Gottlieb S, Lippold OCJ (1983) The 4–6 Hz tremor during sustained contraction in normal human subjects. J Physiol (Lond) 336:499–509

Gottlieb S, Taylor A (1983) Interpretation of fusimotor activity in cat masseter nerve during reflex jaw movements. J Physiol (Lond) 345:423–438

Granit R (1970) The basis of motor control. Academic, London

Granit R, Pascoe JE, Steg G (1957) The behaviour of tonic α and γ motoneurones during stimulation of recurrent collaterals. J Physiol (Lond) 138:381–400

Granit R, Haase J, Rutledge LT (1960) Recurrent inhibition in relation to frequency of firing and limitation of discharge rate of extensor motoneurones. J Physiol (Lond) 154:308–328

Grillner S, Wallén P (1985) Central pattern generators for locomotion, with special reference to verebrates. Annu Rev Neurosci 8:233–261

Gustafsson B, McCrea D (1984) Influence of stretch-evoked synaptic potentials on firing probability of cat spinal motoneurones. J Physiol (Lond) 34:431–451

Haase J, van der Meulen JP (1961 a) Effects of supraspinal stimulation on Renshaw cells belonging to extensor motoneurones. J Neurophysiol 24:510–520

Haase J, van der Meulen JP (1961 b) Die spezifische Wirkung der Chloralose auf die recurrente Inhibition tonischer Motoneurone. Pflügers Arch 274:272–280

Haase J, Vogel B (1971) Direkte und indirekte Wirkungen supraspinaler Reizungen auf Renshaw-Zellen. Pflügers Arch 325:334–346

Haase J, Cleveland S, Ross HG (1975) Problems of postsynaptic autogenous and recurrent inhibition in the mammalian spinal cord. Rev Physiol Biochem Pharmacol 73:73–129

Hagbarth K-E (1973) The effect of muscle vibration in normal man and in patients with motor disorders. In: Desmedt JE (ed) New developments in electromyography and clinical neurophysiology. Karger, Basel, pp 428–443

Hagbarth K-E, Young RR (1979) Participation of the stretch reflex in human physiological tremor. Brain 102:509–526

Hagbarth K-E, Kunesch EJ, Nordin M, Schmidt R, Wallin EU (1986) γ loop contributing to maximal voluntary contractions in man. J Physiol (Lond) 380:575–591

Hamm TM, Koehler W, Stuart DG, Vanden-Noven S (1985a) Partitioning of monosynaptic Ia excitatory postsynaptic potentials in the motor nucleus of the cat semimembranosus muscle. J Physiol (Lond) 369:379–398

Hamm TM, Reinking RM, Roscoe DD, Stuart DG (1985b) Synchronous afferent discharge from a passive muscle of the cat. J Physiol (Lond) 365:77–102

Hamm TM, Sasaki S, Stuart DG, Windhorst U, Yuan C-S (1987a) The measurement of single motor-axon recurrent inhibitory post-synaptic potentials in the cat. J Physiol (Lond) 388:631–651

Hamm TM, Sasaki S, Stuart DG, Windhorst U, Yuan C-S (1987b) Distribution of single-axon recurrent inhibitory post-synaptic potentials in a single spinal motor nucleus in the cat. J Physiol (Lond) 388:653–664

Harrison PJ, Jankowska E (1985a) Sources of input to interneurones mediating group I non-reciprocal inhibition of motoneurones in the cat. J Physiol (Lond) 361:379–401

Harrison PJ, Jankowska E (1985b) Organization of input to the interneurones mediating group I non-reciprocal inhibition of motoneurones in the cat. J Physiol (Lond) 361:403–418

Harrison PJ, Taylor A (1981) Individual excitatory post-synaptic potentials due to muscle spindle Ia afferents in cat triceps surae motoneurones. J Physiol (Lond) 312:455–470

Harrison PJ, Jankowska E, Johannisson T (1983) Shared reflex pathways of group I afferents of different cat hindlimb muscles. J Physiol (Lond) 338:113–127

Hasan Z (1986) Optimized movement trajectories and joint stiffness in unperturbed, inertially loaded movements. Biol Cybern 53:373–382

Hatze H, Buys JD (1977) Energy-optimal controls in the mammalian neuromuscular system. Biol Cybern 27:9–20

Hebb DO (1949) The organization of behaviour. Wiley, New York

Hellweg C, Meyer-Lohmann J, Benecke R, Windhorst U (1974) Response of Renshaw cells to muscle ramp stretch. Exp Brain Res 21:353–360

Henatsch H-D (1983) Facts and hypotheses on the supraspinal control of the spinal Renshaw cell-motoneurone-system. In: Speckmann E-J, Elger CE (eds) Epilepsy and motor system. Urban und Schwarzenberg, München, pp 78–99

Henatsch H-D, Windhorst U (1981) Die Relevanz spinaler Interneuronen-Systeme und ihrer supraspinalen Kontrollen für die Pathogenese von Spastik. In: Bauer HJ, Koella WP, Struppler A (eds) Therapie der Spastik. Verlag f. angew. Wissenschaften, München, pp 39–52

Henatsch H-D, Kaese H-J, Langrehr D, Meyer-Lohmann J (1961) Einflüsse des motorischen Cortex der Katze auf die Renshaw-Rückkopplungshemmung der Motoneurone. Pflügers Arch 274:51

Henatsch H-D, Meyer-Lohmann J, Windhorst U, Schmidt J (1986) Differential effects of stimulation of the cat's red nucleus on lumbar alpha motoneurones and their Renshaw cells. Exp Brain Res 62:161–174

Henneman E, Mendell LM (1981) Functional organization of motoneuron pool and its inputs. In: Brooks VB (ed) The nervous system. Am Physiol Soc, Bethesda, pp 423–507 (Handbook of physiology, vol. II, part 1)

Henneman E, Lüscher H-R, Mathis JO (1984) Simultaneously active and inactive synapses of single Ia fibres on cat spinal motoneurones. J Physiol (Lond) 352:147–161

Hermanson JW, Lennard PR, Takamoto RL (1986) Morphology and histochemistry of the ambiens muscle of the red-eared turtle (Pseudomys scripta). J Morphol 187:39–49

Herring SW, Grimm AF, Grimm BR (1979) Functional heterogeneity in a multipinnate muscle. Am J Anat 154:563–575

Hilaire G, Khatib M, Monteau R (1983) Spontaneous respiratory activity of phrenic and intercostal Renshaw cells. Neurosci Lett 43:97–101

Hilaire G, Monteau R, Bianchi AL (1984) A cross-correlation study of interactions among respiratory neurons of dorsal, ventral and retrofacial groups in cat medulla. Brain Res 302:19–31

Hill DK (1968) Tension due to interaction between sliding filaments in resting striated muscle. The effect of stimulation. J Physiol (Lond) 199:637–684

Hoffer JA, Sugano N, Loeb GE, Marks WB, O'Donovan MJ, Pratt CA (1987a) Cat hindlimb motoneurons during locomotion. II. Normal activity patterns. J Neurophysiol 57:530–553

Hoffer JA, Loeb GE, Sugano N, Marks WB, O'Donovan MJ, Pratt CA (1987b) Cat hindlimb motoneurons during locomotion. III. Functional segregation in sartorius. J Neurophysiol 57:554–562

Hogan N (1984) An organizing principle for a class of voluntary movements. J Neurosci 4:2745–2754

Holmes O, Houchin J (1966) Units in the cerebral cortex of the anaesthetized rat and the correlations between their discharges. J Physiol (Lond) 197:651–671

Honig MG, Collins WF III, Mendell LM (1983) α-motoneuron EPSPs exhibit different frequency sensitivities to single Ia-afferent fiber stimulation. J Neurophysiol 49:886–901

Hore J, Flament D (1986) Evidence that a disordered servo-like mechanism contributes to tremor in movements during cerebellar dysfunction. J Neurophysiol 56:123–136

Hore J, Vilis T (1984) Loss of set in muscle responses to limb perturbations during cerebellar dysfunction. J Neurophysiol 51:1137–1148

Houk JC (1976) An assessment of stretch reflex function. In: Homma S (ed) Understanding the stretch reflex. Elsevier, Amsterdam, pp 193–215 (Progress in brain research, vol. 44)

Houk JC, Rymer WZ (1981) Neural control of muscle length an tension. In: Brooks VB (ed) The nervous system. Am Physiol Soc, Bethesda, pp 257–323 (Handbook of physiology, vol. II, part 1)

Houk JC, Crago PE, Rymer WZ (1981a) Function of the spindle dynamic response in stiffness regulation: a predictive mechanism provided by non-linear feedback. In: Taylor A, Prochazka A (eds) Muscle receptors and movement. Macmillan, London, pp 299–309

Houk JC, Rymer WZ, Crago PE (1981b) Dependence of dynamic response of spindle receptors on muscle length and velocity. J Neurophysiol 46:143–166

Hounsgaard J, Hultborn H, Jespersen B, Kiehn O (1984) Intrinsic membrane properties causing bistable behaviour of α-motoneurones. Exp Brain Res 55:391–394

Huhle R (1985) Topographic studies relating distribution of Ia- and γ-fibres in spinal cord and position of muscle spindles in cat tibialis anterior muscle. Brain Res 333:299–304

Hulliger M (1984) The mammalian muscle spindle and its central control. Rev Physiol Biochem Pharmacol 101:1–110

Hulliger M, Sonnenberg R (1985) Does the paradoxical γ_D-mediated reduction of Ia sensitivity to small stretches persist during larger background movements? Neurosci Lett Suppl 22:S595

Hulliger M, Matthews PBC, Noth J (1977) Static and dynamic fusimotor action on the response of Ia fibres to low-frequency sinusoidal stretching of widely ranging amplitude. J Physiol (Lond) 267:811–838

Hulliger M, Nordh E, Vallbo AB (1982) The absence of position response in spindle afferent units from human finger muscles during accurate position holding. J Physiol (Lond) 322:167–179

Hulliger M, Nordh E, Vallbo AB (1985a) Discharge in muscle spindle related to direction of slow precision movements in man. J Physiol (Lond) 362:437–453

Hulliger M, Zangger P, Prochazka A, Appenteng K (1985b) Fusimotor "set" vs. α-γ linkage in voluntary movement in cats. In: Struppler A, Weindl A (eds) Electromyography and evoked potentials. Springer, Berlin Heidelberg New York, pp 56–63

Hultborn H (1972) Convergence on interneurones in the reciprocal Ia inhibitory pathway to motoneurones. Acta Physiol Scand 375:3–42

Hultborn H (1976) Transmission in the pathway of reciprocal Ia inhibition to motoneurones and its control during the tonic stretch reflex. In: Homma S (ed) Understanding the stretch reflex. Elsevier, Amsterdam, pp 235–255 (Progress in brain research, vol. 44)

Hultborn H, Pierrot-Deseilligny E (1979) Changes in recurrent inhibition during voluntary soleus contractions in man studied by an H-reflex technique. J Physiol (Lond) 297:229–251

Hultborn H, Wigström H (1980) Motor response with long latency and maintained duration evoked by activity in Ia afferents. In: Desmedt JE (ed) Spinal and supraspinal mechanisms of voluntary motor control and locomotion. Karger, Basel pp 99–116 (Progress in clinical neurophysiology, vol. 8)

Hultborn H, Wigström H, Wängberg B (1975) Prolonged activation of soleus motoneurones following a conditioning train in soleus Ia afferents – a case for a reverberating loop? Neurosci Lett 1:147–152

Hultborn H, Lindström S, Wigström H (1979) On the function of recurrent inhibition in the spinal cord of the cat. Exp Brain Res 37:399–403

Hutton RS, Enoka RM (1986) Kinematic assessment of a functional role for recurrent inhibition and selective recruitment. Exp Neurol 93:369–379

Iansek R, Redman SJ (1973) The amplitude, time course and charge of unitary post-synaptic potentials evoked in spinal motoneurone dendrites. J Physiol (Lond) 234:665–688

Iles JF, Roberts RC (1986) Presynaptic inhibition of monosynaptic reflexes in the lower limbs of subjects with upper motoneuron disease. J Neurophysiol Neurosurg Psychiat 49:937–944

Iliya AR, Dum RP (1984) Somatotopic relations between the motor nucleus and its innervated muscle fibers in the cat tibialis anterior. Exp Neurol 86:272–292

Inbar GF (1972) Muscle spindles in muscle control. 3. Analysis of adaptive system model. Kybernetik 11:130–141

Inbar G, Madrid J, Rudomín P (1979) The influence of the gamma system on cross-correlated activity of Ia muscle spindles and its relation to information transmission. Neurosci Lett 13:73–78

Ishizuka N, Mannen H, Hongo T, Sasaki S (1979) Trajectory of group Ia afferent fibers stained with horseradish peroxidase in the lumbosacral spinal cord of the cat: three dimensional reconstructions from serial sections. J Comp Neurol 186:189–212

Ito M (1984) The cerebellum and neural control. Raven, New York

Ito M, Kano M (1982) Long-lasting depression of parallel fiber-Purkinje cell transmission induced by conjunctive stimulation of parallel fibers and climbing fibers in the cerebellar cortex. Neurosci Lett 33:253–258

Ito M, Sakurai M, Tongroach P (1982) Climbing fibre induced depression of both mossy fibre responsiveness and glutamate sensitivity of cerebellar Purkinje cells. J Physiol (Lond) 324:113–134

Ivarsson C, Schmied A, Dick TE, Fetz EE (1986) Cross-correlations between human extensor digitorum communis motor units. Neurosci Lett Suppl 26:S423

Jami L, Murthy KSK, Petit J (1982) A quantitative study of skeleto-fusimotor innervation in the cat peroneus tertius muscle. J Physiol (Lond) 32:125–144

Jami L, Petit J, Proske U, Zytnicki D (1985a) Responses of tendon organs to unfused contractions of single motor units. J Neurophysiol 53:32–42

Jami L, Petit J, Scott JJA (1985 b) Activation of cat muscle spindles by static skeletofusimotor axons. J Physiol (Lond) 369:323–335

Jankowska E (1984) On neuronal pathways of presynaptic depolarization of group I muscle afferents. In: Creutzfeldt O, Schmidt RF, Willis WD (eds) Sensory-motor integration in the nervous system. Springer, Berlin Heidelberg New York Tokyo, pp 72–85

Jankowska E, Lundberg A (1981) Interneurones in the spinal cord. Trends Neurosci 4:230–233

Jankowska E, McCrea DA (1983) Shared reflex pathways from Ib tendon organ afferents and Ia muscle spindle afferents in the cat. J Physiol (Lond) 338:99–111

Jankowska E, Smith DO (1973) Antidromic activation of Renshaw cells and their axonal projections. Acta Physiol Scand 88:198–214

Jankowska E, McCrea DA, Mackel R (1981 a) Pattern of 'non-reciprocal' inhibition of moto-neurones by impulses in group Ia muscle spindle afferents in the cat. J Physiol (Lond) 316:393–409

Jankowska E, McCrea DA, Mackel R (1981 b) Oligosynaptic excitation of motoneurones by impulses in group Ia muscle spindle afferents in the cat. J Physiol (Lond) 316:411–425

Jenkins GM, Watts DG (1968) Spectral analysis and its applications. Holden Day, San Francisco

Johansson H (1985) Reflex integration in the γ-motor system. In: Boyd IA, Gladden MH (eds) The muscle spindle. Stockton, New York, pp 297–301

Joyce GC, Rack PMH, Westbury DR (1969) The mechanical properties of cat soleus muscle during controlled lengthening and shortening movements. J Physiol (Lond) 204:461–474

Kandou TWA, Kernell D (1986) Localization of intramuscular activity during centrally evoked contractions of the cat's peroneus longus. Neurosci Lett Suppl 26:S81

Kato M, Fukushima K (1974) Effect of differential blocking of motor axons on antidromic activation of Renshaw cells in the cat. Exp Brain Res 20:135–143

Katz R, Pierrot-Deseilligny E (1984) Facilitation of soleus-coupled Renshaw cells during voluntary contraction of pretibial flexor muscles in man. J Physiol (Lond) 355:587–603

Keirstead SA, Rose PK, Vanner SJ (1982) Frequency and distribution of axon collaterals from upper cervical spinal motoneurons. Neurosci Soc Abstr 8:724

Khatib M, Hilaire G, Monteau R (1986) Excitatory interactions between phrenic motoneurons in the cat. Exp Brain Res 62:273–280

Kim JH, King GW, Ebner TJ, Bloedel JR (1982) Evidence for motoneuron collateral input to VSCT neurons. Neurosci Soc Abstr 8:725

Kirkwood PA (1979) On the use and interpretation of cross-correlation measurements in the mammalian central nervous system. J Neurosci Methods 1:107–132

Kirkwood PA, Sears TA (1978) The synaptic connexions to intercostal motoneurones as revealed by the average common excitation potential. J Physiol (Lond) 275:103–134

Kirkwood PA, Sears TA (1982 a) Excitatory post-synaptic potentials from single muscle spindle afferents in external intercostal motoneurones of the cat. J Physiol (Lond) 322:287–314

Kirkwood PA, Sears TA (1982 b) The effects of single afferent impulses on the probability of firing of external intercostal motoneurones in the cat. J Physiol (Lond) 322:315–336

Kirkwood PA, Sears TA, Westgaard RH (1981) Recurrent inhibition of intercostal motoneurones in the cat. J Physiol (Lond) 319:111–130

Kirkwood PA, Sears TA, Stagg D, Westgaard RH (1982 a) The spatial distribution of synchronization of intercostal motoneurones in the cat. J Physiol (Lond) 327:137–155

Kirkwood PA, Sears TA, Tuck DL, Westgaard RH (1982 b) Variations in the time course of the synchronization of intercostal motoneurones in the cat. J Physiol (Lond) 327:105–135

Kirkwood PA, Sears TA, Westgaard RH (1984) Restoration of function in external motoneurones of the cat following partial central deafferentation. J Physiol (Lond) 350:225–251

Kniffki KD, Schomburg ED, Steffens H (1981) Effects of fine muscle and cutaneous afferents on spinal locomotion in cats. J Physiol (Lond) 31:543–554

Knox CK (1974) Cross-correlation functions for a neuronal model. Biophys J 14:567–582

Knox CK, Poppele RE (1977) Correlation analysis of stimulus-evoked changes in excitability of spontaneously firing neurons. J Neurophysiol 40:616–625

Koehler W, Windhorst U (1980) Multi-loop representation of the segmental muscle stretch reflex. Its risk of instability. Biol Cybern 38:51–61

Koehler W, Windhorst U (1981) Frequency response characteristics of a multi-loop representation of the segmental stretch reflex. Biol Cybern 40:59–70

Koehler W, Windhorst U (1985) Responses of the spinal α-motoneurone-Renshaw cell system to various differentially distributed segmental afferent and descending inputs. Biol Cybern 51:417–426

Koehler W, Windhorst U, Schmidt J, Meyer-Lohmann J, Henatsch H-D (1978) Diverging influences on Renshaw cell responses and monosynaptic reflexes from stimulation of capsula interna. Neurosci Lett 8:35–39

Koehler W, Hamm TM, Enoka RM, Stuart DG, Windhorst U (1984a) Contractions of single motor units are reflected in membrane potential changes of homonymous α-motoneurons. Brain Res 296:379–384

Koehler W, Hamm TM, Enoka RM, Stuart DG, Windhorst U (1984b) Linear and non-linear summation of α-motoneuron potential changes elicited by contractions of homonymous motor units in cat medial gastrocnemius. Brain Res 296:385–388

Koehler WJ, Schomburg ED, Steffens H (1984) Activity and excitability of Renshaw cells during fictive locomotion. Neurosci Lett Suppl 18:S387

Kohonen T (1984) Self-organization and associative memory. Springer, Berlin Heidelberg New York Tokyo

Kornhuber HH (1974) Cerebral cortex, cerebellum, and basal ganglia: an introduction to their motor functions. In: Schmitt FO, Worden FG(eds) The neurosciences, third study program. MIT Press, Cambridge (Mass), pp 267–280

Kranz H, Baumgartner G (1974) Human alpha motoneurone discharge, a statistical analysis. Brain Res 67:324–329

Krausz HI (1975) Identification of nonlinear systems using random impulse train inputs. Biol Cybern 19:217–230

Kretz R, Shapiro E, Connor J, Kandel ER (1984) Posttetanic potentiation, presynaptic inhibition, and the modulation of the free Ca^{2+} level in the presynaptic terminals. In: Creutzfeldt O, Schmidt RF, Willis WD (eds) Sensory-motor integration in the nervous system. Springer, Berlin Heidelberg New York Tokyo, pp 240–283

Kröller J, Grüsser O-J (1982) Convergence of muscle spindle afferents on single neurons of the cat dorsal spino-cerebellar tract and their synaptic efficacy. Brain Res 253:65–80

Krüger J (1983) Simultaneous individual recordings from many cerebral neurons: techniques and results. Rev Biochem Physiol Pharmacol 98:177–233

Kuipers U, Meyer-Lohmann J, Windhorst U (1986) Responses of cutaneous afferents and spinal interneurones to mechanical activity elicited by motor unit stimulation. Neurosci Lett Suppl 26:S363

Kuipers U, Laouris Y, Meyer-Lohmann J, Windhorst U (1987) Facilitation and depression of the responses of cat dorsal horn neurones to random stimulation of cutaneous afferents. In: Elsner N, Creutzfeldt O (eds) New frontiers in brain research. Thieme, Stuttgart New York, p 196

Kuno M (1959) Excitability following antidromic activation in spinal motoneurones supplying red muscles. J Physiol (Lond) 149:374–393

Kuno M, Miyahara JT (1969) Non-linear summation of unit potentials in spinal motoneurones of the cat. J Physiol (Lond) 201:465–477

Lagerbäck P-A (1983) An ultrastructural study of serially sectioned Renshaw cells. III. Quantitative distribution of synaptic boutons. Brain Res 264:215–223

Lagerbäck P-A, Kellerth J-O (1985a) Light microscopic observations on cat Renshaw cells after intracellular staining with horseradish peroxidase. I. The axonal systems. J Comp Neurol 240:359–367

Lagerbäck P-A, Kellerth J-O (1985b) Light microscopic observations on cat Renshaw cells after intracellular staining with horseradish peroxidase. II. The cell bodies and dendrites. J Comp Neurol 240:368–376

Lagerbäck P-A, Ronnevi L-O (1982) An ultrastructural study of serially sectioned Renshaw cells. II. Synaptic types. Brain Res 246:181–192

Lagerbäck P-A, Ronnevi L-O, Cullheim S, Kellerth J-O (1981a) An ultrastructural study of the synaptic contacts of α-motoneurone axon collaterals. I. Contacts in lamina IX and with identified α-motoneurone dendrites in lamina VII. Brain Res 207:247–266

Lagerbäck P-A, Ronnevi L-O, Cullheim S, Kellerth J-O (1981b) An ultrastructural study of the synaptic contacts of α-motoneurone axon collaterals. II. Contacts in lamina VII. Brain Res 222:29–41

Landmesser LT (1980) The generation of neuromuscular specificity. Annu Rev Neurosci 3:279–302

Langhorst P, Schulz B, Schulz G, Lambertz M, Krienke B (1983) Reticular formation of the lower brainstem. A common system for cardiorespiratory and somatomotor functions: discharge patterns of neighboring neurons influenced by cardiovascular and respiratory afferents. J Auton Nerv Syst 9:411–432

Laporte Y, Emonet-Dénand F, Jami L (1981) The skeleto-fusimotor or β-innervation of mammalian muscle spindles. Trends Neurosci 4:97–99

Lee KS (1983) Cooperativity among afferents for the induction of long-term potentiation in the CA1 region of the hippocampus. J Neurosci 3:1369–1372

Lennerstrand G (1968) Position and velocity sensitivity of muscle spindles in the cat. I. Primary and secondary endings deprived of fusimotor activation. Acta Physiol Scand 73:281–299

Letbetter WD (1974) Influence of intramuscular nerve branching on motor unit organization in medial gastrocnemius muscle. Anat Rec 178:402

Lev-Tov A, Pinter MJ, Burke RE (1983) Posttetanic potentiation of group Ia EPSPs: possible mechanisms for differential distribution among medial gastrocnemius motoneurons. J Neurophysiol 50:379–398

Levy WB, Steward O (1983) Temporal contiguity requirements for long-term associative potentiation/depression in the hippocampus. Neuroscience 8:791–797

Levy WB, Brassel SE, Moore SD (1983) Partial quantification of the associative synaptic learning rule of the dentate gyrus. Neuroscience 8:799–808

Liddell EGT, Sherrington CS (1924) Reflexes in response to stretch (Myotatic reflexes). Proc R Soc B96:212–242

Lindström S (1973) Recurrent control from motor axon collaterals of Ia inhibitory pathways in the spinal cord of the cat. Acta Physiol Scand, Suppl 392:1–43

Lindström S, Schomburg ED (1973) Recurrent inhibition from motor axon collaterals of ventral spinocerebellar tract neurons. Acta Physiol Scand 88:505–515

Lippold OCJ (1970) Oscillation in the stretch reflex arc and the origin of the rhythmical 8–12 c/s component of physiological tremor. J Physiol (Lond) 206:359–382

Lipski J, Fyffe REW, Jodkowski J (1985) Recurrent inhibition of cat phrenic motoneurons. J Neurosci 5:1545–1555

Llinás R, Baker R, Sotelo C (1974) Electrotonic coupling between neurons in cat inferior olive. J Neurophysiol 37:560–571

Lloyd DPC (1949) Post-tetanic potentiation of response in monosynaptic reflex pathways of the spinal cord. J Gen Physiol 33:147–170

Loeb GE (1984) The control and responses of mammalian muscle spindles during normally executed motor tasks. Exercise Sport Sci Rev 12:157–204

Loeb GE (1987) Hard lessons in motor control from the mammalian spinal cord. Trends Neurosci 10:108–113

Loeb GE, Hoffer JA (1985) Activity of spindle afferents from cat anterior thigh muscles. II. Effects of fusimotor blockade. J Neurophysiol 54:565–577

Loeb GE, Marks WB (1985) Optimal control principles for sensory transducers. In: Boyd IA, Gladden MH (eds) The muscle spindle. Stockton, New York, pp 409–415

Loeb GE, White MW, Merzenich MM (1983) Spatial cross–correlation. A proposed mechanism for acoustic pitch perception. Biol Cybern 47:149–163

Loeb GE, Hoffer JA, Marks WB (1985a) Activity of spindle afferents from cat anterior thigh muscles: III. Effects of external stimuli. J Neurophysiol 54:578–591

Loeb GE, Hoffer JA, Pratt CA (1985b) Activity of spindle afferents from cat anterior thigh muscles. I. Identification and patterns during normal locomotion. J Neurophysiol 54:549–564

Loeb GE, Pratt CA, Chanaud CM, Richmond FJR (1987) Distribution and innervation of short, interdigitated muscle fibers in parallel-fibered muscles of the cat hindlimb. J Morphol 191:1–15

Lucas SM, Binder MD (1984) Topographic factors in distribution of homonymous group Ia-afferent input to cat medial gastrocnemius motoneurons. J Neurophysiol 51:50–63

Lucas SM, Cope TC, Binder MD (1984) Analysis of individual Ia-afferent EPSPs in a homonymous motoneuron pool with respect to muscle topography. J Neurophysiol 51:64–74

Lüscher H-R, Ruenzel P, Fetz E, Henneman E (1979) Postsynaptic population potentials recorded from ventral roots perfused with isotonic sucrose: Connections of group Ia and II spindle afferent fibres with large populations of motoneurons. J Neurophysiol 42:1146–1164

Lüscher H-R, Ruenzel P, Henneman E (1980) Topographic distribution of terminals of Ia and group II fibers in spinal cord, as revealed by postsynaptic population potentials. J Neurophysiol 43:968–985

Lüscher H-R, Ruenzel PW, Henneman E (1983) Effects of impulse frequency, PTP, and temperature on responses elicited in large populations of motoneurons by impulses in single Ia-fibers. J Neurophysiol 50:1045–1058

Lundberg A (1971) Function of the ventral spinocerebellar tract. A new hypothesis. Exp Brain Res 12:317–330

Lundberg A (1979) Multisensory control of spinal reflex pathways. In: Granit R, Pompeiano O (eds) Reflex control of posture and movement. Elsevier, Amsterdam, pp 11–28 (Progress in brain research, vol. 50)

Lundberg A, Malmgren K, Schomburg ED (1987) Reflex pathways from group II muscle afferents. 1. Distribution and linkage of reflex actions to α-motoneurones. Exp Brain Res 65:271–281

Lynch G, Baudry M (1984) The biochemistry of memory: a new and specific hypothesis. Science 224:1057–1063

Lynch GS, Dunwiddie T, Gribkoff V (1977) Heterosynaptic depression: a postsynaptic correlate of long-term potentiation. Nature 266:737–739

MacGregor RJ, Lewis ER (1977) Neural modelling. Electrical signal processing in the nervous system. Plenum, New York

MacKay DM (1966) Cerebral organization and the conscious control of action. In: Eccles JC (ed) Brain and conscious experience. Springer, Berlin Heidelberg New York, pp 422–445

MacLean JB, Leffman H (1967) Supraspinal control of Renshaw cells. Exp Neurol 18:94–104

Madden KP, Remmers JE (1982) Short time scale correlations between spike activity of neighbouring respiratory neurons of nucleus tractus solitarius. J Neurophysiol 48:749–760

Magleby KL (1973) The effect of repetitive stimulation on facilitation of transmitter release at the frog neuromuscular junction. J Physiol (Lond) 234:327–352

Magleby KL, Zengel JE (1975) A quantitative description of tetanic and post-tetanic potentiation of transmitter release at the frog neuromuscular junction. J Physiol (Lond) 245:183–208

Magleby KL, Zengel JE (1976a) Augmentation: a process that acts to increase transmitter release at the frog neuromuscular junction. J Physiol (Lond) 257:449–470

Magleby KL, Zengel JE (1976b) Long-term changes in augmentation, potentiation, and depression of transmitter release as a function of repeated synaptic activity at the frog neuromuscular junction. J Physiol (Lond) 257:471–494

Maier A (1979) Occurrence and distribution of muscle spindles in masticatory and suprahyoid muscles of the rat. Am J Anat 155:483–506

Maier A (1981) Characteristics of pigeon gastrocnemius and its muscle spindle supply. Exp Neurol 74:892–906

Mallart A, Martin AR (1967) An analysis of facilitation of transmitter release at the neuromuscular junction of the frog. J Physiol (Lond) 193:679–694

Marchand R, Bridgman CF, Shumpert E, Eldred E (1971) Association of tendon organs with spindles in muscles of the cat's leg. Anat Rec 169:23–32

Marmarelis PZ, Marmarelis VZ (1978) Analysis of physiological systems: the white noise approach. Plenum, New York

Marr D (1969) A theory of cerebellar cortex. J Physiol (Lond) 202:437–470

Marr D (1982) Vision. A computational investigation into the human representation and processing of visual information. Freeman, New York

Marshall J, Walsh EG (1956) Physiological tremor. J Neurol Neurosurg Psychiatry 19:260–267

Mastronarde DN (1983a) Correlated firing of cat retinal ganglion cells. I. Spontaneously active inputs to X- and Y-cells. J Neurophysiol 49:303–324

Mastronarde DN (1983b) Correlated firing of cat retinal ganglion cells. II. Responses of X- and Y-cells to single quantal events. J Neurophysiol 49:325–349

Mastronarde DN (1983c) Interactions between ganglion cells in cat retina. J Neurophysiol 49:350–365

Matthews PBC (1963) The response of de-efferented muscle spindle receptors to stretching at different velocities. J Physiol (Lond) 168:660–678

Matthews PBC (1972) Mammalian muscle receptors and their central actions. Arnold, London

Matthews PBC (1981) Evolving views on the internal operation and functional role of the muscle spindle. J Physiol (Lond) 320:1–30

Matthews PBC (1982) Where does Sherrington's "muscular sense" originate? Muscles, joints, corollary discharges? Annu Rev Neurosci 5:189–218

Matthews PBC (1984) Evidence from the use of vibration that the human long-latency stretch reflex depends upon spindle secondary afferents. J Physiol (Lond) 348:383–415

Matthews PBC, Stein RB (1969) The sensitivity of muscle spindle afferents to small sinusoidal changes in length. J Physiol (Lond) 200:723–743

McCloskey DI (1978) Kinesthetic sensibility. Physiol Rev 58:763–820

McCrea DA, Pratt CA, Jordan LM (1980) Renshaw cell activity and recurrent effects on motoneurons during fictive locomotion. J Neurophysiol 44:475–488

McKeon B, Burke D (1983) Muscle spindle discharge in response to contraction of single motor units. J Neurophysiol 49:291–302

McKeon B, Gandevia S, Burke D (1984) Absence of somatotopic projection of muscle afferents onto motoneurons of same muscle. J Neurophysiol 51:185–194

McMahon TA (1984) Muscles, reflexes, and locomotion. Princeton University Press, Princeton

McNaughton BL (1978) Evidence for two physiologically distinct perforant pathways to the fascia dentata. Brain Res 199:1–19

McNaughton BL (1980) Long-term synaptic enhancement and short-term potentiation in rat fascia dentata act through different mechanisms. J Physiol (Lond) 324:249–262

McNaughton BL, Douglas RM, Goddard GV (1978) Synaptic enhancement in fascia dentata: cooperativity among coactive afferents. Brain Res 157:277–293

McNaughton BL, Barnes CA, Andersen P (1981) Synaptic efficacy and EPSP summation in granule cells of rat fascia dentata studied in vitro. J Neurophysiol 46:952–966

McNaughton N, Miller JJ (1986) Collateral specific long term potentiation of the output of field CA3 of the hippocampus of the rat. Exp Brain Res 62:250–258

Mendell LM (1984) Modifiability of spinal synapses. Physiol Rev 64:260–324

Mendell LM, Henneman E (1971) Terminals of single Ia fibers: location, density and distribution within a pool of 300 homonymous motoneurons. J Neurophysiol 34:171–187

Mendell LM, Weiner R (1976) Analysis of pairs of individual Ia-EPSPs in single motoneurones. J Physiol (Lond) 255:81–104

Merton PA (1951) The silent period in a muscle of the human hand. J Physiol (Lond) 114:183–198

Merton PA (1953) Speculations on the servo-control of movement. In: Wolstenholme GEW (ed) The spinal cord. Churchill, London, pp 247–255.

Mesarovic MD (1960) The control of multivariable systems. Wiley, New York

Metherate R, Dykes RW (1985) Simultaneous recordings from pairs of cat somatosensory cortical neurons with overlapping peripheral receptive fields. Brain Res 341:119–129

Meyer-Lohmann J (1974) Respiratory influence upon the lumbar extensor motor system of decerebrated cats. In: Umbach W, Koepchen HP (eds) Central rhythmic and regulation. Hippokrates, Stuttgart, pp 334–340.

Meyer-Lohmann J, Riebold W, Robrecht D (1974) Mechanical influence of the extrafusal muscle on the static behaviour of deefferented primary muscle spindle endings in cat. Pflügers Arch 352:267–278

Milgram P, Inbar GF (1976) Distortion suppression in neuromuscular information transmission due to interchannel dispersion in muscle spindle firing thresholds. IEEE Trans Biomed Eng BME-23:1–15

Miller S, Scott PD (1977) The spinal locomotor generator. Exp Brain Res 30:387–403

Milner-Brown HS, Stein RB, Yemm R (1973) Changes in firing rate of human motor units during linearly changing voluntary contractions. J Physiol (Lond) 230:371–390

Milner-Brown HS, Stein RB, Lee RG (1975) Synchronization of human motor units: possible roles of exercise and supraspinal reflexes. Electroencephalogr Clin Neurophysiol 38:245–254

Misulis KE, Durkovic RG (1984) Conditioned stimulus intensity: role of cutaneous fiber size in classical conditioning of the flexion reflex in the spinal cat. Exp Neurol 86:81–92

Mittelstaedt H (1971) Reafferenzprinzip – Apologie und Kritik. Vorträge der Erlanger Physiologentagung 1970. Springer, Berlin Heidelberg New York, pp 161–171

Moore GP, Perkel DH, Segundo JP (1966) Statistical analysis and functional interpretation of neuronal spike data. Annu Rev Physiol 28:493–522

Moore GP, Segundo JP, Perkel DH, Levitan H (1970) Statistical signs of synaptic interactions in neurons. Biophys J 10:876–900

Mori S (1973) Discharge patterns of soleus motor units with associated changes in force exerted by foot during quiet stance in man. J Neurophysiol 36:458–471

Mori S (1975) Entrainment of motor-unit discharges as a neuronal mechanism of synchronization. J Neurophysiol 38:859–870

Mori S (1987) Integration of posture and locomotion in acute decerebrate cats and in awake, freely moving cats. Prog Neurobiol 28:161–195

Mori S, Kawahara K, Sakamoto T, Aoki M, Tomiyama T (1982) Setting and resetting of level of postural tone in decerebrate cat by stimulation of brain stem. J Neurophysiol 48:737–748

Mountcastle VB (1979) An organizing principle for cerebral function: The unit module and the distributed system. In: Schmitt FO, Worden FG (eds) The neurosciences, fourth study program. MIT Press, Cambridge (Mass), pp 21–42

Muhl ZF (1982) Active length-tension relation and the effect of muscle pinnation on fiber lengthening. J Morphol 173:285–292

Munson JB, Sypert GW, Zengel JE, Lofton SA, Fleshman JW (1982) Monosynaptic projections of individual spindle group II afferents to type-identified medial gastrocnemius motoneurons in the cat. J Neurophysiol 48:1164–1174

Munson JB, Fleshman JW, Zengel JE, Sypert GW (1984) Synaptic and mechanical coupling between type-identified motor units and individual spindle afferents of medial gastrocnemius muscle of the cat. J Neurophysiol 51:1268–1283

Murphy PR, Stein RB, Taylor J (1984) Phasic and tonic modulation of impulse rates in γ-motoneurons during locomotion in premammillary cats. J Neurophysiol 52:228–243

Murthy KSK (1978) Vertebrate fusimotor neurones and their influences on motor behaviour. Prog Neurobiol 11:249–307

Murthy KSK (1983) Physiological identification of static ß axons in primate muscle. Exp Brain Res 52:6–8

Murthy KSK, Yoon Y (1979) Possible evidence for synchronization amongst fusimotor neurones. J Physiol (Lond) 293:45P

Mutai M, Shibata H, Suzuki T (1986) Somatotopic organization of motoneurons innervating the pronators, carpal and digital flexors and forepaw muscles in the dog: a retrograde horseradish peroxidase study. Brain Res 371:90–95

Nelson PG (1966) Interaction between spinal motoneurons of the cat. J Neurophysiol 29:275–287

Nelson SG, Mendell LM (1978) Projection of single knee flexor Ia fibers to homonymous and heteronymous motoneurons. J Neurophysiol 41:778–787

Nelson SG, Mendell LM (1979) Enhancement in Ia-motoneuron synaptic transmission caudal to chronic spinal cord transection. J Neurophysiol 42:642–654

Nelson SG, Collatos TC, Niechaj A, Mendell LM (1979) Immediate increase in Ia-motoneuron synaptic transmission caudal to spinal cord transmission. J Neurophysiol 42:655–664

Nichols TR, Houk JC (1976) Improvement in linearity and regulation of stiffness that results from actions of stretch reflex. J Neurophysiol 39:119–142

Niemann U, Windhorst U, Meyer-Lohmann J (1986) Linear and nonlinear effects in the interactions of motor units and muscle spindle afferents. Exp Brain Res 63:639–649

Noth J (1983) Autogenetic inhibition of extensor γ-motoneurones revealed by electrical stimulation of group I fibres in the cat. J Physiol (Lond) 342:51–65

Noth J, Thilmann A (1980) Autogenetic excitation of extensor γ-motoneurones by group II muscle afferents in the cat. Neurosci Lett 17:23–26

Oğuztöreli MN, Stein RB (1976) The effects of multiple reflex pathways on the oscillations in neuro-muscular systems. J Math Biol 3:87–101

Osborn CE, Binder MD (1987) Correlation analysis of muscle receptor discharge during active contractions of the cat medial gastrocnemius muscle. J Neurophysiol 57:343–356

Ottoson D (1976) Morphology and physiology of muscle spindles. In: Llinás R, Precht W (eds) Frog neurobiology. Springer, Berlin Heidelberg New York, pp 643–675

Palm G (1982) Neural assemblies. An alternative approach to artificial intelligence. Springer, Berlin Heidelberg New York

Palm G, Aertsen A (eds) (1986) Brain theory. Springer-Verlag, Berlin Heidelberg New York Tokyo

Parmiggiani F, Stein RB, Rolf R (1982) Slow changes and Wiener analysis of nonlinear summation in contractions of cat muscles. Biol Cybern 4:177–188

Partridge LD (1982) How was movement controlled before Newton? Behav Brain Sci 5:561

Partridge LD, Benton LA (1981) Muscle, the motor. In: Brooks, VB (ed) The nervous system. Am Physiol Soc, Bethesda, pp 43–106 (Handbook of physiology, vol. II, part 1)

Pellionisz A (1979) Modeling of neurons and neuronal networks. In: Schmitt FO, Worden FG (eds) The neurosciences: IVth study program. MIT Press, Boston, pp 525–546

Pellionisz AJ (1986) David Marr: A theory of the cerebellar cortex. A model in brain theory for the "Galilean combination of simplication, unification and mathematization". In: Palm G, Aertsen A (eds) Brain theory. Springer, Berlin Heidelberg New York Tokyo, pp 253–257

Pellionisz A, Llinás R (1979) Brain modeling by tensor network theory and computer simulation. The cerebellum: distributed processor for predictive coordination. Neuroscience 4:323–248

Pellionisz A, Llinás R (1980) Tensorial approach to the geometry of brain function: cerebellar coordination via a metric tensor. Neuroscience 5:1125–1136

Pellionisz A, Llinás R (1982) Space-time representation in the brain. The cerebellum as a predictive space-time metric tensor. Neuroscience 7:2949–2970

Perkel DH, Bullock TH (1969) Neural coding. In: Schmitt FO, Melnechuk T, Quarton GC, Adelmann G (eds) Neurosci Res Program Bull. MIT Press, Cambridge (Mass), pp 221–348

Perkel DH, Gerstein GL, Moore GP (1967a) Neuronal spike trains and stochastic point processes. I. The single spike train. Biophys J 7:391–418

Perkel DH, Gerstein GL, Moore GP (1967b) Neuronal spike trains and stochastic point processes. II. Simultaneous spike trains. Biophys J 7:419–440

Perkel DH, Gerstein GL, Smith MS, Tatton WG (1975) Nerve impulse patterns: A quantitative display technique for three neurons. Brain Res 100:271–296

Person RS, Kudina LP (1968) Cross-correlation of electromyogram showing interference patterns. Electroencephalogr Clin Neurophysiol 25:58–68

Phillips CG (1969) Motor apparatus of the baboon's hand. Proc R Soc (Lond) Ser B 173:141–174

Pinter MJ, Curtis RL, Hosko MJ (1983) Voltage threshold and excitability among variously sized cat hindlimb motoneurons. J Neurophysiol 50:644–657

Polc P, Haefely W (1982) Benzodiazepines enhance the bicuculline-sensitive part of recurrent Renshaw inhibition in the cat spinal cord. Neurosci Lett 28:193–197

Pompeiano O (1984) Recurrent inhibition. In: Davidoff A (ed) Anatomy and physiology. Dekker, New York, pp 461–557 (Handbook of the spinal cord, vol 2+3)

Pompeiano O, Wand P (1980) Critical firing level of gastrocnemius-soleus motoneurons showing a prolonged discharge following vibration of the homonymous muscle. Pflügers Arch 385:263–268

Pompeiano O, Wand P, Sontag KH (1975a) The sensitivity of Renshaw cells to velocity of sinusoidal stretches of the triceps surae muscle. Arch Ital Biol 113:280–294

Pompeiano O, Wand P, Sontag KH (1975b) Response of Renshaw cells to sinusoidal stretch of hindlimb extensor muscles. Arch Ital Biol 113:92–117

Pompeiano O, Wand P, Srivastava UC (1985a) Responses of Renshaw cells coupled with hindlimb extensor motoneurons to sinusoidal stimulation of labyrinth receptors in the decerebrate cat. Pflügers Arch 403:245–257

Pompeiano O, Wand P, Srivastava UC (1985b) Influence of Renshaw cells on the response gain of hindlimb extensor muscles to sinusoidal labyrinth stimulation. Pflügers Arch 404:107–118

Poppele RE (1981) An analysis of muscle spindle behaviour using randomly applied stretches. Neuroscience 6:1157–1165

Poppele RE, Bowman RJ (1970) Quantitative description of the linear behavior of mammalian muscle spindles. J Neurophysiol 33:59–72

Poppele RE, Quick DC (1981) Stretch-induced contraction of intrafusal muscle in cat muscle spindle. J Neurosci 1:1069–1074

Poppele RE, Terzuolo CA (1968) Myotatic reflex: its input-output relation. Science 159:743–745

Poppele RE, Kennedy WR, Quick DC (1979) A determination of static mechanical properties of intrafusal muscle in isolated cat muscle spindles. Neuroscience 4:401–411

Powers RK, Binder MD (1985a) Distribution of oligosynaptic group I input to the cat medial gastrocnemius motoneuron pool. J Neurophysiol 53:497–517

Powers RK, Binder MD (1985b) Determination of afferent fibers mediating oligosynaptic group I input to cat medial gastrocnemius motoneurons. J Neurophysiol 53:518–529

Pratt CA, Jordan LM (1987) Ia inhibitory interneurons and Renshaw cells as contributors to the spinal mechanisms of fictive locomotion. J Neurophysiol 57:56–71

Prochazka A (1985) Introduction. In: Barnes WJP, Gladden MH (eds) Feedback and motor control in invertebrates and vertebrates. Croom Helm, London, pp 115–121

Prochazka A (1986) Proprioception during voluntary movement. Can J Physiol Pharmacol 64:499–504

Prochazka A, Hulliger M (1983) Muscle afferent function and its significance for motor control mechanisms during voluntary movements in cat, monkey and man. In: Desmedt JE (ed) Motor control mechanisms in health and disease. Raven, New York, pp 93–132

Prochazka A, Hulliger M, Zangger P, Appenteng K (1985) "Fusimotor set": new evidence for α-independent control of γ-motoneurones during movement in the awake cat. Brain Res 339:136–140

Proske U, Morgan DL (1984) Stiffness of cat soleus muscle and tendon during activation of part of muscle. J Neurophysiol 52:459–468

Racine RJ, Wilson DA, Gingell R, Sunderland D (1986) Long-term potentiation in the interpositus and vestibular nuclei in the rat. Exp Brain Res 63:158–162

Rack PMH (1981a) Limitations of somatosensory feedback in control of posture and movement. In: Brooks VB (ed) The nervous system. Am Physiol Soc, Bethesda, pp 229–256 (Handbook of physiology, vol. II, part 1)

Rack PMH (1981b) A critique of the papers by Houk, Crago and Rymer, Hoffer and Andreassen and Dietz. In: Taylor A, Prochazka A (eds) Muscle receptors and movement. Macmillan, London, pp 347–353

Rack PMH (1985) Stretch reflexes in man: the significance of tendon compliance. In: Barnes WJP, Gladden MH (eds) Feedback and motor control in invertebrates and vertebrates. Croom Helm, London, pp 217–229

Rack PMH, Ross HG (1984) The tendon of flexor pollicis longus: its effects on the muscular control of force and position at the human thumb. J Physiol (Lond) 351:99–110

Rack PMH, Westbury DR (1969) The effects of length and stimulus rate on tension in the isometric cat soleus muscle. J Physiol (Lond) 204:443–460

Rack PMH, Westbury DR (1984) Elastic properties of the cat soleus tendon and their functional importance. J Physiol (Lond) 347:479–495

Rahamimoff R, Lev-Tov A, Meiri A (1980) Primary and secondary regulation of quantal transmitter release: calcium and sodium. J Exp Biol 89:5–18

Rainey WT, Buahin KG, Rymer WZ (1984) Characteristics of spinal interneurons that respond to primary spindle afferent activation. Soc Neurosci Abstr 10, Part 1:329

Rall W (1959) Branching dendritic trees and motoneuron membrane resistivity. Exp Neurol 1:491–527

Rall W, Hunt CC (1956) Analysis of reflex variability in terms of partially correlated excitability fluctuations in a population of motoneurones. J Gen Physiol 39:397–422

Rapoport S (1979) Reflex connexions of motoneurones of muscles involved in head movement in the cat. J Physiol (Lond) 289:311–327

Rayport SG, Kandel ER (1986) Development of plastic mechanisms related to learning at identified chemical synaptic connections in Aplysia. Neuroscience 17:283–294

Redman SJ (1973) The attenuation of passively propagating dendritic potentials in a motoneurone cable model. J Physiol (Lond) 234:637–664

Renshaw B (1941) Influence of discharge of motoneurons upon excitation of neighboring motoneurons. J Neurophysiol 4:167–183

Ribot E, Roll J-P, Vedel J-P (1986) Efferent discharges recorded from single skeletomotor and fusimotor fibres in man. J Physiol (Lond) 375:251–268

Richmond FJR, Abrahams VC (1975a) Morphology and enzyme histochemistry of dorsal muscles of the cat neck. J Neurophysiol 38:1312–1321

Richmond FJR, Abrahams VC (1975b) Morphology and distribution of muscle spindles in dorsal muscles of the cat neck. J Neurophysiol 38:1322–1339

Richmond FJR, Bakker DA (1982) Muscle spindle complexes in muscles around upper cervical vertebrae in the cat. J Neurophysiol 48:62–74

Richmond FJR, Stuart DG (1985) Distribution of sensory receptors in the flexor carpi radialis muscle of the cat. J Neurophysiol 183:1–13

Richmond FJR, MacGillis DRR, Scott DA (1985) Muscle-fiber compartmentalization in cat splenius muscles. J Neurophysiol 53:868–885

Robrecht D (1971) Untersuchungen an primären Muskelspindelendigungen aus dem M. ext. dig. long. der Katze. Ihr Verhalten bei Dehnung und Kontraktion in Abhängigkeit von der Lage der Rezeptoren im extrafusalen Muskelgefüge. Med Dissertation, Universität Göttingen

Roland PE, Ladegaard-Pedersen H (1977) Sensations of tension and kinaesthesia from musculotendinous receptors in man. Evidence for a muscular sense and sense of effort. Brain 100:671–692

Romanes GJ (1951) The motor cell columns of the lumbosacral spinal cord of the cat. J Comp Neurol 94:313–363

Rose PK (1982) Branching structure of motoneuron stem dendrites: A study of neck muscle motoneurons intracellularly stained with horseradish peroxidase in the cat. J Neurosci 2:1596–1607

Rosenthal NP, McKean TA, Roberts WJ, Terzuolo CA (1970) Frequency analysis of stretch reflex and its main subsystems in triceps surae muscles of the cat. J Neurophysiol 33:713–749

Ross HG (1976) Experimentelle Untersuchungen und Modellvorstellungen zur quantitativen Charakterisierung der rekurrenten Inhibition spinaler Alpha-Motoneurone. Med Habilitationsschrift, Universität Düsseldorf

Ross HG, Thewissen M (1987) Inhibitory connections of ipsilateral semicircular canal afferents onto Renshaw cells in the lumbar spinal cord of the cat. J Physiol (Lond) 388:83–99

Ross H-G, Wittrock C (1987) Renshaw cell activity in falling cats. Pflügers Arch 408:R 54

Ross HG, Cleveland S, Haase J (1975) Contribution of single motoneurons to Renshaw cell activity. Neurosci Lett 1:105–108

Ross HG, Cleveland S, Kuschmierz A (1982) Dynamic properties of Renshaw cells: equivalence of responses to step changes in recruitment and discharge frequency of motor axons. Pflügers Arch 394:239–242

Rudomín P (1980) Information processing at synapses in the vertebrate spinal cord: presynaptic control of information transfer in monosynaptic pathways. In: Pinsker HM, Willis WD (eds) Information processing in the nervous system. Raven, New York, pp 125–155

Rudomín P, Madrid J (1972) Changes in correlation between monosynaptic responses of single motoneurons and in information transmission produced by conditioning volleys to cutaneous nerves. J Neurophysiol 35:44–64

Rudomín P, Dutton H, Muñoz-Martinez J (1969) Changes in correlation between monosynaptic reflexes produced by conditioning afferent volleys. J Neurophysiol 32:759–772

Rudomín P, Burke RE, Nuñez R, Madrid J, Dutton H (1975) Control by presynaptic correlation: a mechanism affecting information transmission from Ia fibers to motoneurons. J Neurophysiol 38:267–284

Rudomín P, Jimenez I, Solodkin M, Dueñas S (1983) Sites of action of segmental and descending control of transmission on pathways mediating PAD of Ia- and Ib-afferent fibers in cat spinal cord. J Neurophysiol 50:743–769

Russell CJ, Dunbar DC, Rushmer DS, MacPherson JM, Phillips JO (1982) Differential activity of innervation subcompartments of cat lateral gastrocnemius during natural movements. Soc Neurosci Abstr 8:948

Ryall RW (1970) Renshaw mediated inhibition of Renshaw cells: Patterns of excitation and inhibition from impulses in motor axon collaterals. J Neurophysiol 33:257–270

Ryall RW (1981) Patterns of recurrent excitation and mutual inhibition of cat Renshaw cells. J Physiol (Lond) 316:439–452

Ryall RW, Piercey MF, Polosa C (1971) Intersegmental and intrasegmental distribution of mutual inhibition of Renshaw cells. J Neurophysiol 34:700–707

Ryall RW, Piercey MF, Polosa C, Goldfarb J (1972) Excitation of Renshaw cells in relation to orthodromic and antidromic excitation of motoneurons. J Neurophysiol 35:137–148

Rymer WZ, D'Almeida A (1980) Joint position sense: The effects of muscle contraction. Brain 103:1–22

Rymer WZ, Grill SE (1985) Reflex consequences of muscle afferent input to beta and gamma motoneurones. In: Boyd IA, Gladden MH (eds) The muscle spindle. Stockton, New York, pp 303–307

Rymer WZ, Hasan Z (1981) Prolonged time course for vibratory suppression of stretch reflex in the decerebrate cat. Exp Brain Res 44:101–112

Sacks RD, Roy RR (1982) Architecture of the hind limb muscles of cats: Functional significance. J Morphol 173:185–195

Saggau P, ten Bruggencate G (1987) Local long-term potentiations in hippocampal CA1-populations monitored by optical recording of neural activity. Pflügers Arch 408:R 59

Schmidt RF (1971) Presynaptic inhibition in the vertebrate central nervous system. Ergeb Physiol 63:20–101

Schmidt RF, Kniffki KD, Schomburg ED (1981) Der Einfluß kleinkalibriger Muskelafferenzen auf den Muskeltonus. In: Bauer HJ, Koella WP (ed) Therapie der Spastik. Verlag f. angew. Wiss., München, pp 71–84

Schneider J, Eckhorn R, Reitböck H (1983) Evaluation of neuronal coupling dynamics. Biol Cybern 46:129–134

Schomburg ED, Behrends HB (1978) The possibility of phase-dependent monosynaptic and polysynaptic Ia excitation to homonymous motoneurones during fictive locomotion. Brain Res 143:533–537

Schomburg ED, Steffens H (1985) Convergence in segmental reflex pathways from group II muscle afferents to α-motoneurones. In: Boyd IA, Gladden MH (eds) The muscle spindle. Stockton, New York, pp 273–278

Schulte FJ, Busch G, Henatsch H-D (1959a) Antriebssteigerung lumbaler Extensor-Motoneurone bei Aktivierung der Chemorezeptoren im Glomus caroticum. Pflügers Arch 269:580–592

Schulte FJ, Henatsch H-D, Busch W (1959b) Über den Einfluß der Carotissinus-Sensibilität auf die spinalmotorischen Systeme. Pflügers Arch 259:248–263

Schulz B, Lambertz M, Schulz G, Langhorst P, Krienke B (1983) Reticular formation of the lower brainstem. A common system for cardio-respiratory and somatomotor functions: discharge patterns of neighboring neurons influenced by somatosensory afferents. J Auton Nerv Syst 9:433–449

Schulz G, Lambertz M, Schulz B, Langhorst P, Krienke B (1985) Reticular formation of the lower brainstem. A common system for cardio-respiratory and somatomotor functions. Cross-correlation analysis of discharge patterns of neighbouring neurones. J Auton Nerv Syst 12:35–62

Schwestka R (1981) Interaktionen zwischen mehreren motorischen Einheiten und mehreren Muskelspindelafferenzen. – Ein Beispiel für vielkanalige Informationsübertragung im Nervensystem. Med Dissertation, Universität Göttingen

Schwestka R, Windhorst U, Schaumberg R (1981) Patterns of parallel signal transmission between multiple α-efferents and multiple Ia afferents in the cat semitendinosus muscle. Exp Brain Res 43:34–46

Schwindt PC (1973) Membrane-potential trajectories underlying motoneuron rhythmic firing at high rates. J Neurophysiol 36:434–449

Schwindt PC (1976) Electrical properties of spinal motoneurons. In: Llinás R, Precht W (eds) Frog neurobiology. Springer, Berlin Heidelberg New York, pp 750–764

Schwindt PC, Calvin H (1972) Membrane-potential trajectories between spikes underlying motoneuron rhythmic firing. J Neurophysiol 35:311–325

Scott JG, Mendell LM (1976) Individual EPSPs produced by single triceps surae Ia afferent fibers in homonymous and heteronymous motoneurons. J Neurophysiol 39:679–692

Sears TA, Stagg D (1976) Short-term synchronization of intercostal motoneurone activity. J Physiol (Lond) 263:357–387

Segundo JP, Perkel DH, Wyman H, Hegstad H, Moore GP (1968) Input-output relations in computer-simulated nerve cells. Influence of the statistical properties, strength, number and interdependence of excitatory presynaptic terminals. Kybernetik 4:157–171

Sejnowski TJ (1976) On the stochastic dynamics of neuronal interaction. Biol Cybern 22:203–211

Sherrington CS (1904) Correlation of reflexes and the principle of the common path. Rep Br Ass:728

Shiavi R, Negin M (1975) Multivariate analysis of simultaneously active motor units in human skeletal muscle. Biol Cybern 20:9–16

Simpson JI (1976) Functional synaptology of the spinal cord. In: Llinás R, Precht W (eds) Frog neurobiology. Springer, Berlin Heidelberg New York, pp 728–749

Singer W (1983) Neuronal activity as a shaping factor in the self-organization of neuron assemblies. In: Basar E, Flohr H, Haken H, Mandell AJ (eds) Synergetics of the brain. Springer, Berlin Heidelberg New York, pp 89–101

Sjöström A, Zangger P (1976) Muscle spindle control during locomotor movements generated by the deafferented spinal cord. Acta Physiol Scand 97:281–291

Smith DV, van Buskirk RL, Travers JB, Bieber SL (1983) Coding of taste stimuli by hamster brain stem neurons. J Neurophysiol 50:541–558

Smith JL, Zernicke RF (1987) Predictions for neural control based on limb dynamics. Trends Neurosci 10:123–128

Smith TG, Wuerker RB, Frank K (1967) Membrane impedance changes during synaptic transmission in cat spinal motoneurons. J Neurophysiol 30:1072–1096

Sojka P, Johansson H, Sjölander P, Wadell I (1986) Reflex effects on fusimotor neurones assessed by multi-afferent recordings from muscle spindle afferents. Neurosci Lett Suppl 26:S163

Sotelo C, Llinás R, Baker R (1974) Structural study of inferior olivary nucleus of the cat: morphological correlates of electrotonic coupling. J Neurophysiol 37:541–559

Stein RB (1974) Peripheral control of movement. Physiol Rev 54:215–243

Stein RB (1982) What muscle variable(s) does the nervous system control in limb movement? Behav Brain Sci 5:535–541

Stein RB, Oğuztöreli MN (1976) Does the velocity sensitivity of muscle spindles stabilize the stretch reflex? Biol Cybern 23:219–228

Stein RB, Oğuztöreli MN (1984) Modification of muscle responses by spinal circuitry. Neuroscience 11:231–240

Stiles RN (1976) Frequency and displacement amplitude relations for normal hand tremor. Am J Physiol 40:44–54

Stiles RN (1980) Mechanical and neural feedback factors in postural hand tremor of normal subjects. J Neurophysiol 44:40–59

Stiles RN, Pozos RS (1976) A mechanical-reflex oscillator hypothesis for parkinsonian hand tremor. J Appl Physiol 40:990–998

Stuart DG, Hamm TM, Vanden Noven S (1988) Partitioning of monosynaptic Ia excitation to motoneurons according to neuromuscular topography: generality and functional implications. Prog Neurobiol (in press)

Stuart GL, Rymer WZ, Schotland JL (1986) Characteristics of reflex excitation in close synergist muscles evoked by muscle vibration. Exp Brain Res 65:127–134

Sutula T, Steward O (1986) Quantitative analysis of synaptic potentiation during kindling of the perforant path. J Neurophysiol 56:732–746

Swadlow HA, Kocsis JD, Waxman SG (1980) Modulation of impulse conduction along the axonal tree. Annu Rev Biophys Bioeng 9:143–179

Swett JE, Eldred E (1959) Relation between spinal level and peripheral location of afferents in calf muscles of the cat. Am J Physiol 196:819–823

Swett JE, Eldred E, Buchwald JS (1970) Somatotopic cord-to-muscle relations in efferent innervation of cat gastrocnemius. Am J Physiol 219:762–766

Szentagothai J (1983) The modular architectonic principle of neural centers. Rev Physiol Biochem Pharmacol 98:11–61

Taylor A (1962) The significance of grouping of motor unit activity. J Physiol (Lond) 162:259–269

Taylor A, Appenteng K (1981) Distinctive modes of static and dynamic fusimotor drive in jaw muscles. In: Taylor A, Prochazka A (eds) Muscle receptors and movement. Macmillan, London, pp 179–192

Taylor A, Prochazka A (eds) (1981) Muscle receptors and movement. Macmillan, London

Taylor WK (1965) A model of learning mechanisms in the brain. In: Wiener N, Schade JP (eds) Cybernetics of the nervous system. Elsevier, Amsterdam, pp 369–397 (Progress in brain research, vol. 17)

ten Hoopen M (1974) Examples of power spectra of uni-variate point processes. Kybernetik 16:145–154

ter Haar Romeny BM, Denier van der Gon JJ, Gielen CCAM (1982) Changes in recruitment order of motor units in the human biceps muscle. Exp Neurol 78:360–368

ter Haar Romeny BM, Denier van der Gon JJ, Gielen CCAM (1984) Relation between location of a motor unit in the human biceps brachii and its critical firing level for different tasks. Exp Neurol 85:631–650

Thomas RC, Wilson VJ (1967) Recurrent interactions between motoneurons of known location in the cervical cord of the cat. J Neurophysiol 30:661–674

Thompson RF, Berger TW, Madden IV J (1983) Cellular processes of learning and memory in the mammalian CNS. Annu Rev Neurosci 6:447–491

Torre V, Poggio T (1978) A synaptic mechanism possibly underlying directional selectivity to motion. Proc R Soc Lond B 202:409–416

Toyama K (1978) Interneuronal connectivity in cat visual cortex: studies by cross-correlation analysis of the response of two simultaneously recorded neurons. In: Ito M (ed) Integrative control function of the brain. Kodamsha Sci, Tokyo, pp 65–72

Toyama K, Kimura M, Tanaka K (1981) Organization of cat visual cortex as investigated by cross-correlation technique. J Neurophysiol 46:202–213

Tönnies JF, Jung R (1948) Über rasch wiederholte Entladungen der Motoneurone und die Hemmungsphase des Beugereflexes. Pflügers Arch 250:667–693

Tracey DJ, Walmsley B (1984) Synaptic input from identified muscle afferents to neurones of the dorsal spinocerebellar tract in the cat. J Physiol (Lond) 350:599–614

Tsukahara N (1981) Synaptic plasticity in the mammalian central nervous system. Annu Rev Neurosci 4:351–379

Subject Index